ACTA PHYSICA AUSTRIACA / SUPPLEMENTUM V

PARTICLES, CURRENTS, SYMMETRIES

PROCEEDINGS OF THE
VII. INTERNATIONALE UNIVERSITÄTSWOCHEN
FÜR KERNPHYSIK 1968 DER KARL-FRANZENS-UNIVERSITÄT
GRAZ, AT SCHLADMING (STEIERMARK, AUSTRIA)
26th FEBRUARY—9th MARCH 1968

SPONSORED BY
BUNDESMINISTERIUM FÜR UNTERRICHT
BUNDESMINISTERIUM FÜR HANDEL, GEWERBE UND INDUSTRIE
THE INTERNATIONAL ATOMIC ENERGY AGENCY
STEIERMÄRKISCHE LANDESREGIERUNG AND
KAMMER DER GEWERBLICHEN WIRTSCHAFT FÜR STEIERMARK

EDITED BY
PAUL URBAN
GRAZ

WITH 36 FIGURES

1968

SPRINGER-VERLAG / WIEN · NEW YORK

Organizing Committee:

Chairman: Prof. Dr. PAUL URBAN
Vorstand des Institutes für Theoretische Physik, Universität Graz

Committee Members: Dr. H. KÜHNELT
Dr. H. LATAL

Proceedings: Dr. R. BAIER
H. J. FAUSTMANN

Secretary: A. SCHMALDIENST

Acta Physica Austriaca / Supplementum I
Weak Interactions and Higher Symmetries
published in 1964

Acta Physica Austriaca / Supplementum II
Quantum Electrodynamics
published in 1965

Acta Physica Austriaca / Supplementum III
Elementary Particle Theories
published in 1966

Acta Physica Austriaca / Supplementum IV
Special Problems in High Energy Physics
published in 1967

Library of Congress Catalog Card Number 68-55816

Printed in Austria

Titel-Nr. 9237

Preface

The subject of this year's conference, as of the six previous meetings, was again elementary particle physics. One of the main topics of current interest here is the understanding of strong interactions; some of the attempts to explain them are discussed at length in these proceedings, including current algebra, Regge pole theory and effective Lagrangians. On the other hand, the processes of weak interactions, too, are not yet fully understood, especially when strong interactions intervene, as in the radiative corrections to weak decays. Several other aspects of elementary particle physics are also treated in this volume, every one of them another step towards a thorough understanding of the subatomic world.

I want to take this opportunity to thank all the members of my staff for their assistance in organizing the meeting and for their help in preparing the manuscripts.

Graz, June 1968 **P. Urban**

Contents

Dear Colleagues, Ladies and Gentlemen:

Once again in the history of our annual International Meeting in Schladming we have started with the lectures before the official opening ceremony, on account of the carnival festivities going on here, which I hope you are enjoying. The word "history" I just used is not an overstatement since we are now meeting for the seventh time in Schladming and to many of you our symposium is quite familiar. These people especially, and all those who are participating for the first time, I would like to welcome most cordially and wish all of you two weeks of success in physics and perhaps also in skiing.

To our guests of honour, who show by their attendance the serious interest of the governmental and provincial authorities in our enterprise, I most respectfully extend my welcome. It is a great privilege for me to have the representative of the Ministry of Education with us; may I take the opportunity here to express our sincerest thanks for the generous support we have received over the years. Other grants which have contributed much to the successful organization of this meeting were given by the International Atomic Energy Agency and the Ministry of Trade and Industry. We are further indebted to the Provincial Government of Styria, whose representative I also welcome here, as well as the Chamber of Commerce. It gives us special pleasure to have among our guests of honour the Dean of the Philosophische Fakultät of the University of Graz. We are certain that through the scientific reputation of this meeting the name of our University is already well known to the scientific community. But also our host city, Schladming, and its beautiful surroundings become more familiar each year to a larger group of people, who meet with us here. May I take the opportunity now to thank especially the Mayor of the City, Director Laurich, both personally and as representative of the authorities of Schladming for their continued help throughout the years.

Science—and possibly also skiing—have again this year drawn to Schladming about 180 participants from 18 nations, a number which impressively shows the value of our meeting as a place for learning about new developments and discussing recent results. We are happy that we have succeeded in obtaining the cooperation of outstanding experts as lecturers and I want to thank them for coming here; I hope they will also benefit from their stay.

As expressed in the title of this year's meeting, "Particles, Currents, Symmetries", we shall be dealing with the world of elementary particle physics. Here especially, in the absence of a theory as powerful as quantum-electrodynamics, the strong and weak interaction processes have successively forced us to change our point of view in order to restore the agreement between experiment and theory. The latest such event, caused by the detection of CP violation processes, has again destroyed another supposed conservation principle. These principles are connected with the role of discrete symmetries in elementary particle physics, a concept which provides a means of predicting various relations between physically observable quantities.

The application of modern group theory led to a classification scheme for the multitude of particles we know today in terms of internal symmetries. Here also remarkable progress has been made in recent years; one outflow of these ideas was the concept of current algebra which has already established itself as a powerful tool in theoretical investigations.

Another approach towards explaining the experimental findings of high-energy physics came from such a distant area as non-relativistic potential scattering, the Regge pole model. Already at our second meeting, here in Schladming in 1963, we devoted the main part of our program to the discussion of the various aspects and predictions of this idea. This year, with interest in the Regge poles being revived, we felt it might be worth while to take another look at the model and see what its successes and shortcomings are now.

I have mentioned only a few prominent features in this survey of the theories we employ in describing strong interaction processes. There are, however, other attempts towards establishing a sound theoretical basis for the multitude of experimental data, some of which will also be discussed in the two weeks ahead of us. So we may expect a fruitful exchange of ideas, and I wish all of you a pleasant stay with the mixture of physics and skiing we have provided for you at our VIIth International Meeting.

P. URBAN, Graz

DISCRETE SYMMETRIES IN ELEMENTARY PARTICLE PHYSICS†

By

J. NILSSON

Institute of Theoretical Physics
University of Göteborg, Sweden

Table of Contents

---

† Lecture given at the VII. Internationale Universitätswochen f. Kernphysik, Schladming, 26 February - 9 March 1968.

1. Introduction

In these lectures we shall review the three discrete transformations P (space reflection), C (charge conjugation) and T (time reversal) applied to elementary particle processes. None of the material presented is new and the survey serves a pedagogical purpose only.

The importance of symmetry principles including the discrete ones was realized already in the early days of quantum mechanics. From that point of view the subject is quite old by now. However, various circumstances have led to a revival of interest in it. Most important in this respect are the recent developments on the experimental side indicating that all the symmetries mentioned above seem to be violated in elementary particle physics. Furthermore, we still lack a complete dynamical scheme for elementary particle processes. In this state of affairs the symmetry principles provide some of the most powerful tools. It has been known for a long time that symmetry arguments can be used to derive predictions which may be subjected to experimental tests, and the deductions do not require any knowledge of the dynamics. Hence, symmetry principles can be verified or they can be disproved even before we have a complete theory. At a time when we still search for a theory to account for elementary particle processes this aspect of symmetry principles is particularly pertinent.

Finally, invariance principles have mostly been discussed within the framework of Lagrangian field theory. Of course, they can equally well be expressed without any reference to field theory. We shall follow this latter procedure.

2. Formalism

To lay the grounds for the discussions in the following sections we begin with a brief account of the basic formalism. This will give us an opportunity to establish notations and to introduce some conventions to be used throughout these lectures. In particular we shall discuss at some length the relativistic description of free particle states. The concept of symmetry will then be defined and some immediate implications derived from symmetry arguments will be outlined. For a more complete discussion of the material of this section we refer to [1].

2.1. Basic postulates of quantum theory

A physical experiment consists of two parts:
(i) the preparation of the initial state of the system

under the study, and
(ii) performing measurements on the system at a later time and possibly also in another region of space. It is the aim of theory to account for or "explain" the correlations between preparations and measurements that one may establish by experiments.

The preparation of the initial state corresponds to performing a series of measurements on the physical system to define its properties. After the preparation the system is represented by a state vector $|\alpha(\Omega)>$, where α denotes the outcome of the various measurements, that is, the state vector is labeled by the results of measurements such as momentum, helicity etc. The preparation extends over a finite space-time region Ω and the state vector will in general depend on Ω as indicated by the notation which is used.

Some time after the preparation the system is once more subjected to measurements and the experiment enters the second phase. The interaction between the system and the measuring apparatus will affect the latter and the registration of how it is affected is the outcome of the experiment. As a result of this second set of measurements we have new information about the system and hence it must be represented by a new state vector $|\beta(\Omega')>$, the final state. Of course, this new state vector will depend on the space-time region Ω' in which the measurements were performed. The properties of the physical system, which initially was in the state $|\alpha(\Omega)>$, at the time and place of the measurements are given by the probability distribution $w(\beta,\alpha)$ over all possible final states $|\beta(\Omega')>$, that is, the probability distribution over all possible outcomes of the measurements in Ω'. The fundamental principle of quantum mechanics now asserts that $w(\beta,\alpha)$ is given by

$$w(\beta,\alpha) = |<\beta(\Omega')|\alpha(\Omega)>|^2 \qquad (2.1)$$

The manifold $\mathcal{H}$ of state vectors representing all possible states of a physical system forms a linear vector space or, more precisely , a separable Hilbert space under the scalar product of the eq. (2.1). Each observable, that is, measurable property A of the system is represented by a linear hermitian operator A acting on the vectors in $\mathcal{H}$. The (real) eigenvalues a of A are the only possible results of an exact (sharp) measurement of the property A. The maximum amount of information one may obtain for the system corresponds to a complete set of commuting (compatible) measurements, that is, measurements which do not perturb or interfere with each other. If $\{A_i\}_{i=1}^N$ is such a complete set of commuting observables, then the space $\mathcal{H}$ is spanned by the basis vectors $\{|a_1,...,a_N>\}$ where a_i is the eigenvalue of A_i. This means that any admissable state $|\psi>$ of the system can always be expanded in terms of the vectors $\{|a_1,...,a_N>\}$, where each a_i runs over the whole spectrum of the corresponding observable A_i.

Of course, one may choose different sets of commuting observables and in that way one obtains different sets of basis vectors, which are related by unitary (norm-preserving) transformations (Clebsch-Gordan type expansions). Which set of basis vectors one chooses is a matter of convenience.

2.2. Relativistic description of free particle states

The classification of relativistic particle states is based on the Poincaré group, that is, the group of proper orthochronous Lorentz transformations. These transformations are originally defined on the space-time manifold. If (a,Λ) is an element of the Poincaré group, then the parameters a and Λ are defined by

$$x_\mu \xrightarrow{(a,\Lambda)} x'_\mu = \Lambda_\mu^\nu \; x_\nu + a_\mu \tag{2.2}$$

and we have the following composition law

$$(a_1,\Lambda_1)(a_2,\Lambda_2) = (a_1 + \Lambda_1 \; a_2, \Lambda_1\Lambda_2) \tag{2.3}$$

The eq. (2.3) implies that the Poincaré group is a semi-direct product of the four-dimensional translation group (the four parameters a_μ) and the homogeneous Lorentz group (the six parameters Λ_μ^{ν}). The local properties of $\mathcal{P}$ are given by the Lie-algebra of the corresponding infinitesimal generators P_μ and $J_{\mu\nu}$

$$[P_\mu, P_\nu] = 0$$

$$[J_{\mu\nu}, P_\rho] = - i(g_{\mu\rho}P_\nu - g_{\nu\rho}P_\mu)$$

$$[J_{\mu\nu}, J_{\rho\sigma}] = - i \, (g_{\mu\rho}J_{\nu\sigma} + g_{\nu\sigma}J_{\mu\rho} - g_{\mu\sigma}J_{\nu\rho} - g_{\nu\rho}J_{\mu\sigma}) \tag{2.4}$$

Alternatively, we may define the angular momentum operators J^m by the relation

$$J^{kl} = \sum_m \varepsilon^{klm} \, J^m ; \quad k,l,m = 1,2,3 \tag{2.5}$$

and the generators K^m of pure Lorentz transformations (accelerations)

$$K^m = J^{om} = - J^{mo} \tag{2.6}$$

for which commutation relations may be derived from the eqs. (2.4).

It is a straightforward task to prove that the following two (Casimir) operators commute with all the generators of $\mathcal{P}$

$$P_\mu P^\mu \sim m^2$$

$$W_\mu W^\mu \sim - m^2 s(s+1) \tag{2.7}$$

with

$$W^\mu \equiv \frac{1}{2} \varepsilon^{\mu\nu\rho\sigma} J_{\nu\rho} P_\sigma$$

and

$$\varepsilon^{0123} = - \varepsilon_{0123} = - 1 .$$

Hence, a unitary irreducible representation (UIR) of $\mathcal{P}$ is characterized by the eigenvalues of these operators or, equivalently, by m and s as defined in the eqs. (2.7). These equations are to be understood in the following way: when $P_\mu P^\mu$ resp. $W_\mu W^\mu$ act on a vector belonging to the UIR (m,s), then the eigenvalues of the operators are given in terms of m and s by the eqs. (2.7). Applied to particle states at rest m and s are readily identified with the mass and the spin of the particle. Thus, the manifold of state vectors representing all the possible states which a particle of mass m and spin s may occupy forms a representation space of the UIR (m,s) of $\mathcal{P}$. We may also express this in the following way. Given a state vector representing any possible state for the particle (m,s) all the other physical state vectors representing the same particle (in different states of motion) are obtained by means of Lorentz transformations acting on the original state vector. This particular property provides us with the means to construct all possible state vectors for the particle out of some standard state vector as we shall demonstrate below.

Consider a particle of mass $m > o$ and spin s at rest.

The state of this particle is completely specified by the spin component m_s along, say, the z-axis, since $P_\mu P^\mu$, $W_\mu W^\mu$, $\bar{p}$ and J_3 form a complete set of commuting observables (we suppress internal properties such as electric charge, isospin etc. and confine our attention to the kinematic properties). Furthermore, in the rest system the angular momentum operator is identical with the spin operator. Altogether there are (2s+1) linearly independent states corresponding to $m_s = -s, -s+1, \ldots, s-1, s$. We denote these state vectors by $|\bar{p} = 0, m_s\rangle$ and we suppress the labels m and s since we have restricted ourselves to one UIR of $\mathcal{P}$. The transformations between the different states $|\bar{p} = 0, m_s\rangle$ are accomplished with the raising and lowering operators $J_\pm \equiv (J_1 \pm iJ_2)$ which satisfy the following relations [1]

$$J_\pm |\bar{p} = 0, m_s\rangle = \sqrt{(s\mp m_s)(s\pm m_s+1)}\; |\bar{p} = 0, m_s \pm 1\rangle \quad (2.8)$$

while

$$J_3 |\bar{p} = 0, m_s\rangle = m_s |\bar{p} = 0, m_s\rangle \quad (2.9)$$

A finite rotation $R(\bar{\Omega})$ is given by

$$R(\bar{\Omega}) = \exp\left[-i\bar{J}.\bar{\Omega}\right] \quad (2.10)$$

and acting on $|\bar{p} = 0, m_s\rangle$ one obtains

$$R(\bar{\Omega})|\bar{p} = 0, m_s\rangle \equiv D^s(\bar{\Omega})|\bar{p} = 0, m_s\rangle =$$

$$= \sum_{m_s'} D^s_{m_s' m_s}(\bar{\Omega})|\bar{p} = 0, m_s'\rangle \quad (2.11)$$

More often the rotation is parametrized by means of the Euler angles α, β, γ . With $s \equiv j$ the eq. (2.10) is replaced by

$$D^j(\alpha,\beta,\gamma) \equiv \exp[-i\, J_3\alpha] \quad \exp[-i\, J_2\beta] \; \exp[-i\, J_3\gamma] \tag{2.12}$$

so that

$$D^j_{m'm}(\alpha,\beta,\gamma) \equiv \langle j,m'|D^j(\alpha,\beta,\gamma)|j,m\rangle =$$

$$= \exp[-i(m'\alpha+m\gamma)] \; d^j_{m'm}(\beta) \tag{2.13}$$

with

$$d^j_{m'm}(\beta) \equiv \langle j,m'|\exp[-i\, J_2\beta]|j,m\rangle \tag{2.14}$$

These d-functions may further be expressed in terms of elementary functions [1]. All the results so far are based on the fact that the (2s+1) state vectors representing a particle at rest span a representation space for a (2s+1)-dimensional UIR of the rotation group SO(3).

Starting from a state at rest we may now proceed to define states of nonvanishing momenta. We define

$$|\bar{p},m_s\rangle \equiv R(\phi,\theta,0)L_z(v) \; R^{-1}(\phi,\theta,0)|\bar{p}=o,m_s\rangle \tag{2.15}$$

and

$$|\bar{p},\lambda\rangle \equiv R(\phi,\theta,0)L_z(v)|\bar{p}=o,\lambda\rangle \tag{2.16}$$

where $\bar{p} = (\phi,\theta,p)$ and $v = |\bar{p}|/p_o$. The operator $L_z(v)$ represents a Lorentz transformation along the z-axis. It can be expressed in terms of the infinitesimal generator K_3:

$$L_z(v) = \exp[-i\, K_3\, u] \tag{2.17}$$

with tanh $u = v$. A closer examination shows that the vectors $|\bar{p},\lambda>$ are eigenvectors of the helicity operator $\bar{J}.\bar{p}/|\bar{p}|$ for $\bar{p} \neq 0$. For this reason they are often called helicity state vectors.

It should be noted that the definitions of $|\bar{p},m_s>$ and $|\bar{p},\lambda>$ are unique for

$$0 < \theta < \pi$$

$$-\pi < \phi \leq \pi$$

while for $\theta=0$ and $\theta=\pi$ as well as for $\phi=-\pi$ in case of half-integer spin the definitions are ambiguous and they must be amended to make them unique. We define

$$|0,0,p,\lambda> \equiv L_z(v)|0,0,p,\lambda> \qquad (2.18)$$

and

$$|0,\pi,p,\lambda> \equiv L_{-z}(v)|0,0,0,-\lambda> =$$

$$= \exp[-i\pi s]\; R(\pi,\pi,0)\; L_z(v)|0,0,0,\lambda> \qquad (2.19)$$

It is well known that the plane wave state vectors cannot be normalized. Instead we impose the following generalized normalization conditions

$$<\bar{p}',m_s'|\bar{p},m_s> = p_0\delta^3(\bar{p} - \bar{p}')\;\delta_{m_s m_s'}$$

$$<\bar{p}',\lambda'|\bar{p},\lambda> = p_0\delta^3(\bar{p} - \bar{p}')\;\delta_{\lambda\lambda'} \qquad (2.20)$$

To determine the transformation properties of, say, the vector $|\bar{p},\lambda>$ under a proper Lorentz transformation $(0,\Lambda)$ we introduce the following abbreviated notation

$$L(\bar{p}) \equiv R(\phi,\theta,0)\; L_z(v) \qquad (2.21)$$

where $\bar{p} = (\phi,\theta,p)$ and $v = |\bar{p}|/p_o$. To determine the action of the operator $L(\Lambda)$ on the state vector we note the following decomposition of $L(\Lambda)$

$$L(\Lambda) = L\,(\Lambda\bar{p})\; R(\bar{p},\Lambda)\; L^{-1}(\bar{p}) \qquad (2.22)$$

where $R(\bar{p},\Lambda)$ is a pure rotation, the Wigner rotation, and $\Lambda\bar{p}$ is defined by

$$p_\mu \xrightarrow{(0,\Lambda)} (\Lambda p)_\mu = \Lambda_\mu^{\ \nu}\, p_\nu$$

Jere $\Lambda_\mu^{\ \nu}$ is the 4 × 4 matrix defining the Lorentz trans-ormation on the space-time manifold. The Wigner rotation, xpressed in terms of the corresponding Euler angles for xample, is determined by the eq. (2.22), which holds for rbitrary representations and hence also for the ordinary ector representation from which the angles are often onveniently determined. Inserting the decomposition (2.22) e obtain

$$L(\Lambda)|\bar{p},\lambda\rangle = L(\Lambda\bar{p})\; R(\bar{p},\Lambda)\; L^{-1}(\bar{p})|\bar{p},\lambda\rangle =$$

$$= \sum_{\lambda'=-s}^{+s} D^{(s)}_{\lambda'\lambda}\,[R(\bar{p},\Lambda)]\;|\Lambda\bar{p},\lambda'\rangle \qquad (2.23)$$

Clearly the decomposition (2.22) corresponds to (i) a transformation to the rest system, (ii) a rotation $R(\bar{p},\Lambda)$ and (iii) a Lorentz transformation from rest to the final momentum $\Lambda\bar{p}$.

Besides the plane wave states we shall also define state vectors of definite angular momentum and establish their relation to the plane wave states. To this end we define

$$|j,m,p,\lambda\rangle \equiv \frac{\sqrt{2j+1}}{\sqrt{4\pi}} \int d\omega\; D^{(j)*}_{m\lambda}(\phi,\theta,0)\,|\phi,\theta,p,\lambda\rangle \qquad (2.24)$$

where $d\omega = d(\cos\theta)d\phi$. It is easy to demonstrate that the vector $|j,m,p,\lambda>$ under a rotation transforms according to the rule

$$R|j,m,p,\lambda> \; = \sum_{m'=-j}^{+j} D^{(j)}_{m'm}\,[R]\;|j,m',p,\lambda> \qquad (2.25)$$

and hence $|j,m,p,\lambda>$ carries angular momentum (j,m), These vectors correspond to the following set of commuting observables J^2, J_3, $p = |\bar{p}|$ and $\bar{J}.\bar{p}/|\bar{p}|$. To invert the eq. (2.24) we note the following orthonormality relation

$$\frac{2j+1}{4\pi}\int d\omega D^{(j)^*}_{m\;n}(\phi,\theta,0)\; D^{(j')}_{m'n'}(\phi,\theta,0) = \delta_{jj'}\delta_{mm'}\delta_{nn'}$$

which provides the means to derive the partial wave expansion for the plane wave state vector $|\bar{p},\lambda>$.

$$|\bar{p},\lambda> \; = \sum_{j=0}^{\infty}\;\sum_{m=-j}^{+j} \frac{\sqrt{2j+1}}{\sqrt{4\pi}}\, D^{(j)}_{m\lambda}(\phi,\theta,0)|j,\, m,p,\lambda> \qquad (2.26)$$

An arbitrary normalizable single particle state vector $|\psi>$ can always be expanded in whichever basis is more convenient. For example, it may be expressed as a plane wave expansion

$$|\psi> \; = \sum_{\lambda}\int \frac{d^3p}{p_o}\,\psi_\lambda(\bar{p})|\bar{p},\lambda> \qquad (2.27)$$

where the weight function $\psi_\lambda(\bar{p})$ must be "square integrable" in order that $|\psi>$ be normalizable. From the eq. (2.20) we find

$$<\psi|\psi> \; = \sum_{\lambda}\int \frac{d^3p}{p_o}\,\psi^*_\lambda(\bar{p})\;\psi_\lambda(\bar{p}) < \infty \qquad (2.28)$$

Before we proceed to two-particle states etc. we once more emphasize that the previous discussion was restricted to the case $m>0$. Of physical importance is also the case m=o since there exist massless particles. For m=o there exists no rest system and the previous analysis fails. Instead we may choose as a standard state the one for which $p^{\mu} = (p,0,0,p)$, that is, the particle travels along the positive z-axis. Starting from this state we may generate all other physical states of a massless particle by means of appropriate Lorentz transformations. We further note that a UIR is characterized by m=o in this case but not any more by the spin invariant $W_{\mu}W^{\mu}$ for the physically important case where $W_{\mu}W^{\mu} = 0$. Instead one finds $W_{\mu} = \lambda p_{\mu}$, where λ is the helicity eigenvalue, and λ is the second label characterizing the UIR's. Thus we denote the standard state vector $|(0,\lambda);0,0,p'\rangle$ and define an arbitrary state by

$$|(0,\lambda);\phi,\theta,p\rangle \equiv R(\phi,\theta,0)L_z(v)|(0,\lambda);0,0,p'\rangle \qquad (2.29)$$

with

$$p^{\mu} \equiv \{R(\phi,\theta,0)\; L_z(v)\}^{\mu}_{\nu}\; p'^{\nu}$$

Since we shall not consider states of massless particles in any detail we do not pursue the subject any further.

Returning to the case of particle states of mass m>o we note that experimental arrangements for the study of particle processes most frequently permit a rather accurate determination of individual momenta. From this point of view it would seem most natural to deal with, say, two-particle state vectors of the type

$$|\bar{p}_1,\lambda_1;\bar{p}_2,\lambda_2\rangle \equiv |\bar{p}_1,\lambda_1\rangle \otimes |\bar{p}_2,\lambda_2\rangle \qquad (2.30)$$

These state vectors span a space for which the eigenvalue of

$$P_{\mu}P^{\mu} \equiv (P_{1\mu} + P_{2\mu})(P_1^{\mu} + P_2^{\mu})$$

$$|J,M,p,\lambda_1,\lambda_2\rangle = \frac{\sqrt{2J+1}}{\sqrt{4\pi}} \int d^2\omega \, D^{(J)*}_{M\lambda}(\phi,\theta,0)|\phi,\theta,p,\lambda_1,\lambda_2\rangle \tag{2.35}$$

which is the analogue of the eq. (2.24). The inverse of the eq. (2.35) reads

$$|\phi,\theta,p,\lambda_1,\lambda_2\rangle =$$

$$= \sum_{J=0}^{\infty} \sum_{M=-J}^{+J} \frac{\sqrt{2J+1}}{\sqrt{4\pi}} D^{(J)}_{M\lambda}(\phi,\theta,0)|J,M,p,\lambda_1,\lambda_2\rangle \tag{2.36}$$

and $\lambda = \lambda_1 - \lambda_2$. This is the partial wave expansion of the two-particle plane wave state vector.

For many discussions of selection rules such as parity conservation a somewhat different angular momentum basis is more convenient. It corresponds to coupling first the two individual spins $\bar{s}_1$ and $\bar{s}_2$ to a resultant spin $\bar{\sigma}$, and then $\bar{\sigma}$ is coupled to the relative orbital angular momentum $\bar{\ell}$. In the center of mass system one then has $\bar{J} = \bar{\ell} + \bar{\sigma}$. Thus, we define the vector $|J,M,p,\ell,\sigma\rangle$ by the relation

$$|J,M,p,\ell,\sigma\rangle =$$

$$= \sum_{\lambda_1\lambda_2} \frac{\sqrt{2\ell+1}}{\sqrt{2J+1}} \langle \ell\sigma 0\lambda|J\lambda\rangle\langle s_1 s_2 \lambda_1 -\lambda_2|\sigma\lambda\rangle|J,M,p,\lambda_1,\lambda_2\rangle \tag{2.37}$$

where $\langle j_1 j_2 m_1 m_2|jm\rangle$ is a Clebsch-Gordan coefficient with phase conventions as given for example in [1].

We end this section with some brief comments on many-particle state vectors. Forming the tensor product of single particle plane wave state vectors we obtain n-particle plane wave state vectors

$$|\bar{p}_1,\lambda_1;\ldots\bar{p}_n,\lambda_n\rangle \equiv |\bar{p}_1,\lambda_1\rangle|\bar{p}_2,\lambda_2\rangle\ldots|\bar{p}_n,\lambda_n\rangle \tag{2.38}$$

out of which we may construct arbitrary normalizable state vectors $|\psi>$

$$|\psi> = \sum_{\lambda_1,\dots,\lambda_n} \int \frac{d^3p_1}{p_1^o} \dots \frac{d^3p_n}{p_n^o} \psi_{\lambda_1\dots\lambda_n}(\bar{p}_1,\dots,\bar{p}_n) \times$$

$$\times |\bar{p}_1,\lambda_1;\dots\bar{p}_n,\lambda_n> \tag{2.39}$$

provided, the weight functions are "square integrable", that is,

$$\sum_{\lambda_1,\dots\lambda_n} \int \frac{d^3p_1}{p_1^o} \dots \frac{d^3p_n}{p_n^o} |\psi_{\lambda_1\dots\lambda_n}(\bar{p}_1,\dots,\bar{p}_n)|^2 < \infty \tag{2.40}$$

Just as in the case of the two-particle Hilbert space the tensor product of n single particle Hilbert spaces may be decomposed into irreducible subspaces so that

$$\mathcal{H}_1(m_1,s_1) \otimes \mathcal{H}_2(m_2,s_2) \dots \otimes \mathcal{H}_n(m_n,s_n) =$$

$$= \int dM \sum_{S,a} \mathcal{H}(M,S;a) \tag{2.41}$$

For a n-particle state there are in general (4n-6) degeneracy labels a, some of which range over a continuous spectrum with the summation in (2.41) then replaced by an appropriate integration. The (4n-6) additional quantum numbers which are necessary to label the state vector can be chosen in many different ways. We just note that the method used in the two-particle case lends itself to an immediate generalization. First one forms the tensor product of the Hilbert space of particle 1 and 2 and decomposes it into its irreducible parts. Each of the components in this reduction is then coupled to the Hilbert space of particle 3 and the corresponding product space is once again decomposed into its irreducible parts. Physically this procedure corresponds to a successive coupling of n particles into states of definite relative angular momentum and invariant masses. Thus, first one considers the particles 1 and 2 in their center of mass system and one forms states of a definite relative angular momentum j_{12} and invariant mass m_{12}. Then one performs a suitable Lorentz transformation to the center of mass of the systems (12) and 3 and there these subsystems are coupled to states of definite relative angular momentum j_{123} and invariant mass m_{123} and so on. Of course, in actual computations this procedure becomes elaborate with increasing value of n. For most applications one has n=3 and it is not too difficult a task to carry out the analysis.

2.3. Invariance Principles

An invariance principle or symmetry operation applied to a physical system is a one to one correspondence (mapping) which assigns to each physically realizable state $|\alpha\rangle$ another state $|\alpha'\rangle$ so that all transition probabilities are preserved, that is

$$w(\beta',\alpha') = w(\beta,\alpha) \tag{2.42}$$

or

$$\langle\alpha'|\beta'\rangle\langle\beta'|\alpha'\rangle = \langle\alpha|\beta\rangle\langle\beta|\alpha\rangle \tag{2.43}$$

The same symmetry operation will map the ring of observables for the system onto itself and in fact it is not until the action on the observables has been established that the symmetry principle has any physical content. It has been shown by Wigner [2] that one can reduce the possible solutions of the eq. (2.43) to two alternatives

(i) operations realized by unitary operators

(ii) operations realized by antiunitary operators.

The case of unitary symmetry operations corresponds to the identification

$$\langle\beta'|\alpha'\rangle = \langle\beta|\alpha\rangle \tag{2.44}$$

and

$$|\alpha'\rangle = U|\alpha\rangle \equiv |U\alpha\rangle \tag{2.45}$$

with

$$U^{+} = U^{-1}$$

$$\langle\beta|U^{\dagger}|\alpha\rangle = \langle U\beta|\alpha\rangle = \langle\alpha|U|\beta\rangle^{\dagger} \tag{2.46}$$

For an arbitrary operator Ω one obtains

$$\langle\beta|\Omega|\alpha\rangle = \langle\beta|U^{-1}U\Omega U^{-1}U|\alpha\rangle = \langle\beta'|\Omega'|\alpha'\rangle \tag{2.47}$$

with

$$\Omega' = U\Omega U^{-1} \tag{2.48}$$

which is the transformation law for the operators. The eq. (2.47) implies that the matrix elements of Ω' between the transformed vectors equal the matrix elements of Ω for the original vectors. The operator is said to be invariant under the symmetry operation if

$$\Omega' = \Omega; \quad [U,\Omega] = 0 \tag{2.49}$$

and hence

$$\langle\beta'|\Omega|\alpha'\rangle = \langle\beta|\Omega|\alpha\rangle \qquad (2.50)$$

so that the matrix elements are the same whether calculated in the primed or the unprimed system.

The case of an antiunitary (≡ unitary and antilinear) symmetry operation corresponds to

$$\langle\beta'|\alpha'\rangle = \langle\alpha|\beta\rangle \qquad (2.51)$$

and

$$|\alpha'\rangle = A|\alpha\rangle \equiv |A\alpha\rangle$$

with

$$A^{\dagger} = A^{-1}$$

$$A\lambda|\alpha\rangle = \lambda^{*}A|\alpha\rangle \;;\; \lambda \equiv \text{complex number}$$

$$\langle\beta|A^{\dagger}|\alpha\rangle = \langle A\beta|\alpha\rangle^{*} = \langle\alpha|A|\beta\rangle \qquad (2.52)$$

Note especially the complex conjugation in the definition of the hermitian conjugate operator in the last equation above. For an arbitrary operator Ω one finds

$$\langle\beta|\Omega|\alpha\rangle = \langle\beta|A^{-1}A\Omega A^{-1}A|\alpha\rangle =$$

$$= \langle\beta'|A\Omega A^{-1}|\alpha'\rangle^{*} = \langle\alpha'|(A\Omega A^{-1})^{\dagger}|\beta'\rangle = \langle\alpha'|\Omega'|\beta'\rangle \qquad (2.53)$$

with

$$\Omega' = (A\Omega A^{-1})^{\dagger} \qquad (2.54)$$

The eq. (2.53) implies that the matrix element of Ω' taken between $|\alpha'\rangle$ and $|\beta'\rangle$ equals the matrix element of Ω between $|\beta\rangle$ and $|\alpha\rangle$ respectively. Note the reversed order of "initial" or"final" states. The operator is said to be invariant if

$$\Omega' = \Omega; \quad \Omega^{\dagger} = A\Omega A^{-1} \tag{2.55}$$

which is the analogue of (2.49) for the unitary case.

The concept of symmetry has been introduced in a somewhat formal way and it may be instructive to consider an explicit example in some detail. We choose the principle of Lorentz invariance, that is, invariance under proper orthochronous Lorentz transformations (Poincaré transformations). This principle asserts that the laws of physics are invariant under Poincaré transformations relating the space-time description of a physical event as observed from two different inertial frames. This is known as the passive formulation of the relativity principle. From the previous discussion it is clear that we may rephrase the principle in the following way. There exists a unitary operator mapping the state vectors and the observables representing the physical system in the first reference frame to those of the second reference frame (observer). There is an alternative point of view, the so-called active formulation of the symmetry principle, in which one instead considers one and the same observer as he observes two physical events which are related by the same Poincaré transformation as the two observers in the passive formulation. Thus, a symmetry principle in the active formulation relates the observations from two different experiments corresponding to changes in the experimental arrangements. The active formulation then implies that if $|\alpha\rangle$ is a possible state of a physical system, then $|\alpha'\rangle = U|\alpha\rangle$, where U is a unitary representation of the Poincaré transformation, is also a possible state for the system as seen by the same observer. The analogous statement holds for arbitrary symmetry operators. We note in passing that only unitary and no antiunitary operator can represent the Poincaré transformation since these are continuously connected to the identity transformation, which must be represented by the unit operator. Since the Poincaré transformations are of geometrical nature it is experimentally feasible to test the relativity principle both in the active and the passive formulation and to establish the equivalence between the two descriptions. For other transformations such as time reversal one may only test the invariance principle in the active formulation since we cannot realize the "inverted" observer. Nevertheless, in the passive manner one may always formulate hypothetical invariance principles involving such "pathological" transformations, but they have no physical relevance unless they represent a true symmetry of the system. If not they are merely mathematical operations. Correspondingly, in the active formulation one finds that the operations, which do not constitute symmetry operations of the system, will transform physically realizable states into states which do not occur in nature, that is, the corresponding state vectors lie outside the physical Hilbert space. Mathematically this implies that the operation cannot be represented by a unitary or an antiunitary operator on the physical Hilbert space, since it leads out of the space. A well known example of this phenomenon occurs in the case of space reflection transformations on states representing a neutrino for which only one of the two conceivable helicity states is realized in nature. A space reflection transforms a state of, say, negative helicity into a state of positive helicity. Of course, one may enlarge the physical Hilbert space to include also these non-physical states and in that way make the symmetry operation unitary. However, as previously stated this is a mathematical construction to which we can attach no physical meaning.

From the discussion above it is clear that the active formulation of symmetry principles is most easily subjected to experimental tests. For this reason we shall consistently adopt the active point of view, and hence we shall use symmetry principles to relate observations, which one may make in separate experiments corresponding to changes in the experimental set-ups.

2.4. The S-operator and invariance properties of the S-operator

In section 2.1. we concluded that the probability distribution $w(\beta,\alpha)$ in general depends on the space-time domains Ω and Ω' for the preparation of the initial state $|\alpha(\Omega)\rangle$ and the later measurement establishing that the system is in the state represented by $|\beta(\Omega')\rangle$. This dependence has been suppressed in the discussion of the relativistic description of particle states to simplify the notations, but it must clearly be incorporated in any scheme dealing with interacting particles.

To do so we note that elementary particle experiments deal with processes where the initial state and the final state consist of free, that is, noninteracting particles. We shall refer to these regions where the particles are free as the asymptotic regions. We may take the asymptotic region for the initial state to be the limit $t \to -\infty$ and for the final state to be the limit $t \to +\infty$. This is an idealization since an experiment only extends over a finite time. In practice the limits $t \to \pm\infty$ are replaced by $|t| \gg |t_o|$, where t_o is the characteristic extension in time for the interaction region. With this in mind we write the fundamental eq.

$$w(\beta,\alpha) = |\langle\beta;\ \text{out}|\alpha;\ \text{in}\rangle|^2 \tag{2.56}$$

with the labels "out" and "in" referring to the asymptotic states.

For the no-particle state, the vacuum state $|0\rangle$, and the single particle states the distinction between in- and out-states is superfluous since nothing can happen; they propagate freely. For states of two or more particles a distinction is necessary, however, since the particles may scatter before they reach the asymptotic out-region. Hence, in general we have

$$|\alpha;\ \text{in}\rangle \neq |\alpha\ ;\ \text{out}\rangle$$

On the other hand in the in-states as well as the out-states span the same physical Hilbert space. We may use the same complete set of commuting observables to label the free particle states out of which the many-particle states are formed. Hence, the in-states and the out-states provide us with two different sets of basis vectors and they must be related by a unitary transformation. We put

$$|\alpha;\ \text{in}\rangle = S|\alpha\ ;\ \text{out}\rangle \tag{2.57}$$

for any α. The unitarity of the S-operator implies that

$$S^{\dagger} = S^{-1}$$

and we shall see below that this property leads to conservation of probability. The eq. (2.57) may be written in the following way

$$|\alpha;\ \text{in}\rangle = \sum_{\beta} |\beta;\ \text{out}\rangle \langle\beta;\ \text{out}|S|\alpha;\ \text{out}\rangle =$$

$$= \sum S_{\beta\alpha} \, |\beta; \text{out}> \tag{2.58}$$

with

$$S_{\beta\alpha} \equiv <\beta; \text{out}|S|\alpha; \text{out}> \tag{2.59}$$

The eq. (2.58) implies that the probability to find the initial state $|\alpha; \text{in}>$ in the final outgoing channel $|\beta; \text{out}>$ is given by

$$w(\beta,\alpha) = |S_{\beta\alpha}|^2 \tag{2.60}$$

This result is clearly consistent with the eq. (2.56), since (2.60) is obtained by inserting the definition (2.57) into (2.56). In passing we note that we may use the S-operator to rewrite the eq. (2.59) in the following way

$$<\beta; \text{out}|S|\alpha; \text{out}> = <\beta; \text{in}|SSS^{-1}|\alpha; \text{in}> =$$

$$=<\beta;\text{in}|S|\alpha; \text{in}> \tag{2.61}$$

Thus we may use either in-states or out-states to evaluate the S-matrix elements. For this reason we shall often suppress the labels "in" and "out" when they are not essential for the discussion. Returning to the question of probability conservation we may express it in the following way: the total probability that the initial state $|\alpha; \text{in}>$ at the end is found in some channel $|\beta;\text{out}>$ must be unity so that

$$\sum_\beta w(\beta,\alpha) = 1 \tag{2.62}$$

or

$$\sum_\beta <\alpha|S^\dagger|\beta><\beta|S|\alpha> = 1 \tag{2.63}$$

The summation over β extends over a complete set and hence

$$\langle\alpha|S^{\dagger}S|\alpha\rangle = 1 \qquad (2.64)$$

and this relation must hold for any state vector $|\alpha\rangle$. Furthermore, the operator $S^{\dagger}S$ is clearly hermitian. It then follows that [3]

$$S\,S^{\dagger} = 1 \qquad (2.65)$$

or

$$S^{\dagger} = S^{-1} \qquad (2.66)$$

that is, the S-operator must be unitary. Clearly the converse is also true; a unitary S-operator implies conservation of probability as expressed by the eq. (2.62).

It is clear from the discussion above that the S-operator really contains all physical information about particle processes. If we knew the S-operator we could proceed to compute the outcome of any experiment, that is, we can predict transition rates and cross sections. The explicit relation between the S-matrix elements and these transition rates is given in all standard textbooks and we shall omit it here since we shall not make use of these expressions.

We next proceed to consider the implications of symmetry properties of the S-operator.

If we adopt the active point of view we know from the previous discussion that symmetry operations for the S-operator yield relations between the appropriate S-matrix elements corresponding to different experimental situations. From this we may then derive relations between observable quantities and these predictions may be tested in experiments. It is important to realize that these predictions are derived without any further information about the S-operator beyond the assumed symmetry property. Of course, if we have an explicit form for the S-operator we can check its symmetry properties explicitly. This is the case if we have field-theoretic models for example. If we do not have the explicit form for the S-operator we may still make the assumption that it is invariant under certain symmetry transformations and we can immediately derive predictions as stated above. Of course, in particle physics we do not know the S-operator and hence we are obliged to adopt this latter attitude towards symmetries.

Under special circumstances symmetry principles provide even more powerful tools than relations between matrix elements of the S-operator. We have in mind selection rules. To see how this comes about we note that a unitary operator U may or may not be hermitian. If it is not, one may always define a hermitian operator H by the relation

$$U = \exp[i\,H] \tag{2.67}$$

For continuous symmetry groups this operator H is a function of the group parameters and

$$\Omega_i = \frac{\partial H(\phi_1, \ldots, \phi_n)}{\partial \phi_i} \tag{2.68}$$

is the infinitesimal generator corresponding to the group parameter ϕ_i. The relevance of considering hermitian operators related to symmetry operations is clearly based on the fact that these may represent observables. Now assume that

(i) H is (related to) a symmetry operation for the S-operator, that is

$$[H, S] = 0$$

(ii) $|\alpha\rangle$ and $|\beta\rangle$ are eigenstates of H so that

$$H|\alpha\rangle = h_\alpha|\alpha\rangle$$

$$H|\beta\rangle = h_\beta|\beta\rangle$$

From these assumptions we obtain

$$h_\alpha\langle\beta|S|\alpha\rangle = \langle\beta|SH|\alpha\rangle = \langle\beta|HS|\alpha\rangle = h_\beta\langle\beta|S|\alpha\rangle \tag{2.69}$$

and hence

$$(h_\alpha - h_\beta)\langle\beta|S|\alpha\rangle = 0 \tag{2.70}$$

so that

$$\langle\beta|S|\alpha\rangle \neq 0 \tag{2.71}$$

requires

$$h_\alpha = h_\beta \tag{2.72}$$

Thus, the matrix element of the S-operator may be non-vanishing only if the initial and the final states correspond to the same eigenvalue of H. This is a conservation law or a selection rule for the quantum number h. Note that both the assumptions (i) and (ii) are essential for the derivation. This is the reason why eigenstates of symmetry operations are particularly important as we shall see in the following sections.

3. Definition of the Discrete Transformations P, C and T

In this section we shall define the operation of space reflection (P), charge conjugation (C), and time reversal (T). We shall further give the transformation rules for the various state vectors which were introduced in section 2.2. Some of these transformation rules will be needed for applications later.

The space reflection and the time reversal operations are defined as transformations on the space time manifold. Therefore, they have classical analogues. The charge conjugation operation is not of geometrical nature. Beyond the basic transformations P, C and T various combined operations such as CP and CPT are of importance for physical applications. Since they consist of successive applications of the fundamental operations we shall not consider them further in this section but defer a discussion of them until later. Of course, the transformation laws of state vectors under, say, CP and CPT can be read off from the transformation rules which will be listed below.

3.1. Definition of P and its action on state vectors

From the passive point of view the space reflection operation is defined by the coordinate transformation

$$(x_o, \bar{x}) \xrightarrow{P} (x'_o, \bar{x}') = (x_o, -\bar{x}) \tag{3.1}$$

It corresponds to a description of the same physical event in a righthanded respectively a lefthanded coordinate system. In quantum theory it is implementable by a unitary and hermitian operator which we shall denote by P. Since the transformation is of classical origin we know its action on various observables. For example it should satisfy the following conditions

$$P x_\mu P^{-1} = \varepsilon(\mu)\, x_\mu$$

$$P P_\mu P^{-1} = \varepsilon(\mu) P_\mu$$

$$P J_{\mu\nu} P^{-1} = \varepsilon(\mu)\, \varepsilon(\nu) J_{\mu\nu} \qquad (3.2)$$

where x_μ is the coordinate operator, P_μ the momentum operator and $J_{\mu\nu}$ the angular momentum operator, or, more precisely, $J_{k\ell}$ for $k,\ell = 1,2,3$ is an angular momentum operator while J_{ok} is the generator of a pure Lorentz transformation along the k-axis. The symbol $\varepsilon(\mu)$ is defined in the following way (cf. appendix 1):

$$\varepsilon(o) = +1 \quad ; \quad \varepsilon(k) = -1$$

It is immediately seen that, provided P is a unitary operator, the mapping (3.2) leaves the commutation rules of the Poincaré algebra invariant, that is, the mapping is an automorphism (an outer automorphism). This is clearly a necessary requirement that any symmetry operation must satisfy. We also note that it is not possible to satisfy this requirement if P is antiunitary. Since it is unitary we further conclude from the eq. (3.2) that

$$P^2 = \lambda I \qquad (3.3)$$

and hence by a suitable phase choice the operator P is hermitian so that

$$P^{-1} = P^\dagger = P \qquad (3.4)$$

and its possible eigenvalues are ± 1 .

In adopting the active point of view it is seen from the eq. (3.2) that P transforms a state vector of momentum $\bar{p}$ and helicity λ into a state vector of momentum $-\bar{p}$ and helicity $-\lambda$ (it reverses momenta and leaves the "spin" unchanged). If we anticipate that P is not a general symmetry operation we must bear in mind that acting with P on certain state vectors (for example a neutrino state vector) leads out of the physical Hilbert space. However, only the weak interactions seem to violate space reflection invariance and for matrix elements representing strong or electromagnetic processes the requirements imposed by P invariance are satisfied to the extent that the very small weak contributions are neglected. This makes P invariance a very useful concept for a large class of processes.

After these introductory remarks we list below the transformation properties of the state vectors which were introduced previously in section 2.2. For the single particle state vectors we have

$$P|\bar{p},\lambda\rangle = \eta(P)\, \exp[-i\pi s]\, |-\bar{p},-\lambda\rangle$$

$$P|j,m,p,\lambda\rangle = \eta(P) \exp[i\pi(j-s)]\, |j,m,p,-\lambda\rangle \qquad (3.5)$$

while the corresponding relations for the two particle state vectors are given by

$$P|p,J,M,\lambda_1,\lambda_2\rangle =$$

$$= \eta^{(1)}(P)\eta^{(2)}(P) \exp[i\pi(J-s_1-s_2)]\, |p,J,M,-\lambda_1,-\lambda_2\rangle$$

$$P|p,J,M,\ell,\sigma\rangle =$$

$$= \eta^{(1)}(P)\eta^{(2)}(P)(-1)^{\ell}\, |p,J,M,\ell,\sigma\rangle \qquad (3.6)$$

The factors $\eta(P)$ in (3.5) and (3.6) denote the intrinsic parities of the particles. We shall only derive the first one of the eqs. (3.5) to demonstrate the procedure. From that particular rule the other ones may be derived in a straightforward fashion.

Consider the state vector $|\bar{p},\lambda>$ at rest. We write it $|\bar{p}=o,\lambda> \equiv |0,0,0,\lambda>$ where λ is the third component of the spin. Thus $|0,0,0,\lambda>$ is an eigenvector of $J^2 = (J_1^2 + J_2^2 + J_3^2)$ and J_3 . Since P commutes with $J_{k\ell} = \varepsilon_{k\ell m} J_m$ we have

$$P|0,0,0,\lambda> = \eta(P)|0,0,0,\lambda> \tag{3.7}$$

Furthermore $[P,J_\pm] = 0$ so that $\eta(P)$ must be independent of λ. The phase factor $\eta(P)$ is the intrinsic parity of the state. Next we consider the following operator

$$R_{xz} \equiv R(0,\pi,0)P \tag{3.8}$$

representing a reflection in the xz plane. Clearly

$$[R_{xz},L_z(v)] = 0 \tag{3.9}$$

and from (3.7) and (3.8) we conclude

$$R_{xz}|0,0,0,\lambda> = \eta(P)R(0,\pi,0)|0,0,0,\lambda> =$$

$$= \eta(P) \sum_{\lambda'} D^{(s)}_{\lambda'\lambda}(0,\pi,0)|0,0,0,\lambda'> \tag{3.10}$$

However

$$D^{(s)}_{\lambda'\lambda}(0,\pi,0) = (-1)^{s+\lambda'} \delta_{\lambda',-\lambda} \tag{3.11}$$

and thus

$$R_{xz}|0,0,0,\lambda\rangle = \eta(P)(-1)^{s-\lambda}|0,0,0,-\lambda\rangle \tag{3.12}$$

so that

$$P|0,0,p,\lambda\rangle = R^{-1}(0,\pi,0)\ R_{xz}L_z(v)|0,0,0,\lambda\rangle =$$

$$= \eta(P)(-1)^{s-\lambda}R^{-1}(0,\pi,0)|0,0,p,-\lambda\rangle \tag{3.13}$$

Since

$$R(\phi,\theta,0)\ R^{-1}(0,\pi,0) = R(\phi+\pi,\pi-\theta,-\pi) \tag{3.14}$$

and since P commutes with $R(\phi,\theta,0)$ we conclude from (3.13) (cf. the definition of $|\phi,\theta,p,\lambda\rangle$)

$$P|\phi,\theta,p,\lambda\rangle = \eta(P)(-1)^{s-\lambda}R(\pi+\phi,\pi-\theta,0)\ \times$$

$$\times\ \exp[+i\pi J_3]|0,0,p,-\lambda\rangle =$$

$$= \eta(P)\exp[-i\pi s]|\phi+\pi,\pi-\theta,p,-\lambda\rangle \tag{3.15}$$

which is the result which was previously quoted in the eq. (3.5).

We end this section with some general remarks. In the ordinary coordinate space a space reflection followed by a rotation $R(2n\pi)$ around an arbitrary axis yields the same result as P provided n is integer. Thus let us define a new space reflection transformation

$$P' = R(2n\pi)P\ ; \qquad n \text{ integer} \tag{3.16}$$

If we apply P' to an arbitrary single particle state, which we for simplicity denote $|(m,s)\rangle$, we obtain

$$P'|(m,s)\rangle = \exp[i2\pi ns]P|(m,s)\rangle \tag{3.17}$$

and hence

$$P' = P \quad \text{for integer } s$$

$$P' = (-1)^n P \quad \text{for half-integer } s$$

This is the explanation why one may define the intrinsic parities for integer-spin particles (bosons) while for half-integer spin particles (fermions) one may only define relative intrinsic parities. In this context we also note that in order to make the Dirac formalism invariant under the space reflection transformation one must choose odd relative intrinsic parity for a particle and its antiparticle, that is

$$\eta^{(1)}(P)\eta^{(2)}(P) = -1 \tag{3.18}$$

if 1 and 2 represent a particle of spin 1/2 and its antiparticle.

3.2. Definition of C and its action on state vectors

The classical theory of electromagnetism can be derived from (i) Coulomb's law for the force between charges and (ii) the theory of special relativity. In Coulomb's law there is full symmetry between positive and negative charges; equal charges repel each other while unequal charges attract each other. The theory of special relativity concerns exclusively space-time concepts which have no immediate relation to the concept of charge and hence it is clear that the classical theory exhibits a basic symmetry between the two kinds of electric charge.

When quantum theory was developed it was found that to each kind of particles there must exist antiparticles. Furthermore, the space-time properties of a particle and its antiparticle must be the same. Electromagnetism was incorporated in this scheme following the classical procedure and it then turned out that a consistent theory required that particle and antiparticle carry opposite charge. In this way one was led to consider the charge symmetry more appropriately as a particle-antiparticle symmetry. For historical reasons one retained the name charge conjugation symmetry also in the quantum theory although particle-antiparticle conjugation would be a more appropriate name for it.

The action of the charge conjugation operation on any state vector is to replace the particles by their antiparticles. Since particles and antiparticles have the same space-time properties it follows that the charge conjugation operator must commute with all the generators of the Poincaré group and with the space reflection and the time reversal operations. On the other hand charge, baryon number, lepton number and strangeness have the opposite sign for particle and antiparticle, and the corresponding operators must then anticommute with C. On the basis of this observation we introduce the following classification: those operators which commute with C are said to be of the first class while operators which anticommute with C are said to be of the second class

Since C commutes with all the generators of the Poincaré algebra it must be represented by a unitary operator in order to leave the commutation relations of P invariant. Furthermore, acting with C on a state $|\alpha\,;\bar{p},\lambda\rangle$, where α denotes all the labels corresponding to second class observables (electric charge etc.), one obtains

$$|\alpha;\bar{p},\lambda\rangle \xrightarrow{C} |\alpha';\bar{p},\lambda\rangle = C|\alpha;\bar{p},\lambda\rangle = \eta(C)\ |-\alpha;\bar{p},\lambda\rangle \qquad (3.19)$$

where $\eta(C)$ is a phase factor since C is a unitary operator. Acting twice with C one finds

$$C^2|\alpha;\bar{p},\lambda\rangle = \kappa|\alpha;\bar{p},\lambda\rangle \tag{3.20}$$

where κ is a phase factor. By adjusting the phase of C conveniently we may then obtain $\kappa = 1$ and hence

$$C^2 = 1 \ ; \quad C = C^{-1} = C^{\dagger} \tag{3.21}$$

Thus, also C may be represented by a unitary and hermitian operator. The eigenvalues of C are then restricted to ± 1.

From the previous discussion it clearly follows that eigenstates of C can only those states be for which all second class charges α are zero. For single particle states this is a very restrictive condition since $Q = B = L = S = 0$ leaves very few candidates of which the photon, the π^{o} meson and the η meson are the most important. To determine the charge parity of the photon we recall that the whole concept of charge conjugation invariance originated from the electromagnetic interactions which display invariance under the operation. This interaction may be expressed as the coupling of a particle current to the electromagnetic field of which the photon is the elementary quantum. Under C the particles are transformed into antiparticles and hence the current changes sign. In order that the interaction should remain invariant also the electromagnetic field must change sign and we conclude

$$C|\gamma\rangle = -|\gamma\rangle$$

and

$$C|n\gamma\rangle = (-1)^{n}|\gamma\rangle$$

Similar considerations yield

$$C|\pi^o> = |\pi^o>$$

$$C|\eta> = |\eta>$$

and hence any n-particle state of π^o mesons or η mesons has positive charge parity.

To obtain two-particle state vectors which are eigenstates of C we must clearly consider particle-antiparticle states such as $|e^+e^->$, $|p\bar{p}>$, $|\pi^+\pi^->$ etc. For states of this kind the generalized Pauli principle must be satisfied. This principle asserts that any fermion-antifermion state must be totally antisymmetric and any boson-antiboson state totally symmetric under the transformation which exchanges the two particles. In field theory this principle expresses the fact that creation and destruction operator satisfy (i) commutation relations for bosons, and (ii) anticommutation relations for fermions. If we denote by $A(\sigma,r)$ the operator which exchanges spin and space coordinates for the two particles, then the generalized Pauli principle may be expressed in the following way

$$C\ A(\sigma,r)|\alpha,\ \bar{\alpha}> = (-1)^{2s}|\alpha,\ \bar{\alpha}> \qquad (3.22)$$

where α and $\bar{\alpha}$ denote any particle of spin s and its antiparticle. It follows from (3.22) that any particle-antiparticle eigenstate of $A(\sigma,r)$ is at the same time an eigenstate of C. Of the two-particle state vectors introduced in section 2.2. $|p,J,M,\ell,\sigma>$ is an eigenvector of $A(\sigma,r)$. A closer examination yields

$$A(\sigma,r)|p,J,M,\ell,\sigma> = (-1)^{\ell+\sigma-2s}|p,J,M,\ell,\sigma> \qquad (3.23)$$

and hence for a particle-antiparticle state we obtain

$$C|p,J,M,\ell,\sigma\rangle = (-1)^{\ell+\sigma}|p,J,M,\ell,\sigma\rangle \quad (3.24)$$

which makes these basis vectors particularly useful for discussions of the charge conjugation operation.

3.3. Definition of T and its action on state vectors

The time reversal transformation is defined by

$$(x_o,\bar{x}) \xrightarrow{T} (x_o',\bar{x}') = (-x_o,\bar{x}) \quad (3.25)$$

In quantum theory it is implemented by an operator T. Since the definition of the transformation involves classical concepts we immediately know its action on observables related to space and time. It must satisfy

$$T\, x_\mu\, T^{-1} = -\varepsilon(\mu)\, x_\mu$$

$$T\, P_\mu\, T^{-1} = \varepsilon(\mu)\, P_\mu$$

$$T\, J_{\mu\nu} T^{-1} = -\varepsilon(\mu)\, \varepsilon(\nu)\, J_{\mu\nu} \quad (3.26)$$

provided the operators x_μ, P_μ and $J_{\mu\nu}$ are chosen to be hermitian. It is easily seen that the mapping (3.26) will not preserve the basic commutation rules of the Poincaré algebra unless T is chosen antiunitary. Consider for example the following commutation rule

$$[J_{k\ell}, P_k] = i\, P_\ell \quad (3.27)$$

Since both $J_{k\ell}$ and P_k change sign under T it follows that we obtain a change in sign for the commutation relation if T is unitary. For an antiunitary T this change of sign is cancelled by the additional complex conjugation of the imaginary unit in (3.27).

Since T is an antiunitary operator it does not represent an observable and we obtain no selection rules from it. It is obvious from the very definition of the time reversal transformation that (i) T^2 is a linear operator and (ii) T^2 applied to any state vector is a multiple of the unit operator. Therefore it commutes with all observables and we have

$$T^2 = \lambda I \qquad (3.28)$$

or

$$T = \lambda T^{-1} = \lambda T^{\dagger} \qquad (3.29)$$

The hermitian conjugate equation reads

$$T^{\dagger} = \lambda^{*} T \qquad (3.30)$$

which inserted in (3.29) yields $|\lambda|^2 = 1$ and we conclude that λ is a phase factor. If we multiply the eq. (3.29) from left respectively right by T and recall the antiunitarity of T we obtain

$$\lambda^{*} I = T^2 = \lambda I \qquad (3.31)$$

and thus $\lambda = \pm 1$. Since T commutes with the S-operator we conclude that λ is a constant of motion; an isolated system in a state of $\lambda = +1$ can never develop into a state of $\lambda = -1$ and vice versa. There is a profound difference between this result and the previous results that P^2 and C^2, by a suitable phase convention for P respectively C, may be identified with the unit operator. These latter results do not yield any selection rules while T^2 does. The selection rule obtained from

T^2 is called a superselection rule to distinguish it from other selection rules. A superselection rule expresses the fact that there are certain physical quantities which always have definite values for any physically realizable state vector, that is, any such state vector is an eigenvector of the corresponding operator. A superselection rule follows from the existence of an observable such as T^2 which is strictly conserved and which commutes with all other observables. Of course it is also assumed that the observable providing a superselection rule has a nontrivial eigenvalue spectrum. An ordinary selection rule follows similarly from the existence of an observable which is strictly conserved but it does not commute with all other observables. Mathematically the existence of a superselection rule leads to a decomposition of the Hilbert space into incoherent subspaces. In the case of T^2 one finds

$$\mathcal{H} = \mathcal{H}(\lambda = +1) \oplus \mathcal{H}(\lambda = -1) \tag{3.32}$$

It is easily seen that no observable can have non-vanishing matrix elements between states which belong to different incoherent subspaces. Consider for example an arbitrary observable Ω and the operator T^2. We have previously stated that $[T^2, \Omega] = 0$ and thus

$$\langle\lambda = +1|\Omega|\lambda = -1\rangle = \langle +|T^{2\dagger} T^2\Omega|-\rangle =$$

$$= \langle +|T^{2\dagger}\Omega T^2|-\rangle = -\langle +|\Omega|-\rangle \tag{3.33}$$

which can only hold provided $\langle +|\Omega|-\rangle = 0$. We shall not discuss this matter further in these lectures and we refer the reader to standard references on this subject [4].

We proceed to list the transformation properties of the state vectors under the time reversal transformation.

For single particle state vectors we have

$$T|\phi,\Theta,p,\lambda\rangle = \exp[-i\pi\lambda]\,|\phi+\pi,\pi-\Theta,p,\lambda\rangle$$

$$T|j,m,p,\lambda\rangle = (-1)^{j-m}|j,-m,p,\lambda\rangle \tag{3.34}$$

corresponding to a change in sign for momentum and spin. Similarly for the two particle states one finds

$$T|p,J,M,\lambda_1,\lambda_2\rangle = (-1)^{J-M}|p,J,-M,\lambda_1,\lambda_2\rangle$$

$$T|p,J,M,\ell,\sigma\rangle = (-1)^{J-M}|p,J,-M,\ell,\sigma\rangle \tag{3.35}$$

It is immediately seen from the eqs. (3.34) and (3.35) that states of angular momentum J are eigenstates of T^2 with the eigenvalue

$$T^2 \sim (-1)^{2J} \tag{3.36}$$

Because of the relation (3.36) one often refers to the superselection rule based on T^2 as conservation of the fermion number (no transitions from a state with an even fermion number to a state of odd fermion number or vice versa are possible).

4. Some Applications of P and C

Both the space reflection and the charge conjugation operations are represented by unitary and hermitian operators on the physical Hilbert space. If we assume that the S-operator or the effective S-operator describing a specific process is invariant under anyone of these transformations, then we obtain (cf. section 2.4.)

$$<\beta|S|\alpha> \; = \; <P\beta|S|P\alpha>$$

$$<\beta|S|\alpha> \; = \; <C\beta|S|C\alpha> \qquad (4.1)$$

These equations relate the matrix elements of the S-operator representing different experimental arrangements, and hence on the basis of the assumed symmetry properties of the S-operator we make the prediction that the corresponding transition rates should be the same. These predictions may clearly be subjected to tests and the outcome of such experiments will lend support to or invalidate the underlying symmetry assumption. We shall give examples below where symmetry principles have been applied in this way.

If the states $|\alpha>$ and $|\beta>$ happen to be eigenstates of, say, P, then

$$P|\alpha> \; = \; \eta^{(\alpha)}(P)|\alpha>$$

$$P|\beta> \; = \; \eta^{(\beta)}(P)|\beta> \qquad (4.2)$$

and hence the eq. (4.1) reads

$$<\beta|S|\alpha> \; = \; \eta^{(\alpha)}(P)\eta^{(\beta)}(P)<\beta|S|\alpha> \qquad (4.3)$$

From this we conclude that $<\beta|S|\alpha>$ may be nonvanishing only if

$$\eta^{(\alpha)}(P) \; \eta^{(\beta)}(P) \; = \; +1 \qquad (4.4)$$

that is, the initial and the final state must have the same parity. Entirely analogous results hold, of course, also for C.

We may very briefly summarize the present situation

with regard to the P and C invariance in the following way. There is no compelling evidence for any P or C violation in strong and electromagnetic processes. Weak interactions indicate maximal violation of both P and C invariance as expressed by the V-A theory. It has been of recent interest to test C invariance of virtual electromagnetic interactions for reasons to be explained in section 7.1.

4.1. P and C applied to the process $\nu_\mu+n\to\mu^-+p$

The process

$$\nu_\mu + n \to \mu^- + p \tag{4.5}$$

is a semi-leptonic weak process. It has been investigated experimentally in connection with the discovery that there exist two kinds of neutrinos denoted ν_μ and ν_e . Since neither the space reflection nor the charge conjugation symmetries are respected by the weak interactions we know that relations based on the eqs. (4.1) and (4.3) do not hold. Rather we shall demonstrate the difference between the matrix elements for the space reflected and the charge conjugated processes as compared to the matrix element for the process (4.5).

Let us first consider the process (4.5) and its space reflected counterpart. For notation we refer to figure 1. The most general form of the matrix element within the framework of the V-A theory has been given in the lecture by Prof. Pietschmann. We write it in the following way

$$\langle p,\mu^-|S|n,\nu_\mu\rangle = -\frac{iG}{\sqrt{2}}\frac{1}{(2\pi)^2}\delta^4(P_f-P_i)\frac{\sqrt{m_n m_p m_\mu m_\nu}}{\sqrt{k_o k'_o p_o p'_o}}h_\mu \ell^\mu \tag{4.6}$$

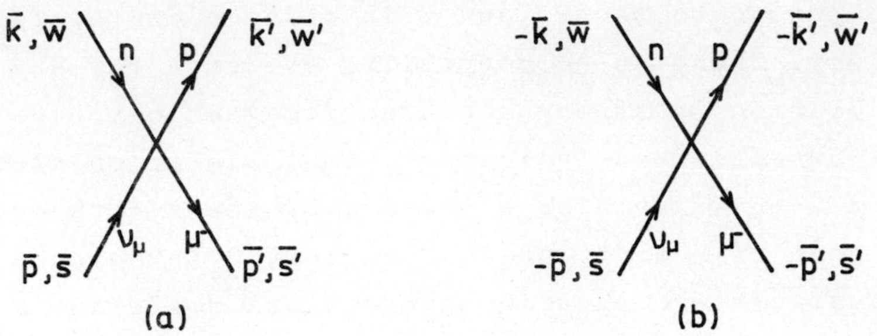

Fig. 1: Two configurations of the process $\nu_\mu + n \to \mu^- + p$ related by a space reflection transformation.

with

$$\frac{1}{(2\pi)^3}\frac{\sqrt{m_n m_p}}{\sqrt{k_o k_o'}}\, h_\mu \equiv \langle p|j_\mu^\dagger(0)|n\rangle$$

$$\frac{1}{(2\pi)^3}\frac{\sqrt{m_\mu m_\nu}}{\sqrt{p_o p_o'}}\, \ell_\mu \equiv \langle \mu^-|\ell_\mu(0)|\nu_\mu\rangle \tag{4.7}$$

so that

$$\begin{aligned} h_\mu &= \bar{u}(\bar{k}',\bar{w}')\{F_1(q^2)\gamma_\mu + iF_2(q^2)\sigma_{\mu\nu}q^\nu + \\ &\quad + F_3(q^2)q_\mu + [G_1(q^2)\gamma_\mu + \\ &\quad + iG_2(q^2)\sigma_{\mu\nu}q^\nu + G_3(q^2)q_\mu]\gamma_5\}u(\bar{k},\bar{w}) \\ \ell_\mu &= \bar{u}(\bar{p}',\bar{s}')\gamma_\mu(1+\gamma_5)u(\bar{p},\bar{s}) \end{aligned} \tag{4.8}$$

and $q_\mu = k_\mu' - k_\mu$. For convenience we introduce the abbreviated notation

$$h_\mu \ell^\mu \equiv \phi(F_i, G_i; \gamma_5) \tag{4.9}$$

We next write down the corresponding expression for the process (b) of figure 1. The relevant parts are

$$h_\mu(P) = \bar{u}(-\bar{k}',\bar{w}')\{F_1(q^2)\gamma_\mu + i\varepsilon(\nu)F_2(q^2)\sigma_{\mu\nu}q^\nu + \\ + \varepsilon(\mu)F_3(q^2)q_\mu + [G_1(q^2)\gamma_\mu + i\varepsilon(\nu) \times \\ \times G_2(q^2)\sigma_{\mu\nu}q^\nu + \varepsilon(\mu)G_3(q^2)q_\mu]\gamma_5\}u(-\bar{k},\bar{w}) \tag{4.10}$$

and

$$\ell_\mu(P) = \bar{u}(-\bar{p}',\bar{s}')\gamma_\mu(1+\gamma_5)\ u(-\bar{p},\bar{s}) \tag{4.11}$$

The factors $\varepsilon(\mu)$ in the eq. (4.10) account for the change in sign of the space components of the momentum transfer q_μ when $\bar{k} \to -\bar{k}$ and $\bar{k}' \to -\bar{k}'$. To rewrite the expression for $h_\mu(P)\ell^\mu(P)$ we make use of the following identities (cf. appendix 2)

$$u(-k,\bar{w}) = \gamma_0\ u(\bar{k},\bar{w})$$

$$\bar{u}(-\bar{k},\bar{w}) = \bar{u}(\bar{k},\bar{w})\gamma_0 \tag{4.12}$$

to obtain

$$\ell_\mu(P) = \varepsilon(\mu)\ \bar{u}(\bar{p}',\bar{s}')\gamma_\mu(1-\gamma_5)\ u(\bar{p},\bar{s}) \tag{4.13}$$

and

$$h_\mu(P) = \varepsilon(\mu)\ \bar{u}(\bar{k}',\bar{w}') \times \\ \times \{F_1(q^2)\gamma_\mu + iF_2(q^2)\sigma_{\mu\nu}q^\nu + F_3(q^2)q_\mu -$$

$$- \left[G_1(q^2)\gamma_\mu + iG_2(q^2)\sigma_{\mu\nu}q^\nu + G_3(q^2)q_\mu\right]\gamma_5\}u(\bar{k},\bar{w}) \tag{4.14}$$

In forming the product $h_\mu(P)\ell^\mu(P)$ the factors $\varepsilon(\mu)$ cancel out and we remain with

$$h_\mu(P)\ell^\mu(P) = \phi(F_i, G_i; -\gamma_5) \tag{4.15}$$

Since an overall change in sign for the S-matrix element does not affect the transition probability we conclude that it is the simultaneous presence of terms with and without γ_5 in both h_μ and ℓ_μ that makes the matrix element for the process (a) differ from that of the process (b).

We next consider the process which is related to the process (a) of figure 1 by a charge conjugation transformation. Using the notations of figure 2 we write

$$\langle\bar{p}|j_\mu(0)|\bar{n}\rangle = \frac{1}{(2\pi)^3}\frac{\sqrt{m_n m_p}}{\sqrt{k_o k'_o}}\, h_\mu(C) \tag{4.16}$$

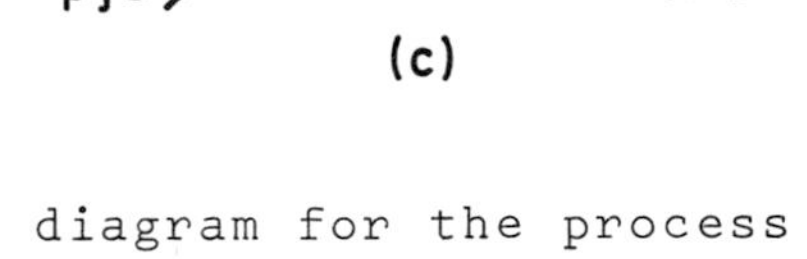

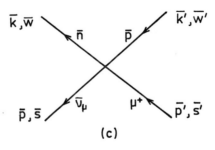

Fig. 2: The diagram for the process $\bar{\nu}_\mu + \bar{n} \to \mu^+ + \bar{p}$. This process is related to the process (a) of figure 1 by a charge conjugation transformation.

and

$$<\mu^+|\ell_\mu(0)|\bar{\nu}_\mu> = \frac{1}{(2\pi)^3}\frac{\sqrt{m_\nu m_\mu}}{\sqrt{p_o p_o'}}\,\ell_\mu(C) \qquad (4.17)$$

From the identity

$$<\bar{p}|j_\mu(0)|\bar{n}> = <\bar{n}|j_\mu^\dagger(0)|\bar{p}>^* \qquad (4.18)$$

we obtain

$$h_\mu(C) = \{\bar{v}(\bar{k}',\bar{w}')[F_1(q^2)\gamma_\mu - i\,F_2(q^2)\sigma_{\mu\nu}q^\nu - F_3(q^2)q_\mu +$$

$$+ (G_1(q^2)\gamma_\mu - iG_2(q^2)\sigma_{\mu\nu}q^\nu - G_3(q^2)q_\mu)\gamma_5]v(\bar{k},\bar{w})\}^* \qquad (4.19)$$

and similarly, for the leptonic part we arrive at

$$\ell_\mu(C) = \bar{v}(\bar{p},\bar{s})\gamma_\mu(1+\gamma_5)\,v(\bar{p}',\bar{s}') \qquad (4.20)$$

To rewrite these expressions we recall the following identities

$$v(\bar{p},\bar{s}) = C\,\bar{u}^T(\bar{p},\bar{s})$$

$$\bar{v}(\bar{p},\bar{s}) = -\,u^T(\bar{p},\bar{s})\,C^{-1} \qquad (4.21)$$

Here C denotes a 4x4 matrix and its properties are listed in the eq. (21) of appendix 2. Inserting (4.21) into (4.20) we obtain

$$\ell_u(C) = -\,u^T(\bar{p},\bar{s})C^{-1}\gamma_\mu(1+\gamma_5)C\,\bar{u}^T(\bar{p}',\bar{s}') =$$

$$= -\,\bar{u}\,(\bar{p}',\bar{s}')[C^{-1}\gamma_\mu(1+\gamma_5)C]^T\,u(\bar{p},\bar{s}) =$$

$$= \bar{u}(\bar{p}',\bar{s}')[\gamma_\mu^T(1+\gamma_5^T)]^T u(\bar{p},\bar{s}) =$$

$$= \bar{u}(\bar{p}',\bar{s}')\gamma_\mu(1-\gamma_5)\, u(\bar{p},\bar{s}) \qquad (4.22)$$

and similarly

$$h_\mu(C) = \bar{u}(\bar{k}',\bar{w}') \times$$

$$\times \{F_1^*(q^2)\gamma_\mu + iF_2^*(q^2)\sigma_{\mu\nu}q^\nu + F_3^*(q^2)q_\mu -$$

$$- [G_1^*(q^2)\gamma_\mu + iG_2^*(q^2)\sigma_{\mu\nu}q^\nu + G_3^*(q^2)q_\mu]\gamma_5\}u(\bar{k},\bar{w}) \qquad (4.23)$$

so that

$$h_\mu(C)\ell^\mu(C) = \phi(F_i^*,\ G_i^*;\ -\gamma_5) \qquad (4.24)$$

Once more we find that it is the simultaneous presence of terms with and without γ_5 in both h_μ and ℓ_μ that makes it impossible to achieve C invariance, that is

$$h_\mu(C)\ \ell^\mu(C) \neq h_\mu\ \ell^\mu$$

Any quantity that depends on the interference between terms with and without γ_5 will change sign when we pass from the process (a) to the process (c).

4.2. The C operation applied to some η-decays

There have been some recent speculations that the electromagnetic interactions violate C invariance. Most of the easily detectable effects of a C violation would not appear, however, unless one is dealing with virtual photons, since the conservation of the electromagnetic current often imposes the same restrictions on matrix elements as invariance under C. This phenomenon that the same predictions may follow from more than one invariance principle will be discussed in section 7.1.

In the case of C non-invariance for virtual electromagnetic processes one has looked for detectable effects in some η-decays. We shall briefly consider the following processes from the point of view of C invariance

(i) $\eta \to \pi^+ + \pi^- + \pi^0$

(ii) $\eta \to \pi^+ + \pi^- + \gamma$

(iii) $\eta \to \pi^0 + e^+ + e^-$

The electromagnetic nature of the first process follows from the fact that the G-parity is not conserved or alternatively, by considering the generalized Pauli principle it is easily seen that the isospin cannot be conserved in the process. If we assume that the interaction is C-invariant then we obtain from the eq. (4.1)

$$<\pi^+(k_+),\ \pi^-(k_-),\ \pi^0(k)|S|\eta> =$$

$$= \eta(C)<\pi^-(k_+),\pi^+(k_-),\pi^0(k)|S|\eta> \qquad (4.25)$$

where $\eta(C)$ is a phase factor which is irrelevant for the present discussion. If we denote by N_+ the number of η-decays of the type (i) in which $E(\pi^+) > E(\pi^-)$ and by N_- the number of decays with $E(\pi^-) > E(\pi^+)$ then it follows from the eq . (4.25) that

$$N_+ = N_- \qquad (4.26)$$

This means that if one determines the sign of the charge for that charged π meson which emerges with the largest velocity (energy) in the final state, then it should be positive and negative equally often. A charge unbalance for the fastest charged π-meson thus implies a C violation. Experimentally one has determined the parameter A defined by

$$A = \frac{N_+ - N_-}{N_+ + N_-} \tag{4.27}$$

and $A \neq 0$ would indicate that the S-matrix element does not satisfy the eq. (4.25) and hence C invariance does not hold. Present experimental results seem to be consistent with $A = 0$.

The argument presented for the $\eta \to 3\pi$ above holds equally well for the process (ii) where the π^0 is replaced by a photon. The experimental results for the parameter A defined just as in the previous case are consistent with C invariance, that is, $A = 0$.

The dominant contribution to the process $\eta \to \pi^0 + e^+ + e^-$ is given by the diagram in figure 3. The relevant part of the transition matrix element is

$$\langle \pi^0 | J_\mu^{EM}(0) | \eta \rangle ,$$

where $J_\mu^{EM}(0)$ is the electromagnetic current operator.

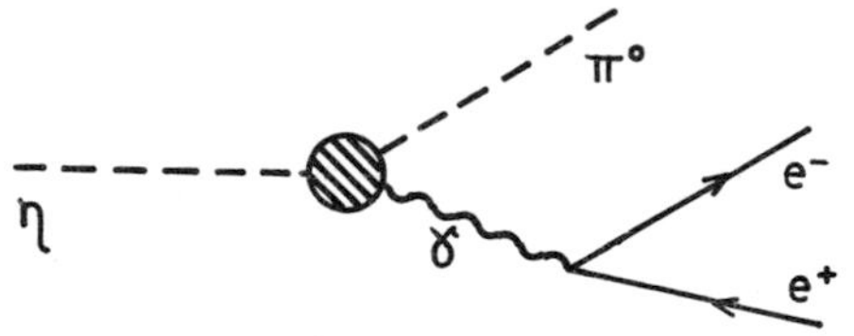

Fig. 3: The lowest order diagram for the decay $\eta \to \pi^0 + e^+ + e^-$

Under C this current operator transforms in the following way

$$C\, J_\mu^{EM}\, C^{-1} = -\, J_\mu^{EM} \tag{4.28}$$

and hence

$$\begin{aligned} <\pi^o|J_\mu^{EM}(0)|\eta> &= <\pi^o|C^{-1}C\, J_\mu^{EM}(0)C^{-1}C|\eta> = \\ &= -<\pi^o|J_\mu^{EM}(0)|\eta> \end{aligned} \tag{4.29}$$

from which it follows that to this order in the electromagnetic coupling the matrix element must vanish unless C is violated. If, for example, the current operator would have a part which is even rather than odd under C then it would in general give a nonvanishing contribution to the matrix element. However, the presence of such a current coupled to the electromagnetic field would make the electromagnetic interaction noninvariant under C. Experimentally one has established that the upper limit for the branching ratio

$$B = \frac{\Gamma(\eta\to\pi^o+e^++e^-)}{\Gamma(\eta\to\pi^o+\pi^++\pi^-)} \tag{4.30}$$

is of the order of 0.01 in agreement with C invariance. It should be noted in this context that higher order (in the electromagnetic coupling) contributions to the process may be nonvanishing even if C is an exact symmetry. Therefore, one expects the process to occur although it should be substantially supressed in comparison with the other electromagnetic decays of the η meson.

5. Some Applications of T

Since the time reversal transformation is represented by an antiunitary operator one obtains the following relation provided the S-operator is invariant under T:

$$\langle\beta|S|\alpha\rangle = \langle T\alpha|S|T\beta\rangle \tag{5.1}$$

We have previously seen that T acting on a state vector reverses momenta and spins leaving the helicities unchanged. For some applications it is more convenient to introduce the t-operator defined by

$$S = 1 + i\,t \tag{5.2}$$

Invariance of the S-operator under time reversal then implies

$$\langle\beta|t|\alpha\rangle = \langle T\alpha|t|T\beta\rangle \tag{5.3}$$

These are the fundamental relations to which we shall return in the applications below.

We may summarize the present experimental status of T invariance in the following way. There are no direct violations of T invariance known. However, indirectly by means of the CPT theorem (cf. section 6) and the observed violation of CP invariance in the decay of the longliving neutral K-meson into two π-mesons we know that it may be violated although the effect appears to be small. These experimental observations will be exhaustively discussed by Prof. Rubbia in his lectures and for that reason we shall not discuss them further.

5.1. T applied to the process $\nu_\mu + n \to \mu^- + p$

We have previously discussed this process in section 4.1. in the context of P and C. Under time reversal the process (cf. figure 1a) is related to the process of figure 4.

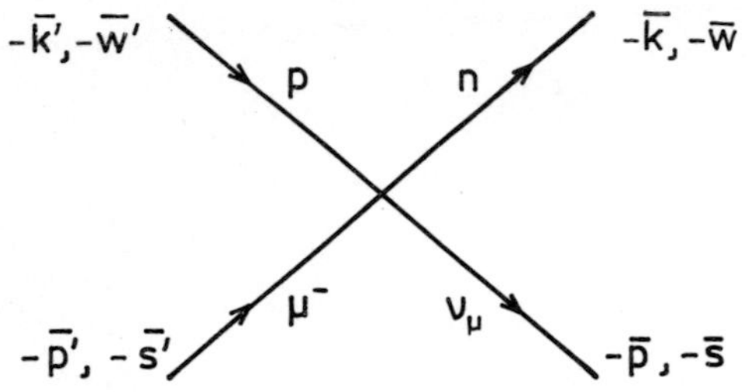

Fig. 4: The diagram describing the process $\mu^- + p \to \nu_\mu + n$ with a configuration representing the time-reversed process of the process (a) of fig.1.

Using the same notations as before we have

$$<T(\nu_\mu, n)|S|T(\mu^-, p)> =$$

$$= - \frac{iG}{\sqrt{2}} \frac{\delta^4(P_f - P_i)}{(2\pi)^2} \frac{\sqrt{m_p m_n m_\mu m_\nu}}{\sqrt{k_o k'_o p_o p'_o}} h_\mu(T)\, \ell^\mu(T) \qquad (5.4)$$

with

$$\frac{1}{(2\pi)^3} \frac{\sqrt{m_n m_p}}{\sqrt{k_o k'_o}} h_\mu(T) \equiv <T(n)|j_\mu(0)|T(p)> =$$

$$= <T(p)|j^\dagger_\mu(0)|T(n)>^*$$

$$\frac{1}{(2\pi)^3} \frac{\sqrt{m_\mu m_\nu}}{\sqrt{p_o p'_o}} \ell_\mu(T) \equiv <T(\nu_\mu)|\ell^\dagger_\mu(0)|T(\mu^-)> =$$

$$= <T(\mu^-)|\ell_\mu(0)|T(\nu_\mu)>^* \qquad (5.5)$$

Comparison with the definition of the matrix element for the original process yields

$$h^*_\mu(T) = \bar{u}(-\bar{k}',-\bar{w}') \times$$

$$\times \{F_1(q^2)\gamma_\mu + i\varepsilon(\nu)F_2(q^2)\sigma_{\mu\nu}q^\nu + \varepsilon(\mu)F_3(q^2)q_\mu +$$

$$+ [G_1(q^2)\gamma_\mu + i\varepsilon(\nu)G_2(q^2)\sigma_{\mu\nu}q^\nu + \varepsilon(\mu)G_3(q^2)q_\mu]\gamma_5\}u(-\bar{k},-\bar{w}) \tag{5.6}$$

To rewrite this expression we recall the following spinor relations

$$u(-\bar{k},-\bar{w}) = T\ \bar{u}^T(\bar{k},\bar{w})$$

$$\bar{u}(-\bar{k},-\bar{w}) = u^T(\bar{k},\bar{w})T^{-1} \tag{5.7}$$

where T denotes a 4 × 4 matrix whose properties are defined by the eq. (22) of appendix 2. After some algebra the eq. (5.6) may then be written

$$h_\mu(T) = \varepsilon(\mu)\ \bar{u}(\bar{k}',\bar{w}')\{F^*_1(q^2)\gamma_\mu + iF^*_2(q^2)\sigma_{\mu\nu}q^\nu + F^*_3(q^2)q_\mu +$$

$$+ [G^*_1(q^2)\gamma_\mu + iG^*_2(q^2)\sigma_{\mu\nu}q^\nu + G^*_3(q^2)q_\mu]\gamma_5\}u(\bar{k},\bar{w}) \tag{5.8}$$

and similarly

$$\ell_\mu(T) = \varepsilon(\mu)\bar{u}(\bar{p}',\bar{s}')\gamma_\mu(1+\gamma_5)u(\bar{p},\bar{s}) \tag{5.9}$$

In the previous notation we may then write

$$h_\mu(T)\ell^\mu(T) = \phi(F^*_i,G^*_i;\gamma_5) \tag{5.10}$$

T invariance implies that

$$h_\mu(T)\ell^\mu(T) = h_\mu \ell^\mu \qquad (5.11)$$

and it follows from (5.10) that (5.11) can be satisfied only if all the form factors are real. Since we may have an overall change in the phase of the matrix element without any detectable change in observable quantities we may express the result somewhat more appropriately by stating that time reversal invariance can only be satisfied if all the six form factors have the same phase. We had anticipated this result in introducing a factor i in the definition of $F_2(q^2)$ and $G_2(q^2)$.

5.2. Final state interactions

In section 2.4. it was shown that the S-operator must be a unitary operator in order to conserve the probability. Thus

$$S^\dagger S = I \qquad (5.12)$$

Inserting the t-operator of the eq. (5.2) we obtain

$$i(t^\dagger - t) = t^\dagger t \qquad (5.13)$$

This relation may be written in terms of matrix elements in the following way

$$i[\langle\beta|t^\dagger|\alpha\rangle - \langle\beta|t|\alpha\rangle] = \sum_\gamma \langle\beta|\ t^\dagger|\gamma\rangle\langle\gamma|t|\alpha\rangle \qquad (5.14)$$

where a complete set of intermediate states $|\gamma\rangle$ has been introduced. It is important to realize that the eq. (5.14) only involves physically realizable state vectors so that the matrix elements correspond to real processes. This implies among other things that the summation over intermediate states is limited by conservation laws such as energy-momentum conservation etc. In many important cases it turns out that there exists no possible intermediate state and we may write the eq. (5.14)

$$<\beta|t^{\dagger}|\alpha> = <\beta|t|\alpha> \tag{5.15}$$

If this situation prevails, then we say that there are no final state interactions. If we further assume that time reversal invariance holds then we may write the eq. (5.3)

$$<\beta|t|\alpha> = <T\beta|t|T\alpha>^{*} \tag{5.16}$$

In the more general case there are final state interactions present and hence, the right hand side of the eq. (5.14) is nonvanishing. It may be, however, that possible intermediate states can only be reached by additional weak or electromagnetic interactions in which case the right hand side is negligible or small and it may be possible to evaluate it. Clearly the most important situation occurs when the intermediate state is reached by an additional strong interaction in which case the "extra" factor on the right hand side is nonnegligible in general. Let us for simplicity consider the case when only one intermediate state is of this type. Let us further assume that the final state $|\beta>$ is a two-particle eigenstate of the strong (and electromagnetic) S-operator, that is

$$S|\beta> = \exp[2i\delta_{\beta}].|\beta> \tag{5.17}$$

where δ_{β} is the physical phase shift. Hence,

$$t|\beta\rangle = -i[\exp(2i\delta_\beta)-1]|\beta\rangle \qquad (5.18)$$

Inserting this in the eq. (5.14) we obtain

$$\langle\beta|t^\dagger|\alpha\rangle = \exp[-2i\delta_\beta]\langle\beta|t|\alpha\rangle \qquad (5.19)$$

If the t-operator is T invariant we may write the eq. (5.19)

$$\langle T\beta|t|T\alpha\rangle^* = \exp[-2i\delta_\beta]\langle\beta|t|\alpha\rangle \qquad (5.20)$$

This is a fundamental relation which is often referred to as the Fermi-Watson theorem and it relates the phase of certain matrix elements to the physical scattering phase shifts of the final state. Of course, if the initial and the final states are eigenstates of T or possibly of $T\Omega$ where Ω is a unitary symmetry operator, then the eq. (5.20) determines the phase of the matrix element $\langle\beta|t|\alpha\rangle$ in terms of δ_β. We shall see an example of this in the subsection 5.4.

5.3. T applied to the process $K^+ \to \pi^0 + \mu^+ + \nu_\mu$

As a first application of the results of the previous subsection we consider the process

$$K^+ \to \pi^0 + \mu^+ + \nu_\mu$$

for which there are no strong or electromagnetic final state interactions. If we assume time reversal invariance it follows from the eq. (5.16) that

$$\langle\pi^0,\mu^+,\nu_\mu|t|K^+\rangle = \langle T(\pi^0,\mu^+,\nu_\mu)|t|T(K^+)\rangle^* \qquad (5.21)$$

We write the matrix element of the original process in

the following way

$$<\pi^0, \mu^+, \nu_\mu|t|K^+> =$$

$$- \frac{G}{(2\pi)^2\sqrt{2}} \frac{\sqrt{m_\mu m_\nu}}{\sqrt{4k_o k'_o q_o q'_o}} \delta^4(P_f - P_i) L_\mu \ell^\mu \qquad (5.22)$$

where

$$\frac{1}{(2\pi)^3} \frac{1}{\sqrt{4k_o k'_o}} L_\mu \equiv <\pi^0|j_\mu(0)|K^+> \qquad (5.23)$$

and

$$\ell_\mu = \bar{u}_\nu(\bar{q},\bar{w})\gamma_\mu(1+\gamma_5)v_\mu(\bar{q}',\bar{w}') \qquad (5.24)$$

From Lorentz invariance we deduce that the most general form for L_μ is given by (cf. the lecture notes by

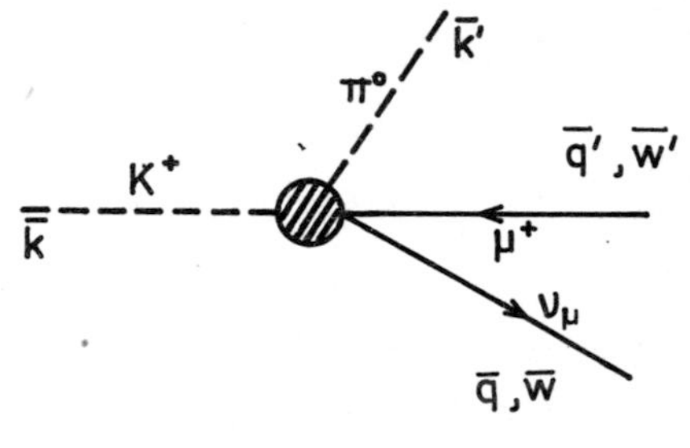

Fig. 5: The decay $K^+ \to \pi^0 + \mu^+ + \nu_\mu$.

Prof. Pietschmann)

$$L_\mu = f_+(p_K + p_\pi)_\mu + f_-(p_K - p_\pi)_\mu \qquad (5.25)$$

where f_+ and f_- are invariant form factors which depend on the only available invariant $(p_K p_\pi)$. From the eq.

(5.21) we conclude that

$$L_\mu \ell^\mu = [L_\mu(T)\ell^\mu(T)]^* \tag{5.26}$$

with

$$L_\mu(T) = \varepsilon(\mu)[f_+(p_K+p_\pi)_\mu + f_-(p_K-p_\pi)_\mu] \tag{5.27}$$

and

$$\ell_\mu(T) = \bar{u}(-\bar{q},-\bar{w})\gamma_\mu(1+\gamma_5)v(-\bar{q}',-\bar{w}') \tag{5.28}$$

From the properties of the spinors (cf. the eq. (5.7)) we deduce that

$$\ell^*_\mu(T) = \varepsilon\,(\mu)\ \bar{u}(\bar{q},\bar{w})\gamma_\mu(1+\gamma_5)v(\bar{q}',\bar{w}') = \varepsilon(\mu)\ell_\mu \tag{5.29}$$

which permits us to conclude from (5.26) that the form factors must be real (have the same phase). This means that $\kappa = \mathrm{Im}[f^*_+ f_-] = 0$. Explicit calculations of time-reversal non-invariant effects always yield results which are proportional to κ . An example of such an effect is the so-called transverse polarization of the muon in the decay $K^+ \to \pi^0 + \mu^+ + \nu_\mu$. The measurement of this polarization effect corresponds to a determination of the correlation $(\bar{p}_\pi \times \bar{p}_\mu)\cdot\langle \bar{\sigma}_\mu \rangle$ where $\bar{\sigma}_\mu$ is the spin of the muon. It is obvious that such a correlation is odd under time reversal. Arguments of this kind referring to the transformation properties of vectors are valid in this example since there are no final state interactions. For the case of non-negligible final state interactions it no longer holds and odd correlations may appear as a consequence of final state interactions even if time reversal is a strict symmetry.

5.4. T applied to the process $\Sigma^- \to n + \pi^-$

The process $\Sigma^- \to n + \pi^-$ is an example of a process with strong final state interactions. We have chosen this particular process since there is only one non-negligible intermediate state in the unitarity relation (5.14), namely the state $|n\ \pi^-\rangle$. All other intermediate states would involve additional electromagnetic or weak interactions and, for that reason, they are neglected.

The Σ^- hyperons are produced by associated production and in general the Σ^- particles have a nonvanishing polarization in a direction perpendicular to the production plane (transverse polarization) as a result of final state interactions. For simplicity we shall consider the decay of a fully polarized Σ^- hyperon whereby we can omit to discuss the production process altogether. The decay configuration in the rest system of the Σ^- is given in figure 6.

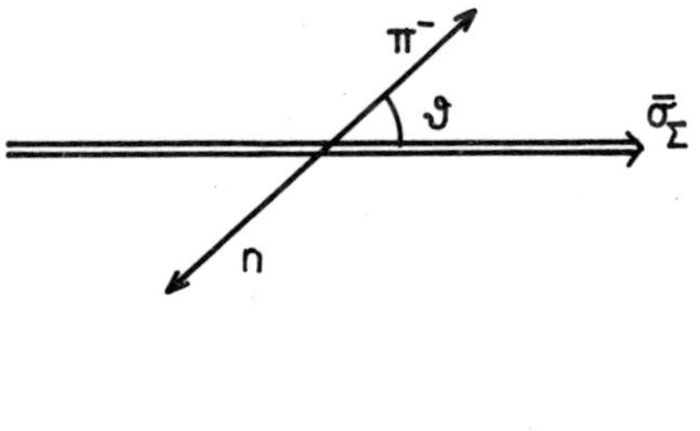

Fig. 6: The decay configuration for the decay of a fully polarized Σ^- into a π^- and a neutron.

Angular momentum conservation requires that $J = M = \frac{1}{2}$ in the final state (the spin quantization axis is chosen along the spin vector of the Σ^-). If we anticipate a P violation since the decay is a weak process, then the final $n\pi^-$ system may be in a S-state or in a P-state.

Thus we have

$$\text{initial state:} \quad |j,m> = |\tfrac{1}{2}, \tfrac{1}{2}>$$

$$\text{final state} \quad : \quad |\phi,\theta,p,\lambda_1,\lambda_2> = |\phi,\theta,p,\pm\tfrac{1}{2}, 0>$$

corresponding to the two possible helicity states for the final neutron. The possible final state vectors may be reexpressed in terms of an orbital angular momentum basis $|p,J,M;\ell,\sigma>$

$$|\phi,\theta,p,-\tfrac{1}{2},0> = -\frac{1}{\sqrt{4\pi}}\sin\theta\,\exp[-i\phi]\,|p,\tfrac{1}{2},\tfrac{1}{2},1,\tfrac{1}{2}>$$

$$|\phi,\theta,p,\tfrac{1}{2},0> = |p,\tfrac{1}{2},\tfrac{1}{2},0,\tfrac{1}{2}> - \frac{1}{\sqrt{4\pi}}\cos\theta\,|p,\tfrac{1}{2},\tfrac{1}{2},1,\tfrac{1}{2}> \tag{5.30}$$

The partial decay rate $d\Gamma/d(\cos\theta)$ is given by

$$\frac{d\Gamma}{d(\cos\theta)} = \text{const} \times \sum_{s=\pm\frac{1}{2}} |<\phi,\theta,p,s,0|t|\tfrac{1}{2},\tfrac{1}{2}>|^2 \tag{5.31}$$

To extract the angular dependence we insert the eq. (5.30) with the following result

$$\frac{d\Gamma}{d(\cos\theta)} = \text{const} \times \{|A_S + A_P\cos\theta|^2 + |A_P\sin\theta|^2\} \tag{5.32}$$

with

$$A_S = \langle p,\tfrac{1}{2},\tfrac{1}{2},0,\tfrac{1}{2}|t|\tfrac{1}{2},\tfrac{1}{2}\rangle \equiv \langle n\pi^-;\ell=0|t|\Sigma^-\rangle \qquad (5.33)$$

and

$$A_P = \langle p,\tfrac{1}{2},\tfrac{1}{2},1,\tfrac{1}{2}|t|\tfrac{1}{2},\tfrac{1}{2}\rangle \equiv \langle n\pi^-;\ell=1|t|\Sigma^-\rangle \qquad (5.34)$$

corresponding to the S and P wave amplitudes. Finally, we may write (5.32) as

$$\frac{d\Gamma}{d(\cos\theta)} = \text{const} \times (1+\alpha\cos\theta) \qquad (5.35)$$

with

$$\alpha = \frac{2\mathrm{Re}(A_S A_P^*)}{|A_S|^2+|A_P|^2} \qquad (5.36)$$

It remains to determine the relative phase of A_S and A_P. To this end we recall the relation (5.20). We have chosen to express the final state in an orbital angular momentum basis since (i) at the relevant energy (corresponding to the energy release) these basis vectors for the π^-n - system are eigenvectors of the strong S-operator corresponding to the phase shifts δ_S respectively δ_P, (ii) for A_S and A_P the unitarity relation contains only one intermediate state each corresponding to S and P wave π^-n scattering, and (iii) assuming rotational invariance for the t-operator one has

$$\langle RT(n\pi^-;\ell)|t|RT(\Sigma^-)\rangle = \langle n\pi^-;\ell|t|\Sigma^-\rangle \qquad (5.37)$$

where $R \equiv R(0,\pi,0)$. Hence,provided T invariance holds, we may use the Fermi-Watson theorem and we conclude from (5.20)

$$A_S = |A_S| \exp[i\delta_S]$$

$$A_P = |A_P| \exp[i\delta_P] \tag{5.38}$$

so that

$$\alpha = \frac{2|A_S|.|A_P|}{|A_S|^2+|A_P|^2} \cos(\delta_S - \delta_P) \tag{5.39}$$

The presence of the cosine factor is a consequence of final state interactions. It is a characteristic feature that correlations which are parity-violating such as $(\bar{\sigma}_\Sigma . \bar{p}_\pi)$ acquire a cosine factor involving the final state scattering phase shifts while correlations odd under time reversal such as for example $(\bar{\sigma}_\Sigma \times \bar{p}_P) . \bar{\sigma}_P$ (the transverse polarization of the proton emitted in the decay of polarized Σ^- hyperons) acquire a sine factor of the same phase shifts. Vanishing final state interactions correspond to vanishing phase shifts and hence, in this limit all odd correlations under time reversal vanish and the simple argument with vectors holds in this case as previously stated. Of course, the phase shifts which enter correspond to the center of mass energy equal to the energy release in the decay.

6. The CPT Theorem and Some Applications

Since we have treated the discrete symmetries C, P and T separately it is a straightforward task to discuss the physical implications of invariance under the combined transformation CPT. We shall do this for the process $\nu_\mu + n \to \mu^- + p$. This leads us to a very fundamental theorem in local field theory, the so-called

CPT theorem. We shall then explore some consequences of CPT invariance in explicit examples. More precisely, assuming CPT invariance we shall prove (i) the equality of mass for a particle and its antiparticle,(ii) the equality of some partial decay rates and for the total lifetime of a particle and its antiparticle,(iii) the equality of electromagnetic form factors (magnetic moment) for a particle and its antiparticle.

6.1. CP and T applied to the process $\nu_\mu + n \to \mu^- + p$

We have previously introduced the notation (cf. fig. 1)

$$<\mu^-, p|S|\nu_\mu, n> = -\frac{iG}{(2\pi)^2\sqrt{2}} \frac{\sqrt{m_n m_p m_\mu m_\nu}}{\sqrt{k_o k'_o p_o p'_o}} h_\mu \ell^\mu \qquad (6.1)$$

with

$$h_\mu \ell^\mu \equiv \phi(F_i, G_i; \gamma_5)$$

Under the transformations P, C and T we have obtained the following relations

$$h_\mu(P)\ell^\mu(P) = \phi(F_i, G_i; -\gamma_5)$$

$$h_\mu(C)\ell^\mu(C) = \phi(F_i^*, G_i^*; -\gamma_5)$$

$$h_\mu(T)\ell^\mu(T) = \phi(F_i^*, G_i^*; \gamma_5) \qquad (6.2)$$

and hence we find

$$h_\mu(CPT)\ell^\mu(CPT) = \phi(F_i, G_i; \gamma_5) = h_\mu \ell^\mu \qquad (6.3)$$

It thus follows that imposing CPT invariance does not restrict the matrix element for the process $\nu_\mu + n \to \mu^- + p$ in any way. The V-A theory applied to this semi-leptonic process is automatically CPT invariant. It is further seen that CP invariance places exactly the same restrictions on the matrix element as T invariance. Thus in this special case we find that CP invariance implies T invariance and a test of either one of the two symmetries is simultaneously a test of the other symmetry. This is just a special case of the CPT theorem which we shall briefly discuss in the next section.

6.2. CPT invariance of the S-operator

So far in these lectures we have made no reference to local field theory although local field theory in many respects provides us with the most natural frame for a description of elementary particle processes. The CPT theorem which was mentioned in the previous subsection emerges in local field theory from very general assumptions. The theorem essentially asserts that any local Lagrangian field theory is invariant under the combined operation CPT, taken in any order whatever, provided the theory is invariant under proper orthochronous Lorentz transformations. For a proof or a more detailed discussion we refer to [5].

Although the CPT theorem strictly speaking has been proved only within field theory it is generally assumed to hold as a strict symmetry obeyed by all the laws of nature. If we assume this to be the case, then clearly the three discrete symmetries C, P and T are no longer independent and invariance under one of them, say, T implies symmetry under CP. In this way the CPT theorem is frequently used and an example of that is the assertion that the CP violating decay $K_L^o \to 2\pi$ implies T non-invariance. As a matter of fact, most tests of T invariance are indirect in the sense that one has tested CP predictions.

The fact that the laws of physics, that is, the S-operator, are assumed to be invariant under CPT implies

$$(CPT)S(CPT)^{-1} = S^{\dagger} \tag{6.4}$$

since $\Theta \equiv CPT$ necessarily is an antiunitary operator due to T. We proceed to derive restrictions on matrix elements based on the eq. (6.4). Consider for this purpose the process

$$|i;\ in\rangle \to |f;\ out\rangle \tag{6.5}$$

where i and f denote arbitrary initial and final states for example characterized by the individual momenta and the helicities of the particles. A CPT transformation clearly takes a particle of momentum $\bar{p}$ and helicity λ into an antiparticle of momentum $\bar{p}$ and helicity $-\lambda$. This follows from the transformation laws given in section 3. We introduce the following notations: the state vector obtained from $|i;in\rangle$ respectively $|f;out\rangle$ by replacing all particles by their antiparticles and vice versa are denoted by $|\bar{i};in\rangle$ and $|\bar{f};out\rangle$, and the state vectors obtained by changing the signs of all the helicities in $|i;in\rangle$ and $|f;out\rangle$ are denoted by $|\tilde{i};in\rangle$ and $|\tilde{f};out\rangle$. Consider next a hermitian operator Ω which is assumed to be CPT invariant, that is,

$$\Theta\ \Omega\ \Theta^{-1} = \Omega \tag{6.6}$$

For the matrix elements of Ω one then finds

$$\langle\bar{f};out|\Omega|\bar{i};in\rangle = \langle\bar{f};out|\Theta^{-1}\Theta\Omega\Theta^{-1}\Theta|\bar{i};in\rangle =$$

$$= \eta\langle\tilde{f};in|\Omega|\tilde{i};out\rangle^{*} \qquad (6.7)$$

since in-states are transformed into out-states etc. The phase factor η depends on the choice of basis vectors and it is known once those are specified. Introducing the S-operator defined by

$$|\alpha;in\rangle = S|\alpha;out\rangle$$

we may write (6.7)

$$\langle\bar{f};out|\Omega|\bar{i};in\rangle = \eta\langle\tilde{f};out|S^{-1}\Omega S|\tilde{i};in\rangle^{*} \qquad (6.8)$$

Let us further make the assumption that $|\tilde{f};out\rangle$ and $|\tilde{i};in\rangle$ are eigenstates of the S-operator. In that case (6.8) takes the form

$$\langle\bar{f};out|\Omega|\bar{i};in\rangle = \eta\exp[2i(\delta_f+\delta_i)]\langle\tilde{f};out|\Omega|\tilde{i};in\rangle^{*} \qquad (6.9)$$

This is the fundamental relation from which we shall deduce a number of predictions, which can be subjected to experimental tests. This has also been done to some extent.

Of course, in those cases when the initial and final states are not eigenstates of the S-operator one may always expand them in such eigenstates. In that case the eq. (6.9) is replaced by an equation where the right hand side consists of a sum of terms of the same type as in (6.9) but each with a different phase factor.

6.3. Some applications of CPT invariance

As a first application of CPT invariance we shall prove that the mass of a particle equals the mass of its antiparticle. To this end we choose $|i;in>$ and $|f;out>$ to represent a stable particle at rest. In this case there is no distinction between in- and out-states and the S-operator acts trivially on the state vectors, that is, $\delta_i = \delta_f = 0$. We further set $\Omega = P_0$ in the relation (6.9) which then reads

$$\bar{m}<\bar{f}|\bar{i}> = m^* <\tilde{f}|\tilde{i}>^* \qquad (6.10)$$

where $\bar{m}$ and m denote the rest masses of the particle and the antiparticle respectively. Since P_0 is hermitian and m represents a diagonal matrix element it follows that m is real and hence

$$\bar{m} = m \qquad (6.11)$$

The same result clearly follows from the more restrictive assumptions of C invariance, but it is important to realize that already the weaker assumption of CPT invariance suffices to derive (6.11). For unstable particles one derives similar relations for the real part of the mass. Experimentally the relation (6.11) has been tested for $\mu^\pm$, $\pi^\pm$ and $K^\pm$ with no sign of disagreement with the prediction. The most accurate test refers to the measured mass difference for K_S^0 and K_L^0 and Prof. Rubbia will discuss this more in detail.

We next consider equalities of partial decay rates and the total life-time for a particle and its antiparticle. Consider for simplicity a particle which decays as a result of the weak interactions. To start with we shall further assume that the final state interactions are negligible and we treat the process to first order in the weak interactions. This implies that the relation (6.9) may be written

$$<\bar{f};out|H_{weak}|\bar{i}> = <\tilde{f};out|H_{weak}|\tilde{i}>^* \qquad (6.12)$$

Since it is the modulus of the matrix element which enters into the expression for the partial decay rates we immediately derive the following equality

$$\Gamma(\bar{i} \to \bar{f}) = \Gamma(\tilde{i} \to \tilde{f}) \qquad (6.13)$$

or after summation over all the possible helicity (spin) states we obtain that

$$\sum_{spins} \Gamma(\bar{i} \to \bar{f}) = \sum_{spins} \Gamma(i \to f) \qquad (6.14)$$

where Γ represents any partial decay rate.

It is important to realize that these results were obtained under the assumption that final state interactions are negligible. For non-negligible final state interactions the result is in general not true. As examples of relations of the type (6.13) and (6.14) we may consider

$$\Gamma(K^+ \to \mu^+ + \nu_\mu) = \Gamma(K^- \to \mu^- + \bar{\nu}_\mu)$$

$$\Gamma(K^+ \to \pi^0 + \mu^+ + \nu_\mu) = \Gamma(K^- \to \pi^0 + \mu^- + \bar{\nu}_\mu) \quad (6.15)$$

For these processes the relations refer to energy spectra, angular correlations etc. after spin summations (or with the appropriate spin orientations). No accurate test of these relations exists so far.

In case the final state interactions are non-negligible it may happen that the final state still is an eigenstate of the S-operator so that the relation (6.9) with only one term on the right hand side still holds. In that case the phase factor $\exp[2i\delta_f]$ is irrelevant in the expression for the partial decay rate since it is proportional to squared modulus of the matrix element and the relations (6.13) and (6.14) still apply. This situation may for example occur as a consequence of selection rules. A specific example of this phenomenon is encountered in the decays $K^{\pm}\rightarrow\pi^{\pm}+\pi^{0}$. Since angular momentum conservation requires that L = 0 (S-state) it follows from the generalized Pauli principle that the final state must have I = 2. Since isospin and parity are both conserved by the strong interactions there is no other state than the 2π state which can be reached by strong final state interactions. Neglecting electromagnetic and weak contributions we conclude that the final state is an eigenstate of the (strong) S-operator and hence we derive from the eq. (6.13) the following relation between partial decay rates

$$\Gamma(K^{+} \rightarrow \pi^{+} + \pi^{0}) = \Gamma(K^{-} \rightarrow \pi^{-} + \pi^{0}) \qquad (6.16)$$

The corrections due to electromagnetic interactions are expected to be of the order 10^{-4}.

In the more general case of non-negligible final state interactions where the final state is not an eigenstate of the S-operator one may, of course, always expand the final state in such eigenstates, and one obtains an equation equivalent to the eq.(6.9) but with a sum of terms on the right hand side, each term with a different phase factor. In this case we do not obtain equalities for partial decay rates corresponding to specific decay channels in general.

The complications which arise in this case are most easily explained by considering an explicit example such as

$$\Lambda \begin{array}{l} \rightarrow p + \pi^{-} \\ \rightarrow n + \pi^{0} \end{array}$$

which both occur due to the weak interactions. Clearly

the final state interactions are non-negligible. As a matter of fact, an event which has been identified as a decay $\Lambda \to p + \pi^-$ may intrinsically have been a $\Lambda \to n + \pi^o$ decay where charge exchange scattering in the final state makes it look like a $\Lambda \to p + \pi^-$ decay. To disentangle these complications one must deal with eigenstates of the strong S-operator. In this particular case there will be four different states to take into account corresponding to $I = \frac{1}{2}, \frac{3}{2}$ and $L = 0,1$, and the corresponding terms are accompanied by the appropriate phase shifts. We shall not pursue the subject further [6] but note that in this case CPT invariance does not suffice to derive

$$\frac{\Gamma(\Lambda \to n + \pi^o)}{\Gamma(\Lambda \to p + \pi^-)} = \frac{\Gamma(\bar{\Lambda} \to \bar{n} + \pi^o)}{\Gamma(\bar{\Lambda} \to \bar{p} + \pi^+)} \tag{6.17}$$

With the assumption of C or T invariance one may derive the equality (6.17).

We finally observe that if we consider the total decay rate into all possible decay channels then we may always analyze the final state in terms of eigenstates of the strong S-operator and since in this case each partial decay rate in this rearranged sum is equal to the particle-conjugate term it follows that the total decay rate and hence the total lifetime for a particle is the same as that of the antiparticle. This prediction has been tested in the decays of $\mu^\pm$, $\pi^\pm$ and $K^\pm$ and no violation has been observed.

What has been stated above for weak decay processes clearly holds also for strong and electromagnetic processes to the extent that appropriate in- and out-states can be defined. The problem how to do this is not too well understood yet, however.

As a last application of CPT invariance we shall

consider the scattering of a charged particle in an external field. The interaction between such a particle and the electromagnetic field is given by the coupling of the appropriate current operator $J_\mu^{EM}(x)$ to the field. Consider for example the scattering of a proton in an external field. To lowest order in the electromagnetic coupling this process is described by the diagram of figure 7, and the relevant part of the matrix element is

$$<p'|J_\mu^{EM}(0)|p> \quad .$$

From Lorentz invariance and current conservation (cf. section 7.1.) it is easily shown that the most general form for this matrix element is given by

$$<f;out|J_\mu^{EM}(0)|i;in> \equiv <\bar{p}',\bar{s}'|J_\mu^{EM}(0)|\bar{p},\bar{s}> =$$

$$= \frac{e}{(2\pi)^3} \sqrt{m_p^2/p_o p_o'} \quad \bar{u}(\bar{p}',\bar{s}')[F_1(q^2)\gamma_\mu + i\ F_2(q^2)\sigma_{\mu\nu}q^\nu] \times$$

$$\times\ u(\bar{p},\bar{s}) \qquad (6.18)$$

where $q_\mu = p_\mu' - p_\mu$. Considering the static limit, that is, $q \to 0$ one easily identifies $e\ F_1(0)$ as the proton charge and $e\ F_2(0)$ as the anomalous magnetic moment of the proton. Since the electric charge of the proton is the same as the charge of the electron it follows that $F_1(0) = 1$. From the fact that $J_\mu^{EM}(x)$ is a hermitian operator it further follows that $F_1(q^2)$ and $F_2(q^2)$ are real functions.

To investigate the consequences of CPT invariance we need the transformation property of the current operator under CPT. It can be derived from the known transformation property of the photon (field) and the requirement that the interaction term should be CPT

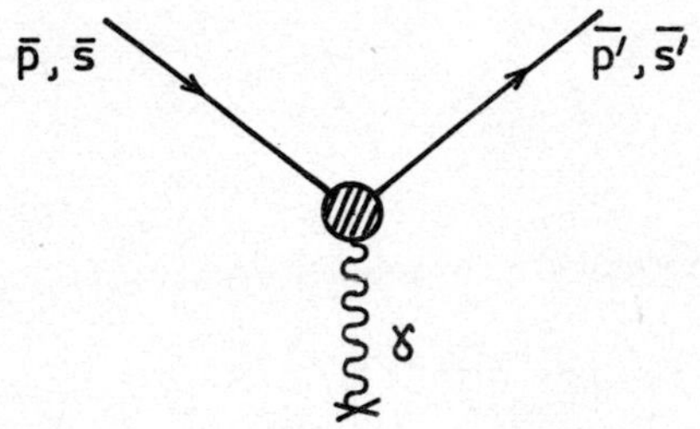

Fig. 7: Proton scattering in an external electromagnetic field.

invariant. This yields

$$\Theta J_\mu^{EM}(0)\, \Theta^{-1} = - J_\mu^{EM}(0) \tag{6.19}$$

and we obtain the following relation, which is the analogue of the eq. (6.9)

$$<\bar{f};out|J_\mu^{EM}(0)|\bar{i};in> = - <\tilde{f};out|J_\mu^{EM}(0)|\tilde{i};in>^* \tag{6.20}$$

This relation permits us to evaluate the antiproton form factors from the proton form factors. Evaluating the right hand side we obtain

$$<\bar{f};out|J_\mu^{EM}(0)|\bar{i};in> =$$

$$= - \frac{e}{(2\pi)^3} \sqrt{m_p^2/p_o p_o'} \{\bar{u}(\bar{p}',-\bar{s}')[F_1(q^2)\gamma_\mu + iF_2(q^2)\sigma_{\mu\nu}q^\nu] \times$$

$$\times\, u(\bar{p}, -\bar{s})\}^* \tag{6.21}$$

If we make use of the following spinor relations

$$u(\bar{p},-\bar{s}) = \gamma_o \; T \; \bar{u}^T(\bar{p},\bar{s})$$

$$\bar{u}(\bar{p},-\bar{s}) = u^T(\bar{p},\bar{s}) \; T^{-1} \; \gamma_o \tag{6.22}$$

it follows that

$$<\bar{f};out|J_\mu^{EM}(0)|\bar{i};in> =$$

$$= - \frac{e}{(2\pi)^3} \sqrt{m_p^2/p_o p_o'} \;\; \bar{u}(\bar{p}',\bar{s}')[F_1(q^2)\gamma_\mu + iF_2(q^2)\sigma_{\mu\nu}q^\nu] \times \times u(\bar{p},\bar{s}) \tag{6.23}$$

since the form factors are real. The change in sign in (6.23) as compared to (6.18) reflects the fact that particle and antiparticle have the opposite charge. From (6.23) and (6.18) it follows immediately that the proton and the antiproton anomalous magnetic moments are the same. The same conclusion clearly holds for any charged particle and it has been tested experimentally for $e^\pm$ and $\mu^\pm$ with no indication of a CPT violation.

7. Some General Remarks on Symmetry Principles

We end this presentation of the discrete symmetries with some brief remarks concerning two important aspects of symmetry arguments, namely (i) a specific prediction may follow from more than one invariance principle and its verification cannot be uniquely attributed to anyone of them, and (ii) the concept of symmetry when symmetry violating interactions are present.

7.1. Predictions which follow from more than one invariance principle

At various occasions we have encountered predictions which can be derived from more than one principle. This is, for example, the case with the vanishing electric dipole moment for the neutron which is a consequence of C or T invariance. A violation of this prediction implies a violation of both C and T invariance. A verification of the prediction lends support to both invariance principles but cannot be taken as evidence for either one. For this reason it is extremely important to clarify under what assumptions a certain prediction may be derived if there are more than one alternative. This inherent weakness of symmetry arguments was once again pointed out very clearly by Bernstein et al. [7] , who noted that very many predictions of C and T invariance in electromagnetic processes can also be derived from other equally fundamental principles. As an example we consider the electromagnetic form factors of a nucleon. In section 6.3. we introduced the concept invoking Lorentz invariance and current conservation. If we only assume Lorentz invariance, then the eq. (6.18) is replaced by

$$<\bar{p}',\bar{s}'|J_\mu^{EM}(0)|\bar{p},\bar{s}> =$$

$$= \frac{e}{(2\pi)^3}\sqrt{m_p^2/p_o p_o'}\;\bar{u}(\bar{p}',\bar{s}')\{F_1(q^2)\gamma_\mu + iF_2(q^2)\sigma_{\mu\nu}q^\nu +$$

$$+ F_3(q^2)q_\mu\}u(\bar{p},\bar{s}) \qquad (7.1)$$

Current conservation implies

$$<\bar{p}',\bar{s}'|q^\mu J_\mu^{EM}(0)|\bar{p},\bar{s}> = 0 \qquad (7.2)$$

Noting that the first term in (7.1) may be written

$$\bar{u}(\bar{p}',\bar{s}')\gamma_\mu q^\mu u(\bar{p},\bar{s}) =$$

$$= \bar{u}(\bar{p}',\bar{s}')[\gamma_\mu p'^\mu - \gamma_\mu p^\mu]u(\bar{p},\bar{s}) \qquad (7.3)$$

it follows from the Dirac equation for the two spinors that this term vanishes. In the second term we obtain $\sigma_{\mu\nu}q^\mu q^\nu$. Since $\sigma_{\mu\nu}$ is antisymmetric in μ and ν while $q^\mu\, q^\nu$ is symmetric also this term vanishes. Thus, from the condition (7.2) we conclude

$$q^2 F_3(q^2) = 0 \qquad (7.4)$$

or

$$F_3(q^2) = 0 \qquad (7.5)$$

except possibly for $q^2 = 0$. This result was anticipated in the eq. (6.18). However, one reaches the same conclusion by invoking (i) that $J_\mu^{EM}(0)$ is hermitian and (ii) that under time reversal $J_\mu^{EM}(0) \to \varepsilon(\mu)J_\mu^{EM}(0)$. From (i) we obtain

$$<\bar{p}',\bar{s}'|J_\mu^{EM}(0)|\bar{p},\bar{s}> = <\bar{p},\bar{s}|J_\mu^{EM}(0)|\bar{p}',\bar{s}'>^* \qquad (7.6)$$

or

$$\bar{u}(\bar{p}',\bar{s}')[F_1\gamma_\mu + iF_2\sigma_{\mu\nu}q^\nu + F_3 q_\mu]u(\bar{p},\bar{s}) =$$

$$= \{\bar{u}(\bar{p},\bar{s})[F_1\gamma_\mu - iF_2\sigma_{\mu\nu}q^\nu - F_3 q_\mu]u(\bar{p}',\bar{s}')\}^* =$$

$$= u^\dagger(\bar{p}',\bar{s}')[F_1^*\gamma_\mu^\dagger + iF_2^*\sigma_{\mu\nu}^\dagger q^\nu - F_3^* q_\mu]\gamma_0 u(\bar{p},\bar{s}) =$$

$$= \bar{u}(\bar{p}',\bar{s}')[F_1^*\gamma_\mu + iF_2^*\sigma_{\mu\nu}q^\nu - F_3^* q_\mu]u(\bar{p},\bar{s}) \qquad (7.7)$$

We conclude that hermiticity requires that F_1 and F_2 are real functions and F_3 a purely imaginary function. With regard to the condition (ii) we have previously considered the case of the weak current. The computations are in this case very similar and we shall not repeat them. The result is that the condition

$$<\bar{p}',\bar{s}'|J_\mu^{EM}(0)|\bar{p},\bar{s}> = \varepsilon(\mu)<-\bar{p}',-\bar{s}'|J_\mu^{EM}(0)|-\bar{p},-\bar{s}>^* \tag{7.8}$$

implies that F_1, F_2 and F_3 must all be real functions. In order to satisfy both conditions we must then choose $F_3=0$ which is the same result as the eq. (7.5).

We shall not pursue the subject further, but we conclude that a verification of a prediction based on a symmetry argument never constitutes an absolute proof of the invariance principle since (i) the same result may follow from an alternative set of principles as discussed above, or (ii) there may be an accidental cancellation for example of dynamical origin. An example of this latter possibility is known from the parity violating asymmetry parameters α in Σ decay. For the $\Sigma_+^+(\Sigma^+\rightarrow n+\pi^+)$ and $\Sigma_-^+(\Sigma^-\rightarrow n+\pi^-)$ decays the asymmetry parameters vanish while for $\Sigma_0^+(\Sigma^+\rightarrow p+\pi^0)$ it is large. Off hand one would say that the two first decays are parity conserving while the third one is parity violating. For the time being these observations can only be understood as accidental cancellations although they neatly conform with the $\Delta I = \frac{1}{2}$ rule.

Finally it should also be stated that although the fundamental laws of physics may satisfy certain symmetry properties the same symmetry properties are reflected in the experimental results only for isolated systems with no "external fields" present. For example, if we neglect the very small CP violation it was found that the two neutral K mesons K_1^o and K_2^o retain their identity until the moment of decay since CP conservation forbids the transitions $K_1^o \rightleftharpoons K_2^o$. How then can one explain that a beam of K_2^o regenerates K_1^o mesons copiously when it passes through matter although all interactions are CP invariant to a high degree of accuracy. The reason is simply that the piece of matter through which the beam passes is not CP invariant and it constitutes an "external field". Quite naturally then the regeneration effect depends on the difference between the scattering amplitudes for KN and $\bar{K}$N scattering. In the same spirit it was first attempted to save the CP invariance in the $K_L^o \to 2\pi$ decay by invoking an external field to account for the apparent CP violation. This possibility has later been ruled out on experimental grounds.

7.2. The concept of symmetry in the presence of symmetry-violating interactions.

We have consistently defined the various symmetry operations by giving their action on certain state vectors. In those cases where there is a classical analogue we have chosen the definitions in such a way that the quantum mechanical operators transform in the same way as their classical counterparts. This procedure may seem very straightforward , but in realistic applications there are some inherent difficulties. For one thing the problem of how to treat unstable states has not been tackled more than superficially, and a completely satisfactory way to do that is not known yet. The other complication arises in the context of approximate symmetries. We have seen that our definition of a space reflection operator requires a Hilbert space which is larger than the physical Hilbert space (≡ the space of physical state vectors) since occasionally non-physical vectors appear (e.g. the neutrino states with positive helicity). In a formal approach strictly based on group theory it is difficult to give an exact meaning to the concept of an approximate symmetry, and yet the concept has turned out to be most useful in elementary particle physics. In a hamiltonian approach approximate symmetries emerge in a rather natural way. Suppose that we can write the total hamiltonian of a physical system in the following way:

$$H = H_{free} + H_{st} + H_{em} + H_{wk} \tag{7.9}$$

where the strong , the electromagnetic and the weak parts depend on the corresponding coupling constants. Suppose further that one may define some limiting procedure so that, say, the weak and the electromagnetic part of the hamiltonian vanish. In this limit, if it exists, it is conceivable that we obtain a model which exhibits full C and P symmetry. This means that the corresponding quantum mechanical operators would not contain any parts which would lead out of the physical Hilbert space when they act on a physical state vector. On the other hand, in this limit the hamiltonian will in general be invariant under entirely new transformations which render the C and P transformations ambiguous, and some of the conventional state labels are no longer relevant. In the example considered above this happens for the electric charge, since it requires the presence of electromagnetism for an operational definition. Since the strong forces are charge independent one arrives at a model which is invariant under the isospin group. Under these circumstances a space reflection or a charge conjugation may just as well contain isospin transformations. Without the charge label we would retain transformation properties for observables analogous to the classical ones for P and the C transformation would, for example, transform a p state into a $\bar{n}$ state etc. In this way one faces the possibility of different C, P and T transformations corresponding to the limits which we conventionally refer to strong, electromagnetic and weak interactions [8]. We shall not develop this subject further, but it is clear that once a symmetry is found not to hold rigorously we must approach the subject with an open mind for entirely new possibilities with drastic changes in old concepts. This is, of course, an observation which is particularly pertinent in these times when old symmetries repeatedly fail under closer scrutiny.

8. Concluding Remarks

It is hoped that these lectures on the discrete symmetries have demonstrated the power of symmetry arguments and at the same time the care that must be exercised in drawing definite conclusions from experimental results.

Only occasionally has reference been made to field theory, since we have based the whole discussion on representations of the Poincaré group. However, all the UIR of the Poincaré group may be realized by means of creation and destruction operators (Fock space representations) and it is essentially a trivial step to define fields once these operators are defined. For symmetry arguments this additional construction of fields is entirely unnecessary, however, and hence it has been omitted here.

APPENDIX 1

Some notations

Throughout the lectures we have used the following metric tensor

$$g_{\mu\nu} = \begin{vmatrix} 1 & 0 & 0 & 0 \\ 0 & -1 & 0 & 0 \\ 0 & 0 & -1 & 0 \\ 0 & 0 & 0 & -1 \end{vmatrix}$$

so that

$$x_\mu x^\mu = x_0^2 - \bar{x}^2$$

etc. For convenience we have further introduced the symbol $\varepsilon(\mu)$ defined by

$$\varepsilon(\mu) = \begin{cases} +1 \text{ for } \mu = 0 \\ -1 \text{ for } \mu \neq 0 \end{cases}$$

Regarding notations, choice of representations for the γ-matrices etc. we refer to appendix 2, where the Dirac equation is briefly discussed.

APPENDIX 2

Brief review of the Dirac equation

Relativistic particles of spin 1/2 are described by a four-component spinor wave function $\psi(x)$. For free particles this wave function satisfies the following equation of motion - the Dirac equation

$$(i\gamma_\mu \frac{\partial}{\partial x_\mu} - m)\,\psi(x) = 0 \tag{A1}$$

The 4×4 matrices $\gamma_\mu (\mu = 0,1,2,3)$ obey the anticommutation rules

$$\gamma_\mu\gamma_\nu + \gamma_\nu\gamma_\mu = 2g_{\mu\nu} \tag{A2}$$

where $g_{\mu\nu}$ is the metric tensor. We shall use the following explicit representation of the γ-matrices

$$\gamma_0 = \begin{vmatrix} I & 0 \\ 0 & -I \end{vmatrix} \quad ; \qquad \gamma_k = \begin{vmatrix} 0 & \sigma_k \\ -\sigma_k & 0 \end{vmatrix} \tag{A3}$$

where I denotes the 2×2 unit matrix and $\sigma_k (k=1,2,3)$ is a Pauli matrix

$$\sigma_1 = \begin{vmatrix} 0 & 1 \\ 1 & 0 \end{vmatrix} \; ; \; \sigma_2 = \begin{vmatrix} 0 & -i \\ i & 0 \end{vmatrix} \; ; \; \sigma_3 = \begin{vmatrix} 1 & 0 \\ 0 & -1 \end{vmatrix} \tag{A4}$$

With these conventions it is easily shown that γ_0 is hermitian and γ_k antihermitian. These properties are summarized by the relation

$$\gamma_\mu^\dagger = \gamma_0\,\gamma_\mu\gamma_0 \tag{A5}$$

If the adjoint wave function $\bar{\psi}(x)$ is defined by

$$\bar{\psi}(x) = \psi^\dagger(x)\,\gamma_0 \tag{A6}$$

then it follows from the eq. (A1) that in the case of free particles it satisfies the equation

$$i \frac{\partial}{\partial x_\mu} \bar{\psi}(x)\gamma_\mu + m\bar{\psi}(x) = 0 \qquad (A7)$$

The following two matrices occur frequently in the Dirac theory

$$\sigma_{\mu\nu} = \frac{1}{2i} (\gamma_\mu\gamma_\nu - \gamma_\nu\gamma_\mu) \qquad (A8)$$

$$\gamma_5 = i \gamma_0 \gamma_1 \gamma_2 \gamma_3$$

The matrix $\sigma_{\mu\nu}$ satisfies the relation

$$\sigma^\dagger_{\mu\nu} = \gamma_0 \sigma_{\mu\nu} \gamma_0 \qquad (A9)$$

while γ_5 is hermitian. Furthermore, the matrix γ_5 anticommutes with the four γ-matrices γ_μ

$$\gamma_\mu\gamma_5 + \gamma_5\gamma_\mu = 0 \qquad (A10)$$

In the standard representation (A3) γ_5 is given by

$$\gamma_5 = \begin{vmatrix} 0 & -I \\ -I & 0 \end{vmatrix} \qquad (A11)$$

The Dirac equation (A1) is satisfied by plane wave solutions of the form

$$\psi(x) = u(\bar{p}) \exp[-i\, p_\mu\, x^\mu]$$

$$\psi(x) = v(\bar{p}) \exp[\, i\, p_\mu\, x^\mu] \qquad (A12)$$

provided the coordinate-independent spinors $u(\bar{p})$ and $v(\bar{p})$ satisfy the following set of equations

$$(\gamma_\mu p^\mu - m)\, u(\bar{p}) = 0$$

$$(\gamma_\mu p^\mu + m)\, v(\bar{p}) = 0 \qquad (A13)$$

The solutions with $u(\bar{p})$ are identified with the particle solutions while those with $v(\bar{p})$ describe antiparticles. It is easily seen that the eqs. (A13) have two independent solutions each corresponding to two possible orientations of the spin. With the standard representation (A3) the solutions are given by

$$u_i(\bar{p}) = \sqrt{(E+m)/2m} \begin{vmatrix} \chi_i \\ \frac{\bar{\sigma}.\bar{p}}{E+m}\,\chi_i \end{vmatrix} \qquad (A14)$$

$$v_i(\bar{p}) = \sqrt{(E+m)/2m} \begin{vmatrix} \frac{\bar{\sigma}.\bar{p}}{E+m}\,\xi_i \\ \xi_i \end{vmatrix} \qquad (A15)$$

with $i = 1,2$ and

$$\chi_1 = \binom{1}{0} = -\,\xi_2 \;; \qquad \chi_2 = \binom{0}{1} = \xi_1 \qquad (A16)$$

The normalization has been chosen in such a way that the following orthonormality conditions hold

$$\bar{u}_i(\bar{p})\, u_j(\bar{p}) = \delta_{ij}$$

$$\bar{v}_i(\bar{p})\, v_j(\bar{p}) = -\,\delta_{ij}$$

$$\bar{u}_i(\bar{p})v_j(\bar{p}) = \bar{v}_i(\bar{p})u_j(\bar{p}) = 0 \qquad (A17)$$

The adjoint spinors $\bar{u}(\bar{p})$ and $\bar{v}(\bar{p})$ are defined by

$$u(\bar{p}) = u^{\dagger}(\bar{p})\,\gamma_0$$

$$v(\bar{p}) = v^{\dagger}(\bar{p})\,\gamma_0 \qquad (A18)$$

and they satisfy equations analogous to (A13)

$$\bar{u}(\bar{p})(\gamma_\mu p^\mu - m) = 0$$

$$\bar{v}(\bar{p})(\gamma_\mu p^\mu + m) = 0 \qquad (A19)$$

In the discussions of charge conjugation and time reversal two more matrices C and T are introduced. They have the following properties

$$C^{-1}\gamma_\mu C = -\gamma_\mu^T$$

$$C^{-1}\gamma_5 C = \gamma_5^T$$

$$C^\dagger = C^{-1}$$

$$C^T = -C \qquad (A20)$$

and

$$T^{-1}\gamma_\mu T = \varepsilon(\mu)\,\gamma_\mu^T$$

$$T^{-1}\gamma_5 T = -\gamma_5^T$$

$$T^\dagger = T^{-1}$$

$$T^T = -T \qquad (A21)$$

By direct inspection it is seen that

$$C = -i\,\gamma_0\,\gamma_2 \qquad (A22)$$

and

$$T = i\,\gamma_2\,\gamma_5 \qquad (A23)$$

satisfy these conditions. In the standard representation

(A3) they are given by

$$C = -i\begin{vmatrix} 0 & \sigma_2 \\ \sigma_2 & 0 \end{vmatrix}$$

$$T = i\begin{vmatrix} -\sigma_2 & 0 \\ 0 & \sigma_2 \end{vmatrix} \qquad \text{(A24)}$$

For applications regarding P, C and T we note the following relations, which can be obtained, for example, by direct inspection from the eqs. (A14) and (A15) with the explicit representation (A3) for the γ-matrices

$$u(-\bar{p},\pm\bar{s}) = \gamma_o u(\bar{p},\pm\bar{s})$$

$$v(-\bar{p},\pm\bar{s}) = -\gamma_o v(\bar{p},\pm\bar{s}) \qquad \text{(A25)}$$

with

$$u(\bar{p},\bar{s}) \equiv u_1(\bar{p})$$

$$u(\bar{p},-\bar{s}) \equiv u_2(\bar{p}) \qquad \text{(A26)}$$

etc. Similarly

$$v(\bar{p},\bar{s}) = C\bar{u}^T(\bar{p},\bar{s})$$

$$\bar{v}(\bar{p},\bar{s}) = -u^T(\bar{p},\bar{s})C^{-1} \qquad \text{(A27)}$$

and

$$u(-\bar{p},-\bar{s}) = T\,\bar{u}^T(\bar{p},\bar{s})$$

$$\bar{u}(-\bar{p},-\bar{s}) = u^T(\bar{p},\bar{s})T^{-1} \qquad \text{(A28)}$$

References

1. J. Werle, Relativistic Theory of Reactions, North Holland Publishing Company, Amsterdam (1966).
2. E. P. Wigner, Gruppentheorie und ihre Anwendung auf die Quantenmechanik der Atomspektren, Friedrich Vieweg Verlag, Braunschweig, (1931).
3. N. I. Achieser and I. M. Glasmann, Theorie der linearen Operatoren im Hilbertraum, Akademieverlag Berlin (1965).
4. See for example R. F. Streater and A. S. Wightman, PCT, Spin and Statistics, and all that, W. A. Benjamin, Inc., New York (1964).
5. W. Pauli, "Niels Bohr and the Development of Physics", McGraw-Hill, New York (1955); G. Lüders, Kongl. Dansk Medd. Fys. 28, No.5(1954); Annals of Physics 2, 1 (1957); R. F. Streater and A. S. Wightman, "PCT,Spin and Statistics, and all that", W. A. Benjamin, New York (1964).
6. T. D. Lee, R. Oehme and C. N. Yang, Phys. Rev. 106, 340 (1957).
7. J. Bernstein, G. Feinberg and T. D. Lee, Phys. Rev. 139, B1650 (1965).
8. T. D. Lee and G. C. Wick, Phys. Rev. 148, 1385 (1966).

SEMI - LEPTONIC WEAK DECAY PROCESSES†

By

H. PIETSCHMANN

Physikalisches Institut, Universität Bonn

and

Institut für Theoretische Physik,

Universität Wien

1. Introduction

These lectures are meant as an introduction to semi-leptonic weak decay processes. Form factors describing the influence of strong interactions on these processes are treated with the main emphasis. The lectures are intimately connected to those of Professor Jan Nilsson and frequent reference is made to his lectures.

The formalism for semi-leptonic weak decay processes is built up with the minimum number of assumptions and thus represents nothing but a unified description of the empirical material without additional assumptions. The consequence of 3 additional assumptions (G-parity properties, conserved vector current and Cabibbo's assumptions) are then derived. Finally some recent developments in $K^{\pm}_{\ell 3}$ form factor physics are presented.

Most of the calculations are given in detail. For further elaboration on some techniques the reader is re-

---

† Lecture given at the VII. Internationale Universitätswochen für Kernphysik, Schladming, February 26 - March 9, 1968.

ferred to ref. 1.

2. Phenomenology of Semi-Leptonic Baryon Decays

The oldest example of a semi-leptonic baryon decay is the decay of the free neutron. After the discovery of hyperons, strangeness changing semi-leptonic decays have been observed, however no decay has been observed so far that changes strangeness by more than 1 unit. Therefore, as an empirical result, we may post the selection rule

$$|\Delta S| \leq 1 \tag{1}$$

The support of this rule from semi-leptonic processes is not too overwhelming but in weak hadronic processes there is one very convincing example, the K_1^o - K_2^o mass difference. It is a $\Delta S = 2$ phenomenon and quantitatively, it is of second order in the weak coupling constant, proving that $\Delta S = 2$ processes are absent to first order in G.

Assuming the selection rule (1), there is a total number of 34 semi-leptonic baryon decays compatible with kinematical constraints. In this number, muonic and electronic modes of otherwise the same decay have not been counted separately. These 34 decay processes are exhibited in tables 1, 2 and 3. We have included Σ^o decays although they have to compete with electromagnetic decay modes and will therefore not be observed in the near future. We did not take into account strongly decaying baryons.

Notice that there are no muonic decay processes with $\Delta S = 0$. If we define a "hadronic charge change" by

$$\Delta Q \equiv (Q_f - Q_i)_{\text{hadrons}} \tag{2}$$

where Q_f and Q_i are the charges of final and initial hadrons, only one of the 6 decays with $\Delta S = 0$ has $\Delta Q = -1$, whereas 5 have $\Delta Q = +1$.

The strangeness change in tables 2 and 3 is defined analogous to eq. (2) by

$$\Delta S = S_f - S_i \quad . \tag{3}$$

Table 1:

Strangeness conserving semi-leptonic baryon decays.

observed:	not yet observed:
$n \to p + e^- + \bar{\nu}_e$	$\Sigma^0 \to \Sigma^+ + e^- + \bar{\nu}_e$
$\Sigma^- \to \Lambda + e^- + \bar{\nu}_e$	$\Sigma^- \to \Sigma^0 + e^- + \bar{\nu}_e$
$\Sigma^+ \to \Lambda + e^+ + \nu_e$	$\Xi^- \to \Xi^0 + e^- + \bar{\nu}_e$

Table 2:

$|\Delta S| = 1$ decays of baryons with 3 final particles.

$\Delta S = \Delta Q$	$\Delta S = -\Delta Q$
$\Lambda \to p + \ell^- + \bar{\nu}_\ell$ observed	$\Sigma^+ \to n + \ell^+ + \nu_\ell$ observed
$\Sigma^0 \to p + \ell^- + \bar{\nu}_\ell$	$\Xi^0 \to \Sigma^- + \ell^+ + \nu_\ell$
$\Sigma^- \to n + \ell^- + \bar{\nu}_\ell$ observed	
$\Xi^0 \to \Sigma^+ + \ell^- + \bar{\nu}_\ell$	

$\Xi^- \to \Lambda + \ell^- + \bar{\nu}_\ell$ observed	
$\Xi^- \to \Sigma^0 + \ell^- + \bar{\nu}_\ell$	
$\Omega^- \to \Xi^0 + \ell^- + \bar{\nu}_\ell$	

Table 3:

$|\Delta S| = 1$ decays of baryons with more than 3 final particles.

$\Delta S = \Delta Q$	$\Delta S = - \Delta Q$
$\Lambda \to p + \pi^0 + e^- + \bar{\nu}_e$	$\Lambda \to n + \pi^- + e^+ + \nu_e$
$\Lambda \to n + \pi^+ + e^- + \bar{\nu}_e$	$\Sigma^0 \to n + \pi^- + \ell^+ + \nu_\ell$
$\Sigma^0 \to p + \pi^0 + \ell^- + \bar{\nu}_\ell$	$\Sigma^+ \to p + \pi^- + \ell^+ + \nu_\ell$
$\Sigma^0 \to n + \pi^+ + \ell^- + \bar{\nu}_\ell$	$\Omega^- \to \Xi^- + \pi^- + \ell^+ + \nu_\ell$
$\Sigma^- \to n + \pi^0 + \ell^- + \bar{\nu}_\ell$	$\Omega^- \to \Xi^- + \pi^- + \pi^0 + e^+ + \nu_e$
$\Sigma^- \to p + \pi^- + \ell^- + \bar{\nu}_\ell$	$\Sigma^+ \to n + \pi^0 + \ell^+ + \nu_\ell$
$\Xi^- \to \Lambda + \pi^0 + e^- + \bar{\nu}_e$	$\Omega^- \to \Xi^0 + \pi^- + \pi^- + e^+ + \nu_e$
$\Omega^- \to \Xi^0 + \pi^0 + \ell^- + \bar{\nu}_\ell$	
$\Omega^- \to \Xi^- + \pi^+ + \ell^- + \bar{\nu}_\ell$	
$\Omega^- \to \Xi^0 + \pi^0 + \pi^0 + e^- + \bar{\nu}_e$	
$\Omega^- \to \Xi^0 + \pi^+ + \pi^- + e^- + \bar{\nu}_e$	
$\Omega^- \to \Xi^- + \pi^+ + \pi^0 + e^- + \bar{\nu}_e$	

There is an inherent asymmetry in the quantum numbers of baryons (no positive strangeness occurs among baryons) so that only 2 of the 9 $|\Delta S| = 1$ decays with 3 final particles have $\Delta S = -\Delta Q$. But there is no kinematical reason why the Σ^+ decay, for example, should not occur with similar frequency to the Σ^- decay. Yet it takes 2 or 3 years before another Σ^+ decay is found and only 3 have been accumulated so far [2]. (They occurred in 1962, 1964 and 1967). On the other hand, several hundred Σ^- decays are collected so that we can set up another empirical selection rule

$$\Delta S = \Delta Q \tag{4}$$

keeping in mind that it is probably not a strict rule but that it holds to a good accuracy.

The Gell-Mann-Nishijima relation

$$Q = I_3 + \frac{1}{2}(N+S) \tag{5}$$

allows for a relation of selection rule (4) to isospin selection rules. Since the baryon number is strictly conserved and Q in eq. (5) refers to hadronic charge only, we have

$$\Delta Q = \Delta I_3 + \frac{1}{2}\Delta S \tag{6}$$

Consequently

$$\Delta Q = \Delta S \Longrightarrow \Delta I_3 = \pm\frac{1}{2} \Longrightarrow \Delta I \geq \frac{1}{2}$$

$$\Delta Q = -\Delta S \Longrightarrow \Delta I_3 = \pm\frac{3}{2} \Longrightarrow \Delta I \geq \frac{3}{2}$$

where ΔI is defined by

$$\Delta I = |\vec{I}_f - \vec{I}_i| \tag{7}$$

Thus $\Delta Q = -\Delta S$ processes require at least $\Delta I = \frac{3}{2}$ and are therefore of more complex structure than $\Delta Q = \Delta S$ processes which are compatible with $\Delta I = \frac{1}{2}$.

Although Ω^- decays have been listed in tables 2 and 3, they shall be excluded from the subsequent discussion because the Ω^- has probably spin 3/2.

3. Description of a General Baryon Decay

In the following we shall set up the formalism to describe a general semi-leptonic baryon decay

$$A \rightarrow B + \ell^- + \bar{\nu}_\ell \tag{8}$$

It is depicted in fig. 1 where the momenta of the particles are also drawn. The "black box" in fig. 1 describes the so-called interaction region where particles are close enough together in space time so that strong interactions come into play. Consequently, we have little knowledge about this region and a phenomenological description has to start from the observation that particle A is annihilated at point x_2 where it enters the interaction region. At point x_1, particle B is created and at x, the lepton pair is emitted. In principle, the 2 leptons could leave the interaction region at different points, but it seems to be justified empirically to assume "local action of the lepton current" which is precisely the assumption that leptons are emitted at one and the same point.

Outside the interaction region, particles can be treated as free particles as long as we neglect electromagnetic corrections.

This can be done because electromagnetic corrections can be attached to experimental results so that a compa-

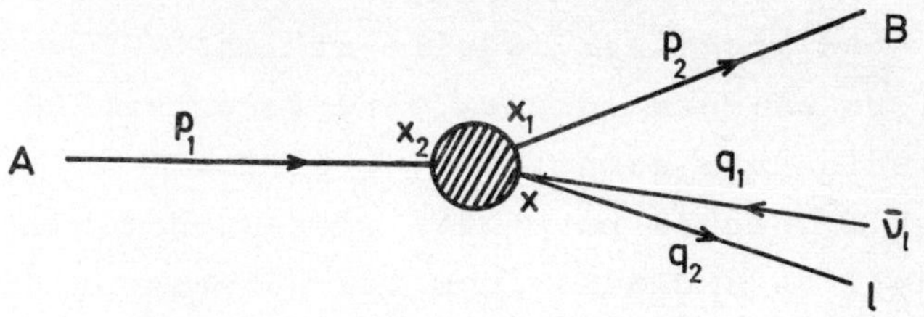

Fig. 1.: The general semi-leptonic baryon decay.

rison of the net effect is possible.

The creation and annihilation of a free fermion is described by the field operator

$$\psi(x) =$$

$$= \frac{1}{(2\pi)^{3/2}} \int d^3p \sqrt{m/p_0} \sum_{r=1}^{2} \{a_r(p) u_r(p) e^{-ipx} +$$

$$+ b_r^+(p) v_r(p) e^{ipx} \} \qquad (9)$$

The index r refers to spin states. $a_r(p)$ is the annihilation operator for a fermion of momentum p and spin r and $b_r^+(p)$ is the corresponding creation operator for the anti-fermion. The fact that one and the same field operator destroys a fermion and creates an anti-fermion is an example for the so-called "substitution law" which states that a particle line can be bent back and forth in a diagram. This gives rise to different processes where a particle on one side of the process equation can be brought onto the other side provided it is changed into the antiparticle. The function which describes the interaction region remains unchanged under these operations. We shall frequently make use of this possibility.

In order to write down an S-operator for processes (8), we shall have to make one more assumption, namely the "V-A" nature of the lepton current. This assumption is well supported by experiments [1]. Before writing down the S-operator, the 3 assumptions we have made on the way shall be collected here.

a) Local action of the lepton current

b) Neglect of electromagnetic corrections

c) V-A nature of the lepton current.

It then follows from Lorentz invariance that the S-operator is of the form

$$S_{eff} = I - \frac{iG}{\sqrt{2}} \int d^4x \, d^4x_1 d^4x_2 \bar{\psi}_B(x_1)\tilde{\Gamma}_\lambda(x_1,x_2,x)\psi_A(x_2)\ell^\lambda(x) \tag{10}$$

with

$$\ell^\lambda(x) = \bar{\psi}_\ell(x) \gamma^\lambda(1 + \gamma_5)\psi_{\nu_\ell}(x) \tag{11}$$

This S-operator is unitary only up to order G and we have therefore denoted it by S_{eff} (for "effective S-operator"). $\tilde{\Gamma}_\lambda(x_1,x_2,x)$ describes the interaction region and is therefore unknown due to our ignorance on strong interactions.

It is now convenient to define a "baryon current" by

$$j^+_\lambda = \int d^4x_1 d^4x_2 \; \bar{\psi}_B(x_1)\tilde{\Gamma}_\lambda(x_1,x_2,x) \; \psi_A(x_2) \tag{12}$$

so that the S-operator becomes

$$S_{eff} = I - \frac{iG}{\sqrt{2}} \int d^4x \; \{j^+_\lambda(x)\ell^\lambda(x) + h.c.\} \tag{13}$$

Here, we have included the hermitian conjugate (h.c.) which describes the corresponding antiparticle process $\bar{A} \to \bar{B} + \ell^+ + \nu_\ell$ The coupling constant G is as yet undefined because the vertex function $\tilde{\Gamma}_\lambda(x_1,x_2,x)$ has not been normalized so far.

The S-matrix element describing a particular kinematical configuration of process (8) is

$$<p_2, q_1, q_2 | S_{eff} | p_1> =$$

$$= \frac{-iG}{\sqrt{2}} \int d^4x \; <p_2 | j^+_\lambda(x_2) | p_1><q_1 q_2 | \ell^\lambda(x) | 0> \qquad (14)$$

The matrix element of the lepton current can directly be worked out to give

$$<q_1 q_2 | \ell^\lambda(x) | 0> =$$

$$= \frac{1}{(2\pi)^3} \sqrt{m_\ell m_\nu / q_1^o q_2^o} \; \bar{u}_\ell(q_2) \gamma^\lambda (1+\gamma_5) v_\nu(q_1) e^{ix(q_1+q_2)} \qquad (15)$$

Note that we have attached a small nonvanishing mass to the neutrino in order to be able to use a common normalization. This mass is put equal to zero in the final results.

Because of translation invariance, the vertex function $\tilde{\Gamma}_\lambda$ can depend only on the difference of space-time points. Moreover, the x-dependence of the field operator is determined, once a definite matrix element is taken. Hence

$$<p_2 | j^+_\lambda(x) | p_1> =$$

$$= \int d^4x_1 d^4x_2 <p_2 | \bar{\psi}_B(o) \tilde{\Gamma}_\lambda(x_1 - x, x_2 - x) \psi_A(o) | p_1> e^{i(x_1 p_2 - x_2 p_1)}$$

A shift of integration variables yields

$$<p_2 | j^+_\lambda(x) | p_1> = \int d^4\xi_1 d^4\xi_2 \; \times$$

$$\times \; <p_2 | \bar{\psi}_B(o) \tilde{\Gamma}_\lambda(\xi_1, \xi_2) \psi_A(o) | p_1> \; e^{ix(p_2 - p_1)} \; e^{i(\xi_1 p_2 - \xi_2 p_1)}$$

so that we conclude

$$\langle p_2 | j^+_\lambda(x) | p_1\rangle = \langle p_2 | j^+_\lambda(o) | p_1\rangle \, e^{ix(p_2 - p_1)} \tag{16}$$

Defining the Fourier transform of the vertex function by

$$\Gamma_\lambda(p_1, p_2) = \int d^4\xi_1 \, d^4\xi_2 \, \tilde{\Gamma}_\lambda(\xi_1, \xi_2) \, e^{i(\xi_1 p_2 - \xi_2 p_1)} \tag{17}$$

we finally arrive at

$$\langle p_2 | j^+_\lambda(x) | p_1\rangle = \langle p_2 | \bar{\psi}_B(x) \Gamma_\lambda(p_1, p_2) \psi_A(x) | p_1\rangle =$$

$$= \frac{1}{\sqrt{(2\pi)^3}} \sqrt{m_A m_B / p_1^o p_2^o} \;\; \bar{u}(p_2) \Gamma_\lambda(p_1, p_2) u(p_1) e^{ix(p_2 - p_1)} \tag{18}$$

The next question is: can we say anything more about the vertex function? Note that we have not assumed V-A nature for the baryonic current so far. In fact, $\tilde{\Gamma}_\lambda(\xi_1, \xi_2)$ may even include derivatives. But we know that it is a Lorentz vector. Therefore, its general form will be a linear combination of all possible vectors that can be built out of momentum and spin vectors. The coefficients are general functions of the only invariant variable that can be formed, $(p_1 p_2)$.

It is convenient to use sum and difference of the momenta p_1 and p_2 instead of p_1 and p_2 themselves

$$p = p_2 + p_1$$

$$q = p_2 - p_1 \tag{19}$$

The vector q has the physical interpretation of the four momentum transfer between the baryons. Likewise, q^2, the invariant four momentum transfer square, shall be used as argument for the coefficient functions instead of $(p_1 p_2)$.

There are 12 vectors and 12 axial vectors to be formed out of γ-matrices and momenta*

p_λ	q_λ	γ_λ
$p_\lambda \not{p}$	$q_\lambda \not{p}$	$i\sigma_{\lambda\nu} \; p^\nu$
$p_\lambda \not{q}$	$q_\lambda \not{q}$	$i \; \sigma_{\lambda\nu} \; q^\nu$
$ip_\lambda \; \sigma_{\nu\mu} \; p^\nu \; q^\mu$	$iq_\lambda \; \sigma_{\mu\nu} \; p^\mu \; q^\nu$	$i \; \gamma_\lambda \; \sigma_{\mu\nu} \; p^\mu \; q^\nu .$

The 12 axial vectors are obtained from the vectors by multiplication with γ_5 . To be sure that the list is complete, we note that

$$\not{p} \; \not{p} = p^2 = \Sigma^2 + \Delta^2 - q^2$$

$$\not{q} \; \not{q} = q^2$$

$$\not{p} \; \not{q} + \not{q} \; \not{p} = - \Sigma\Delta \qquad (20)$$

where

$$\Sigma = m_A + m_B$$

$$\Delta = m_A - m_B \qquad (21)$$

By means of eqs. (20), any vector containing $\not{p}$ or $\not{q}$ twice can be reduced to one of the 12 vectors times a function of q^2, which can be absorbed in the coefficient.

Application of the Dirac equation allows for an elimination of some of the vectors by means of

---

* Our definitions follow those of S.S.Schweber , H.A. Bethe, F.de Hoffmann, Mesons and Fields (Row, Peterson; Evanston 1955) except that our $\gamma_5 = i \; \gamma_0 \gamma_1 \gamma_2 \gamma_3$.

$$\bar{u}(p_2)\, \not{p}\, u(p_1) = \Sigma \bar{u}(p_2)\, u(p_1)$$

$$\bar{u}(p_2)\, \not{q}\, u(p_1) = -\Delta \bar{u}(p_2) u(p_1)\ .$$

$$\bar{u}(p_2) i\sigma_{\mu\nu}\, p^{\mu} q^{\nu}\, u(p_1) = (q^2-\Delta^2)\bar{u}(p_2) u(p_1)$$

$$\bar{u}(p_2) i\, \gamma_{\lambda}\, \sigma_{\mu\nu} p^{\mu} q^{\nu} u(p_1) =$$

$$= (q^2-\Sigma^2)\bar{u}(p_2)\gamma_{\lambda} u(p_1) + 2m_A \bar{u}(p_2)(p+q)_{\lambda} u(p_1)$$

Similar relations hold for the axial vectors. Hence there remain only 5 vectors and 5 axial vectors:

$$\gamma_{\lambda},\ i\sigma_{\lambda\nu} q^{\nu},\ i\sigma_{\lambda\nu} p^{\nu},\ p_{\lambda},\ q_{\lambda}$$

$$\gamma_{\lambda}\gamma_5,\ i\sigma_{\lambda\nu} q^{\nu}\gamma_5, i\sigma_{\lambda\nu} p^{\nu}\gamma_5,\ p_{\lambda}\gamma_5,\ q_{\lambda}\gamma_5$$

Between these 10 vectors, 4 relations hold because of the Dirac equation

$$\bar{u}(p_2)[i\sigma_{\mu\nu} q^{\nu} + p_{\mu} - \gamma_{\mu}\Sigma] u(p_1) = 0$$

$$\bar{u}(p_2)[i\sigma_{\mu\nu} p^{\nu} + q_{\mu} + \gamma_{\mu}\Delta] u(p_1) = 0$$

$$\bar{u}(p_2)[i\sigma_{\mu\nu} q^{\nu} + p_{\mu} + \gamma_{\mu}\Delta]\gamma_5 u(p_1) = 0$$

$$\bar{u}(p_2)[i\sigma_{\mu\nu} p^{\nu} + q_{\mu} - \gamma_{\mu}\Sigma]\gamma_5 u(p_1) = 0 \qquad (22)$$

Therefore, we can randomly drop 2 vectors and 2 axial vectors and write the vertex function in its most general form

$$\Gamma_{\lambda}(p_1, p_2) = F_1(q^2)\gamma_{\lambda} + F_2(q^2) i\sigma_{\lambda\nu} q^{\nu} + F_3(q^2) q_{\lambda} +$$

$$+ \left[G_1(q^2)\gamma_\lambda + G_2(q^2)q_\lambda + G_3(q^2)i\sigma_{\lambda\nu}q^\nu\right]\gamma_5 \qquad (23)$$

Hence we are able to describe the completely unknown interaction region by 6 functions of a single variable In Professor Nilssons lectures it is shown that these so-called form factors are real functions if time reversal invariance holds true.

4. Lifetime and μ - e Ratio

With eqs. (15)and (18), the S-matrix element (14) can be written in the following way:

$$S_{fi} = \frac{-iG}{\sqrt{2}} \frac{1}{(2\pi)^2} \delta^4(p_1-p_2-q_1-q_2) \left[\frac{m_A m_B m_\ell m_\nu}{p_1^0 p_2^0 q_1^0 q_2^0}\right]^{1/2} \times$$

$$\times \bar{u}_\ell(q_2)\gamma^\lambda(1+\gamma_5)v_\nu(q_1)\bar{u}_B(p_2)\Gamma_\lambda(p_1,p_2)u_A(p_1) \qquad (24)$$

The T-matrix element is defined by taking out the δ-function, i times the volume normalization $(2\pi)^{-2}$ and $1/\sqrt{2E}$ for each particle; it thus reads

$$T_{fi} = \frac{G}{\sqrt{2}} \left[16\ m_A m_B m_\ell m_\nu\right]^{1/2} \times$$

$$\times \bar{u}_\ell(q_2)\gamma^\lambda(1+\gamma_5)\ v_\nu(q_1)\bar{u}_B(p_2)\Gamma_\lambda(p_1,p_2)u_A(p_1) \qquad (25)$$

The "induced pseudoscalar and scalar" form factors, $G_2(q^2)$ and $F_3(q^2)$ respectively, are multiplied into the momentum transfer only (no γ-matrices except γ_5). Therefore, they can be taken together with the lepton current and by means of the Dirac equation, their contribution is shown to be of order m_ℓ/Σ.

e.g.:

$$\bar{u}_B(p_2)q_\lambda F_3(q^2)u_A(p_2).\bar{u}_\ell(q_2)\gamma^\lambda(1+\gamma_5)v_\nu(q_1) =$$

$$= - \bar{u}_B(p_2)F_3(q^2)u_A(p_2).\bar{u}_\ell(q_2)(\not{q}_1+\not{q}_2)(1+\gamma_5)v_\nu(q_1) =$$

$$= - \frac{m_\ell}{\Sigma} \bar{u}_B(p_2)\ \Sigma.F_3(q^2)u_A(p_2)\bar{u}_\ell(q_2)(1+\gamma_5)v_\nu(q_1)$$

(Note that $\Sigma F_3(q^2)$ is dimensionless). Hence these form factors can be neglected in a first approximation.

Similarly, the "weak magnetism and induced pseudo-tensor" form factors, $F_2(q^2)$ and $G_3(q^2)$ contribute to order q^λ/Σ or equivalently, Δ/Σ. Hence in a first approximation, only F_1 and G_1 contribute. In the same approximation, these 2 form factors may be taken at their zero momentum transfer value, assuming smooth variation over a range of order $(\Delta/\Sigma)^2$. (In the terminology of nuclear physics, this approximation corresponds to "allowed transitions".)

With the full vertex function, the decay rate is given by

$$\Gamma_{A,B,\ell} =$$

$$= \frac{1}{(2\pi)^5}\frac{1}{2m_A}\int\frac{d^3q_1}{2q_1^o}\frac{d^3q_2}{2q_2^o}\frac{d^3p_2}{2p_2^o}\delta^{(4)}(p_1-p_2-q_1-q_2)\frac{1}{2}\sum_{spins}|T_{fi}|^2 =$$

$$= \frac{1}{(2\pi)^5}\frac{G^2}{m_A}\int\frac{d^3q_1}{2q_1^o}\frac{d^3q_2}{2q_2^o}\frac{d^3p_2}{2p_2^o}\delta^{(4)}(p_1-p_2-q_1-q_2)\ \times$$

$$\times\ N_{\lambda\mu}\{q_1^\lambda q_2^\mu + q_2^\lambda q_1^\mu - g^{\lambda\mu}(q_1q_2) - i\varepsilon^{\lambda\alpha\mu\beta}q_{1\alpha}q_{2\beta}\} \tag{26}$$

where

$$N_{\lambda\mu} = \sum_{\text{spins}} \bar{u}_B(p_2)\Gamma_\lambda u_A(p_1)[\bar{u}_B(p_2)\Gamma_\mu u_A(p_1)]^* \, 4m_A m_B \tag{27}$$

The term with the ε-tensor can be dropped since it vanishes when we integrate over lepton momenta. The integration over lepton momenta is easily carried out by means of the general formulae

$$\int \frac{d^3q_1}{2q_1^o}\frac{d^3q_2}{2q_2^o}\delta^{(4)}(P-q_1-q_2)f(q_1q_2) =$$

$$= \frac{\pi}{2}\frac{1}{P^2}w(P^2;m_1^2,m_2^2)f[\tfrac{1}{2}(P^2-m_1^2-m_2^2)] \tag{28}$$

and

$$\int \frac{d^3q_1}{2\,q_1^o}\frac{d^3q_2}{2\,q_2^o}\delta^{(4)}(P-q_1-q_2)f(q_1q_2)q_1^\mu q_2^\nu =$$

$$= \frac{\pi}{12}\frac{1}{P^4}w(P^2,m_1^2,m_2^2)f[\tfrac{1}{2}(P^2-m_1^2-m_2^2)] \times$$
$$\times \{\frac{g^{\mu\nu}}{2}[w(P^2,m_1^2,m_2^2)]^2 + \frac{P^\mu P^\nu}{P^2}[P^4+P^2(m_1^2+m_2^2)-2(m_1^2-m_2^2)]\} \tag{29}$$

where $f(q_1q_2)$ are arbitrary functions and

$$w(a,b,c) = [a^2+b^2+c^2-2(ab+ac+bc)]^{1/2} \tag{30}$$

The decay rate (26) then becomes

$$\Gamma_{A,B,\ell} = \frac{1}{(2\pi)^5}\frac{G^2}{m_A}\frac{\pi}{12}\int\frac{d^3p_2}{2p_2^o}(1-\frac{m_\ell^2}{q^2})^2 N_{\lambda\mu} \quad \times$$

$$\times \quad \{2 \frac{q^\lambda q^\mu}{q^2} (q^2+2m_\ell^2) - g^{\lambda\mu}(2q^2+m_\ell^2)\} \quad .$$

The remaining integration over the baryon momentum can be cast into an integration over the recoil energy by means of

$$\int \frac{d^3p_2}{2p_2^o} f(q^2) = \int_{m_\ell^2}^{\infty} ds \int \frac{d^3p_2}{2p_2^o} f(s)\,\delta(s-q^2) =$$

$$= \frac{\pi}{2m_A^2} \int_{m_\ell^2}^{\Delta^2} ds \; [(\Sigma^2-s)(\Delta^2-s)]^{1/2} \; f(s)$$

s is related to the kinetic recoil energy T by

$$s = \Delta^2 - 2m_A T \tag{31}$$

If we now neglect all terms of order m_ℓ/Σ and Δ/Σ , the vertex function becomes

$$\Gamma_\lambda(p_1,p_2) = F_1(o)\,\gamma_\lambda + G_1(O)\gamma_\lambda\gamma_5 + O(\Delta/\Sigma, m_\ell/\Sigma) \tag{32}$$

and $N_{\lambda\mu}$ is easily calculated by means of the usual trace technique. It is convenient to define a "vector" and an "axial vector coupling constant" by

$$G_V = G\,F_1(o)$$

$$G_A = G\,G_1(o)$$

$$\lambda = G_A/G_V \tag{33}$$

The decay rate then becomes

$$\Gamma_{A,B,\ell} = \frac{G_V^2}{384\pi^3} \frac{1}{m_A^3} \int_{m_\ell^2}^{\Delta^2} ds (1-\frac{m_\ell^2}{s})^2 [(\Sigma^2-s)(\Delta^2-s)]^{1/2} \times$$

$$\times \left\{ \left[\Delta^2 (4s - m_\ell^2) + 2\Sigma^2\Delta^2 \frac{s+2m_\ell^2}{s} - (\Sigma^2+2s)(2s+m_\ell^2) \right] + \right.$$

$$\left. + \lambda^2 \left[\Sigma^2 (4s - m_\ell^2) + 2\ \Sigma^2\Delta^2 \frac{s+2m_\ell^2}{s} - (\Delta^2+2s)(2s+m_\ell^2) \right] \right\} \quad (34)$$

If s is expressed through eq. (31) and if the integral is dropped eq. (34) represents the recoil energy spectrum for process (8).

Defining dimensionless variables by

$$x = m_\ell / \Delta$$

$$\sigma = s/\Delta^2$$

$$\delta = \Delta/\Sigma \quad (35)$$

the decay rate becomes

$$\Gamma_{A,B,\ell} = \frac{G_V^2 \Delta^5}{48\pi^3} \left(\frac{\Sigma}{2m_A}\right)^3 \int_{x^2}^{1} d\sigma \left(1 - \frac{x^2}{\sigma}\right)^2 \sqrt{1-\sigma} \ \times$$

$$\times \left\{ x^2 \left(\frac{4}{\sigma} - 1\right) + 2(1-\sigma) + \lambda^2 \left[x^2 \left(\frac{4}{\sigma} - 1\right) + 2(1+2\sigma) \right] \right\}$$

$$+ O(\delta^2)$$

or, after integration

$$\Gamma_{A,B,\ell} = \frac{G_V^2\ \Delta^5}{60\pi^3} \left(\frac{\Sigma}{2m_A}\right)^3 (1+3\lambda^2) r(x) + O(\delta^2) \quad (36)$$

where r(x) is the μ-e ratio (note that $x \simeq o$ for the electronic mode!)

$$r(x) = \frac{\Gamma_{A,B,\mu}}{\Gamma_{A,B,e}} = \frac{1}{2}(1-x^2)^{3/2} (2-7x^2) + \frac{15}{4} x^4 \left[\log \frac{1+\sqrt{1-x^2}}{1-\sqrt{1-x^2}} - \right.$$

$$\left. - 2\sqrt{1-x^2} \right] \quad (37)$$

For the 3 observed decays [4], this ratio is exhibited in table 4.

Table 4:

μ-e ratio for some baryon decays.

Mode	$x = m_\mu/\Delta$	$r_{theor.}$	$r_{exp.}$
$\Lambda \to p+\ell^- +\bar{\nu}_\ell$	0.6	0.16	0.15 ± 0.07
$\Sigma^- \to n+\ell^- +\bar{\nu}_\ell$	0.41	0.46	0.50 ± 0.12
$\Xi^- \to \Lambda+\ell^- +\bar{\nu}_\ell$	0.51	0.27	?

The agreement wit experiments is good. Unfortunately, this does not test the V-A theory because the ratio is the same for P and S. It does however test the assumption of local action of the lepton current and improvement on the experimental numbers would thus be welcome. We notice in passing that several alternative models have been proposed in which local action of the lepton current is not incorporated. An example is the theory of S. V. Pepper et al., ref. 3, in which a neutral lepton current may act non-locally to first order in G.

5. First and Second Class Currents

So far, we have made no assumptions about the nature of the hadronic current or the vertex function. The form of eq. (23) is completely general.

One of the guide lines in constructing a physical theory is simplicity as well as universality. If these concepts are applied to the hadronic current, we have to assume that this current is of the V-A form, if strong interactions were absent. Since strong interactions are always present, this "bare current" is never observed. Nevertheless it is possible to draw some conclusions on the bare current by means of symmetry considerations. If the bare current is of V-A nature, the full current has to show the same properties under any symmetry operation that is conserved in strong interactions.

The concept of G-parity has proven to be particularly useful in this connection. G-parity is defined by

$$G = C\, e^{i\pi I_2} , \qquad (38)$$

where C is the charge conjugation operator defined in Professor Nilsson's lectures and $\exp(i\pi I_2)$ describes a rotation of 180° around the second axis in isospin space. G-parity provides an extension of charge conjugation symmetry to charged systems with $N = S = 0$. A charged state with $N = S = 0$ is transformed into a state with opposite charge by C. The rotation in isospin space turns the charge back to its original sign so that eigenstates of G can be constructed provided $N = S = 0$.

In widening the range of applicability, we have to pay the price of non-conservation by electromagnetic interactions. G-parity is a good quantum number only with respect to strong interactions because it includes isospin transformation.

In order to apply G-parity to $\Delta S = 0$ processes, we have to distinguish 2 cases:

case A: Transitions within an iso-multiplet

case B: Transitions between different iso-multiplets.

As an example for case A, let us consider the free neutron decay.

Using the isospin matrices

$$\tau_{\pm} = \frac{1}{\sqrt{2}} (\tau_1 \pm i\tau_2) \qquad (39)$$

the bare vector and axial vector current can be written

$$j_\lambda^V(x) = \bar{\psi}_N(x)\gamma_\lambda \frac{\tau_-}{\sqrt{2}} \psi_N(x)$$

$$j_\lambda^A(x) = \bar{\psi}_N(x)\gamma_\lambda\gamma_5 \frac{\tau_-}{\sqrt{2}} \psi_N(x) \quad . \tag{40}$$

We will show presently that these currents have definite transformation properties under G-parity. To this end, consider

$$G\, j_\lambda^V G^{-1} = \frac{1}{\sqrt{2}}\, G\, \bar{\psi}_N\, G^{-1}\, \gamma_\lambda \tau_-\, G\, \psi_N\, G^{-1} \tag{41}$$

Using

$$e^{iA} B\, e^{-iA} = B + i[A,B] + \frac{i^2}{2!}[A,[A,B]] + \ldots \tag{42}$$

and

$$[I_\ell, \psi_N(x)] = -\frac{1}{2}\tau_\ell\, \psi_N(x) \tag{43}$$

we obtain

$$G\, \psi_N\, G^{-1} = \exp\{-i\, \frac{\pi}{2}\, \tau_2\} C\, \psi_N\, C^{-1} = -i\, \tau_2 \mathcal{C} \bar{\psi}^T(x)$$

where $\mathcal{C}$ is the 4 × 4 matrix defined in Professor Nilsson's lectures and the transposition sign "T" does not include transposition in the Hilbert space of creation and destruction operators.

Similarly,

$$G\, \bar{\psi}_N G^{-1} = i\, \psi^T\, \mathcal{C}^{-1}\, \tau_2$$

so that with

$$\tau_2 \, \tau_- \, \tau_2 = - \, \tau_-^T$$

we end up with

$$G \, j_\lambda^V \, G^{-1} = \frac{1}{\sqrt{2}} \, \psi_N^T \, \mathcal{C}^{-1} \, \tau_-^T \, \gamma_\lambda \, \mathcal{C} \, \bar{\psi}_N^T =$$

$$= - \frac{1}{\sqrt{2}} \, \psi_N^T \, \tau_-^T \, \gamma_\lambda^T \, \bar{\psi}_N^T = \frac{1}{\sqrt{2}} \, \bar{\psi}_N \, \gamma_\lambda \, \tau_- \, \psi_N$$

The last minus sign stems from the anticommutativity of fermion field operators. The axial vector current acquires an additional minus sign from the anti-commutativity of γ_λ and γ_5. Therefore, the transformation properties under G are

$$G \, j_\lambda^V \, G^{-1} = j_\lambda^V$$

$$G \, j_\lambda^A \, G^{-1} = - \, j_\lambda^A \qquad (44)$$

It follows, that the full vector and axial vector current after switching on of strong interactions have to show the same transformation properties (44) because strong interactions cannot induce terms with wrong transformation properties unless they violate G-parity.

For the matrix element follows

$$\langle a | j_\lambda^i | b \rangle = \langle a | G^{-1} \, G \, j_\lambda^i \, G^{-1} \, G | b \rangle =$$

$$= \varepsilon_i \langle Ga | j_\lambda^i | Gb \rangle \qquad i = V, A \qquad (45)$$

with

$$\varepsilon_i = \left\{ \begin{matrix} +1 \\ -1 \end{matrix} \right. \text{ for } i = \left\{ \begin{matrix} V \\ A \end{matrix} \right. \qquad (46)$$

so that we can connect 2 processes by a G-transformation

in much the same way as it was done for C,P and T in Prof. Nilsson's lectures. For nucleons, the G transformation yields [1]

$$G|p> = -|\bar{n}>$$

$$G|n> = |\bar{p}> \tag{47}$$

and 2 processes connected by a G-transformation are depicted in fig. 2.

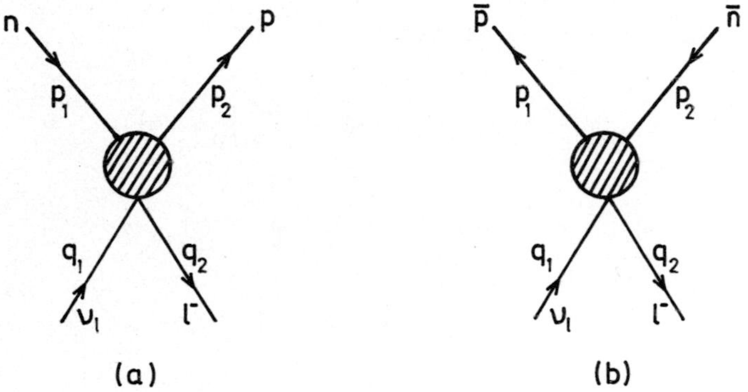

Fig. 2 : Two processes related by a G-transformation
(a) $\nu_\ell + n \rightarrow \ell^- + p$ and
(b) $\nu_\ell + \bar{p} \rightarrow \ell^- + \bar{n}$

Process (b) can be converted into process (a) with the help of the substitution law. Hence G-parity will yield restrictions on matrix elements for transitions within an isomultiplet.

The matrix element for process (b) is given by (recall eq.(12))

$$<\bar{n}(p_2)|j^+_\lambda(x)|\bar{p}(p_1)> =$$

$$= \frac{-1}{(2\pi)^3} \frac{M}{\sqrt{p_1^o p_2^o}} \int d^4x_1 d^4x_2 \; \bar{v}(p_1)\tilde{\Gamma}_\lambda(x_1-x,x_2-x) \; v(p_2)\times$$

$$\times \; \exp\{i(p_2x_2-p_1x_1)\} \tag{48}$$

where the minus sign stems from the interchange of antiparticle creation and destruction operators. M is the common nucleon mass. Shifting the integration and using eq. (17) yields

$$<\bar{n}(p_2)|j^+_\lambda(x)|\bar{p}(p_1)> =$$

$$= \frac{-1}{(2\pi)^3} \frac{M}{\sqrt{p^o_1 p^o_2}} e^{ix(p_2-p_1)} \bar{v}(p_1)\Gamma_\lambda(-p_2,-p_1)v(p_2)$$

Remember now the properties of the matrix $\mathcal{C}$ as given in Prof. Nilsson's lectures

$$\bar{v}(p_1)\Gamma_\lambda\, v(p_2) = - u^T(p_1)\, \mathcal{C}^{-1}\, \Gamma_\lambda \mathcal{C}\, \bar{u}^T(p_2) =$$

$$= - \bar{u}(p_2)\, \mathcal{C}\, \Gamma^T_\lambda\, \mathcal{C}^{-1}\, u(p_1)$$

Comparison with eq. (45) yields Weinberg's equation [5]

$$\Gamma^i_\lambda(p_1,p_2) = - \varepsilon_i\, \mathcal{C}\, \Gamma^{iT}_\lambda(-p_2,-p_1)\, \mathcal{C}^{-1}; \qquad i=V,A \qquad (49)$$

ε_i is defined by eq. (46).

If we start with a pure V-A theory, only those terms can be induced by strong interactions that obey eq. (49). Table 5 exhibits the transformation properties of the various terms in the vertex function.

It is seen that $F_3(q^2)$ and $G_3(q^2)$ must vanish in a pure V-A theory. Those contributions are called "second class currents".

A word of warning is here in order. So far we have shown the consequences of G-parity only for transitions within an isomultiplet. In a quark model with a bare V-A current no second class currents can occur. However, in other models, one may have V-A type transitions with $\Delta S = 0$ between different isomultiplets that do not show the required G-parity properties. By strong interactions, all these transitions are interconnected and thus second class currents may also be induced in n-decay for example. Hence the relation of first and second class currents to V-A type interactions is not model independent. The only model independent definition of first and second class currents is by means of eq. (49). A recent survey on the experimental side of the question has been made by P. Hertel [6].

$\Gamma_\lambda(p_1,p_2)$	$-\mathcal{C}\Gamma_\lambda^T(-p_2,-p_1)\mathcal{C}^{-1}$	parity	G-parity
γ_λ	γ_λ	V	even
$i\,\sigma_{\lambda\nu}\,q^\nu$	$i\,\sigma_{\lambda\nu}\,q^\nu$	V	even
q_λ	$-q_\lambda$	V	odd
$\gamma_\lambda\,\gamma_5$	$-\gamma_\lambda\,\gamma_5$	A	odd
$q_\lambda\,\gamma_5$	$-q_\lambda\,\gamma_5$	A	odd
$i\,\sigma_{\lambda\nu}\,q^\nu\gamma_5$	$i\,\sigma_{\lambda\nu}\,q^\nu\gamma_5$	A	even

Table 5:

Transformation properties of vertex terms under G-parity

Let us now turn to case B, i.e. transitions between different isomultiplets. The processes

$$\Sigma^+ \to \Lambda + e^+ + \nu_e$$

$$\Sigma^- \to \Lambda + e^- + \bar{\nu}_e \qquad (50)$$

may serve as example. Two processes related by a G-transformation are depicted in fig. 3.

We have

$$G|\Lambda> = |\bar{\Lambda}>$$

$$G|\Sigma^+> = -\,|\bar{\Sigma}^-> \qquad (51)$$

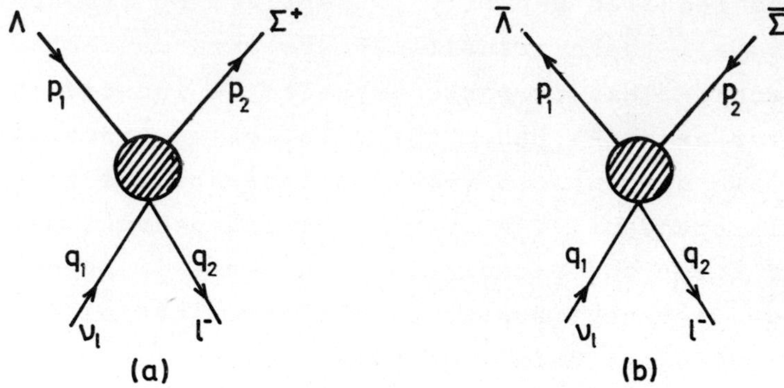

Fig. 3: Two processes related by a G-transformation.
(a) $\nu_\ell + \Lambda \to \ell^- + \Sigma^+$ and
(b) $\nu_\ell + \bar{\Lambda} \to \ell^- + \overline{\Sigma^-}$

Here, the substitution law allows for a transformation of process (b) into the Σ^- - Λ decay. Therefore, G-parity establishes a relation between the 2 processes (50). If only first class currents are present, we have

$$\langle\Lambda|j^i_\lambda|\Sigma^+\rangle = - \varepsilon_i \langle\bar{\Lambda}|j^i_\lambda|\overline{\Sigma^-}\rangle \qquad (52)$$

The second matrix element can be recast into the following form

$$\langle\bar{\Lambda}|j_\lambda(o)|\overline{\Sigma^-}\rangle = - \mathcal{N}\, \bar{v}_{\Sigma^-} \Gamma_\lambda v_\Lambda = + \mathcal{N}\, \bar{u}_\Lambda \mathcal{C}\, \Gamma_\lambda{}^T \mathcal{C}^{-1} u_{\Sigma^-}$$

$$\mathcal{N} = \frac{1}{(2\pi)^3} \left|\frac{m_\Sigma m_\Lambda}{E_\Sigma E_\Lambda}\right|^{1/2}$$

so that eq. (52) requires

$$\Gamma^i_\lambda(\Sigma^+\Lambda) = - \varepsilon_i\, \mathcal{C}\, \Gamma^{iT}_\lambda(\Sigma^-\Lambda)\, \mathcal{C}^{-1} \; ; \qquad i=V,A \qquad (53)$$

The known transformation properties of Γ_λ (table 5) then require

$$F_i^{\Sigma^+} = F_i^{\Sigma^-}$$

$$G_i^{\Sigma^+} = G_i^{\Sigma^-} \qquad i=1,2 \tag{54}$$

and

$$F_3^{\Sigma^+} = - F_3^{\Sigma^-}$$

$$G_3^{\Sigma^+} = - G_3^{\Sigma^-} \tag{55}$$

Since F_i and G_i can be neglected for $i = 2,3$ to a very good accuracy, eq. (36) yields

$$\frac{\Gamma_{\Sigma^-\Lambda}}{\Gamma_{\Sigma^+\Lambda}} = \left[\frac{m_{\Sigma^-} - m_\Lambda}{m_{\Sigma^+} - m_\Lambda}\right]^5 = 1.64 \tag{56}$$

The experimental value [4] for this ratio is

$$(\Gamma_{\Sigma^-\Lambda}/\Gamma_{\Sigma^+\Lambda})_{exp} = 1.5 \pm 0.6 \tag{57}$$

The agreement is good, but errors are still far too large to exclude second class currents.

6. Consequences of CVC for $\Delta S = 0$ Processes

There exists a broad literature on the conserved vector current (CVC) theory [7]. We can therefore restrict the discussion to some parts of the theory that fit particularly well into this discussion.

The CVC theory consists of 2 distinct hypotheses

i) the weak vector current is conserved

ii) the electromagnetic isovector current and the weak vector currents (divided by $\sqrt{2}$) are components of one and the same isovector current.

Historically, hypothesis i) has been suggested by Gershtein and Zel'dovich [8] while the full theory was proposed by Feynman and Gell-Mann [9]. It is interesting to observe that neither of the 2 hypotheses is a pre-requisite of the other one. A conserved vector current that does not coincide with the rotated electromagnetic isovector current is easily conceivable. However, there could be a situation in which the vector current contains in addition to the rotated electromagnetic isovector current a second class current that breaks the conservation equation.

CVC requires for the vector part of the vertex function

$$q^{\lambda}\bar{u}_B(p_2)\Gamma_{\lambda}{}^{V}(p_1,p_2)\; u_A(p_1) =$$

$$= \bar{u}_B(p_2)\{\not{q}\; F_1(q^2) + q^2F_3(q^2)\}u_A(p_1) = 0 \qquad (58)$$

or, equivalently,

$$\Delta F_1(q^2) = q^2F_3(q^2) \qquad (59)$$

From now on, we have to distinguish the 2 cases again.

Case A

transitions within an isomultiplet:
In this case, we have by definition

$$\Delta = 0 \quad \Longrightarrow \quad F_3(q^2) = 0 \qquad (60)$$

Hence CVC excludes the second class vector current in case A.

Case B

transitions between different iso-multiplets:

$$\Delta \neq 0 \implies F_1(o) = 0 \tag{61}$$

Hence in the allowed approximation, the transition is a pure axial vector transition. In the case of Σ-Λ decay, 2 recent measurements have been reported

$$\text{ref.10:} \quad |V/A| = 0.31 \pm 0.30$$

$$\text{ref.11:} \quad V/A = -0.3 \pm 0.4 \tag{62}$$

Improvement on the errors would be extremely welcome.

7. Cabibbo's Assumptions

So far we have mainly concentrated on $\Delta S = 0$ transitions. An elegant unification of all semi-leptonic baryon decays is provided by Cabibbo [12] in the framework of unitary symmetry SU(3). Cabibbo's theory is based on 3 assumptions:

i) the total hadronic current J_μ transforms as a component of an octet of SU(3),

ii) the vector part of J_μ is in the same octet as is the electromagnetic current,

iii) the assumption of Gell-Mann and Lévy [13]

$$J_\mu = \cos\theta \; j_\mu + \sin\theta \; \hat{J}_\mu$$

where j_μ is the $\Delta S = 0$ current and $\hat{J}_\mu$ is the $|\Delta S| = 1$ current. θ is called "Cabibbo angle".

Ass. (i) excludes $|\Delta S| > 1$ transitions as well as $\Delta S = - \Delta Q$ decays. Ass.(ii) is a generalization of the second hypothesis of CVC to $|\Delta S| = 1$ transitions and ass. (iii) tells us that the coupling constants for semi-leptonic processes are

$$G' = G \cos\theta \qquad \text{for} \quad \Delta S = 0$$

$$G'' = G \sin\theta \qquad \text{for} \quad |\Delta S| = 1 \qquad (63)$$

where G is the leptonic coupling constant from μ- decay. Only now is the coupling of eq. (10) defined. Without this definition (that replaces old universal fermi interaction which put G = G' = G") even the result of CVC, $F_1(o) = 1$, is an empty statement because only the product $G_V = G' F_1(o)$ is measurable.

In the allowed approximation, the vertex function is given by eq. (32). Ass. (i) of Cabibbo allows us to write

$$\left.\begin{aligned} F_1(o) &= c_f\, F_1^f + c_d\, F_1^d \\ G_1(o) &= c_f\, G_1^f + c_d\, G_1^d \end{aligned}\right\} \quad \Delta S = 0 \qquad (64)$$

$$\left.\begin{aligned} F_1(o) &= C_f\, F_1^f + C_d\, F_1^d \\ G_1(o) &= C_f\, G_1^f + C_d\, G_1^d \end{aligned}\right\} \quad |\Delta S| = 1 \qquad (65)$$

where $c_{f,d}$ and $C_{f,d}$ are Clebsch-Gordan coefficients of SU(3). They can be read off the following traces

$$c_f \; : \; \mathrm{Tr}\; \bar{B}[E_1, B] = \bar{p}n - \sqrt{2}(\bar{\Sigma}^+\Sigma^o - \bar{\Sigma}^o\Sigma^-) - \bar{\Xi}^o\Xi^-$$

$$c_d \; : \; \mathrm{Tr}\; \bar{B}\{E_1, B\} = \bar{p}n + \sqrt{2/3}(\bar{\Sigma}^+\Lambda + \bar{\Lambda}\,\Sigma^-) + \bar{\Xi}^o\Xi^- \qquad (66)$$

$$C_f \; : \; \mathrm{Tr}\; \bar{B}[E_2, B] = -\sqrt{3/2}\;\bar{p}\Lambda - \frac{1}{\sqrt{2}}\,\bar{p}\Sigma^o - \bar{n}\Sigma^- + \frac{1}{\sqrt{2}}\,\bar{\Sigma}^o\Sigma^- + \bar{\Sigma}^+\Xi^o + \sqrt{3/2}\;\bar{\Lambda}\Xi^-$$

$$C_d : \mathrm{Tr}\, \bar{B}\{E_2, B\} = -\frac{1}{\sqrt{6}}\bar{p}\Lambda + \frac{1}{\sqrt{2}}\bar{p}\Sigma^0 + \bar{n}\Sigma^- + \frac{1}{\sqrt{2}}\bar{\Sigma}^0\Xi^- + $$

$$+ \bar{\Sigma}^+\Xi^0 - \frac{1}{\sqrt{6}}\bar{\Lambda}\Xi^- \qquad (67)$$

where B and $\bar{B}$ are the familiar 3 × 3 baryon matrices of SU(3) [14] and E_1 and E_2 are the $\Delta S = 0$ and $\Delta S = \Delta Q$ transition matrices of SU(3).

There remain 4 unknown parameters, $F_1^{f,d}$ and $G_1^{f,d}$ (the Cabibbo angle can be determined from meson decays to be about 15°). Ass. (ii) determines 2 of them because the electromagnetic current is a pure f-type current

$$F_1^f = 1$$

$$F_1^d = 0 \qquad (68)$$

From n-decay, we know

$$G_1^d + G_2^f = G_1(o)\Big|_{n,p} = 1{,}2 \qquad (69)$$

In section 6 we have learned that the Σ-Λ decay is a pure axial transition. In fact, c_f is zero for $(\Sigma\Lambda)$, so that

$$\Gamma_\lambda(\Sigma\Lambda) = \sqrt{2/3}\, G_1^d\, \gamma_\lambda\gamma_5 \qquad (70)$$

From the experimental lifetime [4], G_1^d is determined to be

$$G_1^d = 0.75 \qquad (71)$$

so that, from eq. (69)

$$G_1^f = 0.45 \qquad (72)$$

It is now possible to write down all semi-leptonic baryon decays without further parameters. As an example, we choose the Λ-p decay. From eqs. (65), (67), (68),(71) and (72), we find

$$\Gamma_\lambda(\Lambda p) = - \sqrt{3/2}\, F_1^f \gamma_\lambda - [\sqrt{3/2}\;\; G_1^f + \frac{1}{\sqrt{6}} G_1^d]\gamma_\lambda\gamma_5 =$$

$$= - \sqrt{3/2}\,(\gamma_\lambda + 0.7\,\gamma_\lambda\gamma_5) \qquad (73)$$

This result is also borne out in the quark model without appeal to ΣΛ decay [15].

A better way to determine G_1^f and G_1^d is of course to make a least square fit to all known data. This has been done by Brene et al., ref. 16.

8. Phenomenology of Semi-Leptonic Meson Decays

A convenient notation for a semi-leptonic decay of a pseudoscalar meson P is $P_{\ell n}$, where ℓ specifies the type of leptons (μ or e) and n is the total number of particles in the final state. All kinematically allowed semi-leptonic meson decays are exhibited in tables 6 and 7. Just as in the case of baryon decays the $\Delta S = \Delta Q$ rule is empirically well satisfied, that is to say $\Delta S = -\Delta Q$ processes are strongly suppressed. We shall restrict ourselves to $P_{\ell 2}$ and $P_{\ell 3}$ decays in the following sections.

Table 6:

Semi-leptonic pion decays ($\Delta S = 0$)

$\pi_{\ell 2}$	$\pi^+ \to \ell^+ + \nu_\ell$
π_{e3}	$\pi^+ \to \pi^0 + e^+ + \nu_e$

Table 7:

Semi-leptonic kaon decay

	$\Delta Q = \Delta S$	$\Delta Q = -\Delta S$
$K_{\ell 2}$	$K^+ \to \ell^+ + \nu_\ell$	-
$K_{\ell 3}$	$K^+ \to \pi^0 + \ell^+ + \nu_\ell$	-
$K_{\ell 4}$	$K^+ \to \pi^0 + \pi^0 + \ell^+ + \nu_\ell$ $K^+ \to \pi^+ + \pi^- + \ell^- + \bar{\nu}_\ell$	$K^+ \to \pi^+ + \pi^+ + \ell^- + \bar{\nu}_\ell$
$K_{\ell 5}$	$K^+ \to 3\pi^0 + e^- + \bar{\nu}_e$ $K^+ \to \pi^+ + \pi^- + \pi^0 + e^+ + \nu_e$	$K^+ \to \pi^+ + \pi^+ + \pi^0 + e^+ + \nu_e$

9. $P_{\ell 2}$ - Decays

The matrix element of the hadron current between a meson and the vacuum can be analyzed in exactly the same way as the baryonic matrix element in section 3. However, since there is no spin involved and since only one momentum vector is available, the matrix element has a particularly simple structure:

$$<0|J_\lambda(x)|P(k)> = \frac{1}{(2\pi)^{3/2}} \frac{1}{\sqrt{2k^0}} f_p \, k_\lambda \, e^{-ikx} \tag{74}$$

It is immedieately seen, that a conserved current would lead to a vanishing f_p and the $P_{\ell 2}$ decay would be forbidden. It follows that the axial vector cannot be conserved, because $P_{\ell 2}$ decays are the dominant decay channels of both π and K. By means of the general formulae of section 4, the decay rate is easily worked out to be

$$\Gamma(P_{\ell 2}) = \frac{(Gf_p)^2}{8\pi} \left(\frac{m_\ell}{m_p}\right)^2 \frac{(m_p^2 - m_\ell^2)^2}{m_p} \tag{75}$$

From eq. (75), the μ-e ratio for $P_{\ell 2}$ decays is obtained to be

$$R_P \equiv \frac{\Gamma(P_{e2})}{\Gamma(P_{\mu 2})} = \left(\frac{m_e}{m_\mu}\right)^2 \left(\frac{m_P^2-m_e^2}{m_P^2-m_\mu^2}\right)^2 \qquad (76)$$

The famous factor $(m_e/m_\mu)^2$ is typical for the V-A interaction and hence eq. (76) tests the presence of other interactions with high accuracy. Notice however, that this factor has nothing to do with parity violation, it would also be present in a parity conserving pure vector interaction. (In fact a similar factor shows up in the electromagnetic decay $\pi^0 \rightarrow e^+ + e^-$) .

Inserting mass values into eq. (76), one obtains

$$R_\pi = 1{,}283 \times 10^{-4}$$

$$R_K = 2{,}1 \times 10^{-5} \qquad (77)$$

in excellent agreement with experiments:

$$\text{ref. 17: } R_\pi = (1{,}247 \pm 0{,}028) . 10^{-4} \qquad (78)$$

$$\text{ref. 18: } R_K = (3{,}8\ {}^{+3{,}2}_{-2{,}3}) \ . \ 10^{-5}$$

$$\text{ref. 19: } R_K = (1{,}9\ {}^{+0{,}7}_{-0{,}5}) \ . \ 10^{-5} \qquad (79)$$

10. $P_{\ell 3}$ - Decays

The matrix element of the hadron current between a π^0 and a meson P contains 2 form factors because 2 momentum vectors are available.

$$<\pi^o(k')|J_\lambda(x)|P(k)> =$$

$$= \frac{1}{(2\pi)^3}\frac{1}{\sqrt{4k_o k_o'}}\left\{(k+k')_\lambda\ f_+(q^2) + (k-k')_\lambda f_-(q^2)\right\} \times$$

$$\times\ e^{ix(k'-k)} \qquad (80)$$

where

$$q^2 = (k-k')^2$$

First let us note in passing that CVC requires

$$(k^2-k'^2)\ f_+(q^2) + q^2 f_-(q^2) = 0$$

and hence

$$f_-(q^2) = 0 \qquad \text{if } k^2 = k'^2$$

For this reason, $\pi_{\ell 3}$ decay is described by a single form factor which is $\sqrt{2}$ times the electromagnetic form factor. Consequently, $\pi_{\ell 3}$ decay is completely specified and can be computed without ambiguity. The result agrees with experiments [7].

For $K_{\ell 3}$ decay, the variation of the form factors cannot be neglected since the energy release is of the same order as the hadron masses. We shall restrict to charged kaons and write

$$f_\pm(q^2) = f_\pm(o)\left\{1 + \lambda_\pm \frac{q^2}{\mu^2} + \ldots\right\} \qquad (81)$$

With the techniques described in section 4, one can now compute the recoil energy spectrum (ω is the total pion energy) for $K_{\ell 3}$ decay, where we can neglect $f_-(q^2)$ because it contributes to order m_ℓ/m_K

$$n(\omega)d\omega = \frac{G^2}{12\pi^3} m_K |f_+(s)|^2 P_\pi^3 d\omega \tag{82}$$

with

$$s = m_K^2 + m_\pi^2 - 2m_K\omega$$

For constant form factors, one defines

$$\xi \equiv f_-/f_+ \tag{83}$$

and computes the μ-e ratio

$$\Gamma(K_{\mu 3})/(K_{e3}) = 0.646+0.126\xi+0.0192\xi^2 \geq 0.44 \tag{84}$$

as well as the lepton recoil spectrum

$$n(\varepsilon)d\varepsilon = \frac{G^2 f_+^2}{32\pi^3} m_K^4 \; p_\ell \; \left(1-\frac{m_\pi^2}{t}\right)^2 \{4\varepsilon(1-2\varepsilon) + 5\alpha^2\varepsilon - \alpha^4 +$$

$$+ \xi\alpha^2(4-6\varepsilon+2\alpha^2) + \xi^2\alpha^2(\varepsilon-\alpha^2)\} \, d\varepsilon \tag{85}$$

where

$$t = m_K^2 + m_\ell^2 - 2m_K E_\ell$$

$$\alpha = m_\ell/m_K$$

$$\varepsilon = E_\ell/m_K \tag{86}$$

Experimentally, one obtains from these equations [20]

$$f_+ = 0.16 \pm 0.01$$

$$\xi = 0.3 \pm 0.4$$

$$\lambda_+ = 0.023 \pm 0.008 \tag{87}$$

On the other hand the longitudinal lepton polarization allows for a determination of ξ via

$$P_\ell(\varepsilon) = \frac{p_\ell}{m_K} \frac{-4(1-2\varepsilon) - 3\alpha^2 + 2\alpha^2\xi + \alpha^2\xi^2}{4\varepsilon(1-2\varepsilon)+5\alpha^2\varepsilon-\alpha^4+\xi\alpha^2(4-6\varepsilon+2\alpha^2)+\xi^2\alpha^2(\varepsilon-\alpha^2)} \tag{88}$$

The experimental value from eq. (88) is

$$\xi = -1.25 \pm 0.32 \tag{89}$$

in contradiction to eq. (87). It should be kept in mind, that these measurements are extremely delicate and that a mistake cannot be completely excluded. If both values are taken seriously, it means that the neglection of variation of the form factors in eqs. (84), (85) and (88) is unjustified. In fact, a recent very elegant investigation of d'Espagnat and Gaillard [21] , based on the method of Fubini and Furlan [22], suggests that $f_-(q^2)$ shows anomalously large variation with q^2, in other words that λ_- is much larger than λ_+ . It seems that the near future will provide interesting developments in this particular area.

Acknowledgment

This work was supported in part by Bundesministerium für wissenschaftliche Forschung.

References

1. J. Nilsson and H. Pietschmann, An Introduction to

Weak Interaction Physics (McGraw Hill Publ. Comp. N. Y.), in preparation.

2. A. Barbaro-Galtieri, W. H. Barkas, H. H. Heckman, J. W. Patrick, F. M. Smith, Phys.Rev.Lett.9,26(1962); U. Nauenberg, P. Schmidt, J. Steinberger, S. Marateck, R. J. Plano, H. Blumenfeld, L. Seidlitz, Phys. Rev. Lett. 12, 679 (1964); F. Eisele, R. Engelmann, H. Filthuth, W. Föhlisch, V. Hepp, E. Kluge, E. Leitner, P. Lexa, P. Mokry, W. Presser, H. Schneider, M. L. Stevenson, G. Zech, Z. f. Physik 205, 409 (1967).
3. S. V. Pepper, C. Ryan, S. Okubo, R. E. Marshak, Phys. Rev. 137, B1259 (1965).
4. A. H. Rosenfeld, A. Barbaro - Galtieri, W. J. Podolsky, L. R. Price, M. Roos, P. Soding, W. J. Willis, C. G. Wohl, Rev. Mod. Phys. 39, 1 (1967).
5. S. Weinberg, Phys. Rev. 112, 1375 (1958).
6. P. Hertel, Z. f. Physik 202, 383 (1967).
7. Cf. for instance C. W. Wu, Rev. Mod. Phys. 36, 618 (1964).
8. S. S. Gershtein and Y. B. Zel'dovich, Soviet Physics, JETP 2, 576 (1956).
9. R. R. Feynman, M. Gell-Mann, Phys. Rev. 109, 193 (1958); M. Gell-Mann, Phys. Rev. 111, 362 (1958).
10. Result from the Maryland group, quoted by G. Zech at the Heidelberg conference 1967.
11. F. Eisele, R. Engelmann, H. Filthuth, W.Föhlisch, V. Hepp, E. Leitner, P. Mokry, W. Pressner, H. Schneider, M. L. Stevenson, G. Zech, Univ. Heidelberg preprint, August 1967.
12. N. Cabibbo, Phys. Rev. Lett. 10, 531 (1963).
13. M. Gell-Mann and M. Lévy, Nuovo Cim. 16, 705 (1958).
14. In Schladming, these matrices were first written down in the lectures of M. E. Mayer, Acta Physica Austr. Suppl. 1, 19 (1964).
15. W. Thirring, Acta Phys. Austriaca, Suppl. 3, 294

(1966).

16. N. Brene, L. Veje, M. Roos, C. Cronström, Phys. Rev. 149, 1288 (1966).
17. E. di Capua, R. Garland, L. Pondrom, A. Strelzoff, Phys. Rev. 133, B1333 (1964).
18. D. R. Bowen, A. K. Mann, W. K. McFarlane ,A. D. Franklin, E. B. Hughes, R. L. Imlay, G. K. O'Neill, D. H. Reading, Phys. Rev. 154, 1314 (1967).
19. D. R. Botterill, R. M. Brown, I. F. Corbett, G. Culligan, J. Mc. L. Emmerson, R. C. Field, J. Garvey, P. B. Jones, N. Middlemas, D. Newton, T. W. Quirk, G. L. Salmon, P. Steinberg, W. S. C. Williams, Heidelberg Int. Conf. El. Particles, Sept. 1967.
20. W. J. Willis, plenary lecture at Heidelberg, Int. Conf. El. Particles, Sept. 1967; cf. also ref. 4.
21. B. d'Espagnat and M. K. Gaillard, Phys. Lett. 25B, 346 (1967); see also A. K. Mann and H. Primakoff, Phys. Rev. Lett. 20, 32 (1968).
22. S. Fubini and G. Furlan, Physics 1, 229 (1965).

INTRODUCTION TO RECENT DEVELOPMENTS IN REGGE POLE THEORY†

By

Y. DOTHAN×

Physics Department, Tel Aviv University,
Ramat Aviv, Israel

INTRODUCTION

Regge poles have their origin in potential scattering. If one assumes two spinless particles interacting through a finite range potential one can get as an outcome the existence of Regge poles which dominate the scene at high values of $\cos\theta$. This means that if we can apply similar results together with the idea of crossing in the analysis of general high energy two body reactions, we shall get a useful way of parametrizing them. Thus in the same way that a few partial wave amplitudes are needed to describe low energy scattering we hope that a few Regge trajectories and residue functions are needed to describe high energy scattering. An

---

† Lecture given at the VII.Internationale Universitätswochen f.Kernphysik,Schladming,February 26-March 9,1968.

× Research sponsored by the Air Force Office of Scientific Research, Office of Aerospace Research, United States Air Force, under AFOSR grant number AF EOAR 66-39, through the European Office of Aerospace Research.

interesting aspect of this parametrization is the fact that particles with the quantum numbers of the crossed channel may lie on the Regge trajectories describing scattering in the channel we are discussing. Thus we shall have correlations between very high energy scattering phenomena and particle spectrum. It therefore seems that the application of Regge pole analysis to high energy scattering is an attractive idea. However, if we actually carry out this program we face and have to overcome various difficulties. First there is the question of branch points or cuts in the complex angular momentum plane on which the progress in understanding is slow. Here we shall develop the subject without considering these complications. Next is the problem of kinematical singularities of the scattering amplitude, which means the specification of the way the scattering amplitude behaves near certain kinematical points. Stated differently it means that there are constraint equations near these points. This has nothing to do with Regge pole theory. These constraints have to be satisfied in any decent theory. The point is that if we try to impose them on amplitudes contributed by Regge poles only we may get nontrivial connections relating different trajectories which are classified under names like daughters and conspirators.

This lecture will deal with the following subjects:

1. Reggeization of helicity amplitudes and kinematical constraints,
2. Van Hove model for spinless external particles, and
3. O(4) analysis in the equal mass case.

1. Reggeization of Helicity Amplitudes

Consider a two body scattering process (fig. 1):

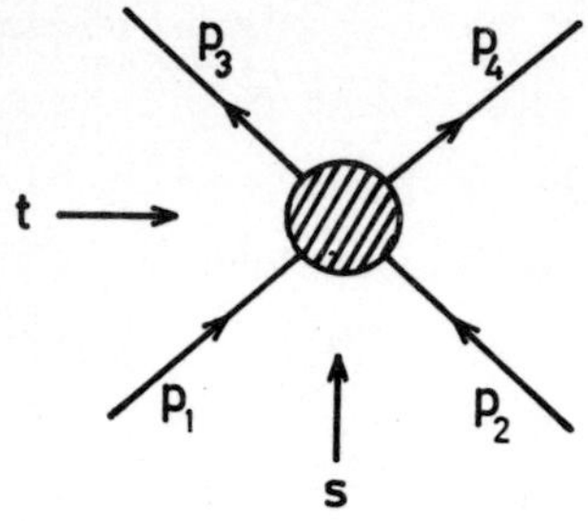

Fig. 1

We define the various channels by

s - channel $1 + 2 \rightarrow 3 + 4$

t - channel $1 + \bar{3} \rightarrow \bar{2} + 4$

u - channel $\bar{3} + 2 \rightarrow \bar{1} + 4$

where we label the channels by the variables which are the square of the respective c. m. energy

$$s = (p_1+p_2)^2 = (p_3+p_4)^2$$

$$t = (p_1-p_3)^2 = (p_4-p_2)^2$$

$$u = (p_2-p_3)^2 = (p_4-p_1)^2 \qquad (1)$$

satisfying $s+t+u = \Sigma m_i^2$ and we assume that all external particles are massive. Then our single particle states in the helicity scheme [1] are defined as

$$|m,s;\vec{p},\lambda> = R(\phi,\Theta,-\phi)\ L_z(\vec{p})|m,s;\vec{0},\lambda> \quad , \qquad (2)$$

where $|m,s;\vec{o},\lambda>$ is the rest state of a particle of mass m and spin s, λ is the component of the spin with respect to the quantization axis z. This state is now boosted along the z-direction by $L_z(\vec{p})$ and then a rotation R is made.

We use the covariant normalization

$$<\vec{p},\lambda|\vec{p}\,',\lambda'> = (2\pi)^3 2E\delta^{(3)}(\vec{p}-\vec{p}\,') \ . \qquad (3)$$

The invariant scattering amplitude F is defined by

$$<f|S|i> = <f|i> - i(2\pi)^4\delta^4(p_f-p_i)<f|F|i> \ . \qquad (4)$$

In our case $|i>$ and $|f>$are two particle states.
The c. m. differential cross section in the s-channel for unpolarized particles is given by

$$\frac{d\sigma}{d\Omega} = \frac{1}{64\pi s(2s_1+1)(2s_2+1)} \frac{|\vec{p}_f^{\,s}|}{|\vec{p}_i^{\,s}|} \sum_{spins} |<f|F|i>|^2 . \qquad (5)$$

In order to make the connection between $|\vec{p}_f|, |\vec{p}_i|$, s and the masses we introduce the following quantities

$$S_{ij}^{+} = [s-(m_i+m_j)^2]^{1/2} \ ,$$

$$S_{ij}^{-} = [s-(m_i-m_j)^2]^{1/2} \ ,$$

$$S_{ij} = S_{ij}^{+} S_{ij}^{-} \ .$$

Then

$$|\vec{p}_f^{\,s}| = \frac{S_{34}}{2\sqrt{s}} \quad \text{and} \quad |\vec{p}_i^{\,s}| = \frac{S_{12}}{2\sqrt{s}} \ .$$

The scattering angle in the s-channel c. m. system is given by

$$\cos\theta_s = \frac{s(t-u)+(m_1^2-m_2^2)(m_3^2-m_4^2)}{S_{12}\,S_{34}} \ . \qquad (6)$$

Following Jacob and Wick [1] we expand the scattering

amplitude into partial waves in the center of mass system

$$\langle f|F|i\rangle = \sum_{J=M}^{\infty} (2J+1) F^J_{\lambda_3\lambda_4;\lambda_1\lambda_2}(s)\, d^J_{\lambda,\mu}(\theta_s) \equiv$$

$$\equiv f^s_{\lambda_3,\lambda_4;\lambda_1,\lambda_2}(s,t,u) \tag{7}$$

with $\lambda = \lambda_1 - \lambda_2$, $\mu = \lambda_3 - \lambda_4$ and $M = \max(|\lambda|,|\mu|)$. (We shall abbreviate later on by $\{\lambda\}$ the set $\lambda_3,\lambda_4;\lambda_1,\lambda_2$). Here we have assumed the scattering to take place in the xz-plane. Parity conservation gives the following relations between the

$$(2s_a+1)(2s_b+1)(2s_c+1)(2s_d+1)$$

matrix elements of F^J

$$F^J_{\lambda_3,\lambda_4;\lambda_1,\lambda_2}(s) =$$

$$= \eta(-1)^{s_3+s_4-s_1-s_2} F^J_{-\lambda_3,-\lambda_4;-\lambda_1,-\lambda_2}(s); \tag{8}$$

η is the relative parity between the initial and final states. The functions $d^J_{\lambda\mu}(\theta)$ are related to the Jacobi polynomials $P_\ell^{(\alpha,\beta)}$

$$P_\ell^{(\alpha,\beta)}(x) = (1-x)^{-\alpha}(1+x)^{\beta}(-1)^{\ell}\, \frac{1}{2^\ell \ell!}\, \frac{d^\ell}{dx^\ell}\{(1-x)^{\ell+\alpha}(1+x)^{\ell+\beta}_{\alpha}$$

in the following form

$$d^J_{\lambda,\mu}(\theta) = (1-\cos\theta)^{1/2|\lambda-\mu|}(1+\cos\theta)^{1/2|\lambda+\mu|}(-1)^{\lambda-\mu} \times \tag{9}$$

$$\times\, 2^{-\lambda}\, \frac{\sqrt{(J+\lambda)!(J-\lambda)!}}{\sqrt{(J+\mu)!(J-\mu)!}}\; P_{J-\lambda}^{(\lambda-\mu,\lambda+\mu)}(\cos\theta)\ ,\quad \lambda > \mu \geq 0\ .$$

We list some useful relations which help us to define $d^J_{\lambda,\mu}(\theta)$ in the other cases

$$d^J_{\mu,\lambda}(\theta) = (-1)^{\lambda-\mu}\, d^J_{\lambda,\mu}(\theta)$$

$$d^J_{-\mu,-\lambda}(\theta) = d^J_{\lambda,\mu}(\theta) \quad . \tag{10}$$

Therefore, assuming parity conservation, (9) and (10) reduce the number of amplitudes by means of

$$f_{\lambda_3,\lambda_4;\lambda_1,\lambda_2}(s,t) = \\ = \eta(-1)^{s_3+s_4-s_1-s_2}(-1)^{\lambda-\mu}\, f_{-\lambda_3,-\lambda_4;-\lambda_1,-\lambda_2}(s,t) \tag{11}$$

As seen in (9), apart from the factors

$$(1 - \cos\theta)^{1/2|\lambda-\mu|} \quad \text{and} \quad (1+\cos\theta)^{1/2|\lambda+\mu|},$$

the $d^J_{\lambda,\mu}$ are polynomials in cosθ of degree J-λ which we denote by $\hat{d}^J_{\lambda,\mu}$. Since these factors do not depend on J we can take them out of the summation over J in (7) and we are left with a Jacobi polynomials series with coefficients depending on s . Since for fixed s, cos θ_s is linear in t, $f_{\{\lambda\}}(s,t)$ is analytic in t whereever the series converges. By a known theorem [2] the region of convergence is an ellipse in the cosθ - plane with foci at ±1. Let us review briefly the arguments [3] leading to the reggeization of helicity amplitudes. One starts by considering the t-channel helicity amplitude

$$f_{\lambda_{\bar{2}},\lambda_4;\lambda_1,\lambda_{\bar{3}}}(t) =$$

$$= \hat{f}_{\{\lambda\}}(t,\theta_t)(\sqrt{2}\cos\theta_t/2)^{|\lambda+\mu|}(\sqrt{2}\sin\theta_t/2)^{|\lambda-\mu|} ,$$

$$\hat{f}_{\{\lambda\}}(t,\theta_t) = \sum_{J=M}^{\infty}(2J+1)F^J_{\{\lambda\}}(t)\hat{d}^J_{\lambda,\mu}(\theta_t) \quad . \qquad (12)$$

By arguments similar to the ones above (s and t are now interchanged), every term in the series (12) is a regular function of s. Therefore the only singularities of $\hat{f}_{\{\lambda\}}(s,t)$ for fixed t arise when this sum does not converge. This motivates the assumption that $\hat{f}_{\{\lambda\}}(s,t)$ satisfies a dispersion relation for fixed t:

$$\hat{f}_{\{\lambda\}}(s,t) = \frac{1}{\pi}\int_{s_o}^{\infty}\frac{A^s_{\{\lambda\}}(s',t)}{s'-s}ds' + \frac{1}{\pi}\int_{u_o}^{\infty}\frac{A^u_{\{\lambda\}}(u',t)}{u'-u}du' \qquad (13)$$

where s_o and u_o are the respective threshold values and $u = \Sigma m_i^2 - s - t$. Our aim is now to define $F^J_{\{\lambda\}}(t)$ for complex J. We start by inverting the expansion (12) of $\hat{f}_{\{\lambda\}}$ and we get with the use of (13)

$$F^J_{\{\lambda\}}(t) =$$

$$= \frac{1}{2}\int_{-1}^{+1} dx\ \hat{f}_{\{\lambda\}}(t,s(t,x))(1-x)^{|\lambda-\mu|}(1+s)^{|\lambda+\mu|}\hat{d}^J_{\lambda,\mu}(x) =$$

$$= \frac{1}{2\pi}\frac{1}{|\vec{p}^{\,t}_i||\vec{p}^{\,t}_f|}\Big\{\int_{s_o}^{\infty} ds'\ A^s_{\{\lambda\}}(s',t)C^J_{\lambda,\mu}\Big(a+\frac{s'}{2|\vec{p}^{\,t}_i||\vec{p}^{\,t}_f|}\Big) +$$

$$+ \int_{u_o}^{\infty} du'\ A^u_{\{\lambda\}}(u',t)\ C^J_{\lambda,\mu}\Big(-b-\frac{u'}{2|\vec{p}^{\,t}_i||\vec{p}^{\,t}_f|}\Big)\Big\} , \qquad (14)$$

where

$$a + \frac{s'}{2|\vec{p}_i^{\,t}||\vec{p}_f^{\,t}|} = \cos\theta_t$$

or

$$4t|\vec{p}_i^{\,t}||\vec{p}_f^{\,t}|a = t^2 - t\Sigma m_i^2 + (m_1^2 - m_3^2)(m_2^2 - m_4^2) \quad ,$$

$$4t|\vec{p}_i^{\,t}||\vec{p}_f^{\,t}|b = t^2 - t\Sigma m_i^2 - (m_1^2 - m_3^2)(m_2^2 - m_4^2) \quad ,$$

and

$$C^J_{\lambda,\mu}(x) = \frac{1}{2}\int_{-1}^{+1} \frac{dx'}{x-x'} (1-x')^{|\lambda-\mu|}(1+x')^{|\lambda+\mu|}\hat{d}^J_{\lambda,\mu}(x') \ .$$

Let us remark about the question of subtractions. Our assumption of fixed t dispersion relations can be rephrased in the form that $\hat{f}_{\{\lambda\}}(s,t)$ is bounded by a power of s (or $\cos\theta_t$) as $|s|$ goes to infinity

$$|f_{\{\lambda\}}(t,\cos\theta_t)(\cos\theta_t)^{-N}| \xrightarrow[|\cos\theta_t|\to\infty]{} 0 \qquad ,$$

which means that we have a finite number of subtractions. If we have to take into account n subtractions in (13) we split the sum (12) into two parts, one going from J=M to J=M+n and the second one going from M+n+1 to infinity. Due to orthogonality relations of the $d^J_{\lambda,\mu}$ the finite sum does not contribute to the integrals in (14) for J large enough, and the analysis is the same as before. The only difference then comes about when we open the contour in our Sommerfeld-Watson transformation, since the background integral is not deformed along the line Re J = M but along Re J = M+n for subtracted dispersion relations. We shall have to remember that when pushing the line to the left as far as possible.

We now consider the expression (14) for $F^J_{\{\lambda\}}(t)$ for $J > n$ only, which guarantees the convergence of the integrals. With the help of the symmetry property

$$C^J_{\lambda,\mu}(-x) = -(-1)^{J+\lambda}\, C^J_{\lambda,-\mu}(x)$$

we can rewrite (14) to get

$$F^J_{\{\lambda\}}(t) =$$

$$= \frac{1}{2\pi|\vec{p}^{\,t}_i||\vec{p}^{\,t}_f|} \Big\{ \int_{s_0}^{\infty} ds'\, A^s_{\{\lambda\}}(s',t)\, C^J_{\lambda,\mu}\Big(a + \frac{s'}{2|\vec{p}^{\,t}_i||\vec{p}^{\,t}_f|}\Big) -$$

$$- (-1)^{J+\lambda} \int_{u_0}^{\infty} du'\, A^u_{\{\lambda\}}(u',t)\, C^J_{\lambda,-\mu}\Big(b + \frac{u'}{2|\vec{p}^{\,t}_i||\vec{p}^{\,t}_f|}\Big)\Big\} \quad . \qquad (15)$$

Our next step is the discussion of the asymptotic behaviour in J for $F^J_{\{\lambda\}}(t)$ which was up to now defined for physical values of J, i.e. for integer or half integer values. From (15) we see that the only dependence on J is carried by the function $C^J_{\lambda,\mu}$. Once we know how to continue analytically the $C^J_{\lambda,\mu}$ in the complex J-plane, we can continue analytically $F^J_{\{\lambda\}}(t)$ provided the integrals converge. Since the $C^J_{\lambda,\mu}$ are related to hypergeometric [2] functions their analytic continuation is well determined and it turns out that the integrals in (15) converge exactly in the region we assumed by taking into account subtractions.

Now the factor $(-1)^{J+\lambda}$ whose analytic continuation is $e^{i\pi(J+\lambda)}$ destroys the argument above, since at $J\to\infty$ it exhibits an essential singularity. We overcome this difficulty by the well known trick of replacing it by ±1 throughout, thereby defining two independent analytic continuations coinciding with $F^J_{\{\lambda\}}(t)$ at alternating physical values of J. With this definition we introduced the concept of signature.

Actually there is some freedom in the above choice since as long as we introduce a J independent phase factor nothing will change. Because of the symmetry properties of the d^J and therefore of the C^J the phase $(-1)^{\lambda+M}$ seems to be convenient. By a change of the integration variables in (15) we then get the two analytic continuations

$$F^{\pm}_{\{\lambda\}}(J,t) = \frac{1}{\pi} \int_{\min(x_s,x_u)}^{\infty} dx' \, \{A^{\pm}_{\{\lambda\}}(t,x') \, C^{J+}_{\lambda,\mu}(x') + A^{\mp}_{\{\lambda\}}(t,x') \, C^{J-}_{\lambda,\mu}(x')\}, \tag{16}$$

where

$$C^{J\pm}_{\lambda,\mu}(x) = \frac{1}{2} \left[C^{J}_{\lambda,\mu}(x) \pm (-1)^{\lambda+M} \, C^{J}_{\lambda,-\mu}(x)\right]$$

and

$$A^{\pm}_{\{\lambda\}}(t,x) = A^{s}_{\{\lambda\}}(2|\vec{p}^{\,t}_{i}||\vec{p}^{\,t}_{f}|(-a+x),t) \pm A^{u}_{\{\lambda\}}(2|\vec{p}^{\,t}_{i}||\vec{p}^{\,t}_{f}|(-b+x),t) \; .$$

Then of course

$$F^{J}_{\{\lambda\}}(t) = \frac{1}{2}[1+(-1)^{J-M}]F^{+}_{\{\lambda\}}(J,t) + \frac{1}{2}[1-(-1)^{J-M}]F^{-}_{\{\lambda\}}(J,t) \; . \tag{17}$$

Therefore it is natural to define

$$\hat{f}_{\{\lambda\}}(t,\cos\theta_t) = \hat{f}^{+}_{\{\lambda\}} + \hat{f}^{-}_{\{\lambda\}} ,$$

with

$$\hat{f}^{\pm}_{\{\lambda\}}(t,\cos\theta_t) =$$

$$= \Sigma(2J+1) \{\frac{1\pm(-1)^{J-M}}{2} F^{\pm}_{\{\lambda\}}(J,t)\ \hat{d}^{J+}_{\lambda,\mu}(\cos\theta_t) +$$

$$+ \frac{1\mp(-1)^{J-M}}{2} F^{\mp}_{\{\lambda\}}(J,t)\ \hat{d}^{J-}_{\lambda,\mu}(\cos\theta_t)\} \quad , (18)$$

where $\hat{d}^{J\pm}_{\lambda,\mu}$ are defined analogous to the $C^{J\pm}_{\lambda,\mu}$ in (16). We now perform the usual Sommerfeld - Watson transformation, where the contour of integration C encircles the physical values of J:

$$\hat{f}^{\pm}_{\{\lambda\}}(t,x = \cos\theta_t) =$$

$$= \frac{1}{2i} \int_C dJ \frac{2J+1}{\sin\pi(J-M)} \{F^{\pm}_{\{\lambda\}}(J,t)\ [e^{J+}_{\lambda,\mu}(x) \pm e^{J+}_{\lambda,\mu}(-x)] +$$

$$+ F^{\mp}_{\{\lambda\}}(J,t)\ [e^{J-}_{\lambda,\mu}(x) \mp e^{J-}_{\lambda,\mu}(-x)]\}.$$

(19)

If we want to deform the contour of integration we need the explicit form of the various functions in (19) which reads for the $e^{J\pm}_{\lambda,\mu}$ [2]

$$e^{J}_{\lambda,\mu}(z) = (-1)^{\lambda-\mu} \frac{2^{-J}\ \Gamma(2J+1)}{[\Gamma(J+\lambda+1)\Gamma(J-\lambda+1)\Gamma(J+\mu+1)\Gamma(J-\mu+1)]^{1/2}}$$

$$\times (z-1)^{J-\lambda}\ {}_2F_1(-J+\lambda,-J+\mu;-2J;\frac{2}{1-z}) \quad .(20)$$

In this deformation process one picks up contributions from moving singularities of $F^{\pm}_{\{\lambda\}}(J,t)$, which we assume to be simple poles and therefore look like

$$F^{\pm}(J,t) = \frac{\beta^{\pm}(t)}{J-\alpha^{\pm}(t)} \qquad (21)$$

in the neigbourhood of $J = \alpha^{\pm}(t)$. Our deformation process includes moving the integration path of the background integral from Re J = M+n to Re J = -1/2. If we first move the integration path to Re J = M we get contributions which cancel the finite sum from J = M to J = M+n in the helicity expansion. Then we pick up more terms, which when $\cos\theta_t \to \infty$ behave like

$$|\cos\theta_t|^{-1},\ |\cos\theta_t|^{-2},\ldots|\cos\theta_t|^{-n}\ ,\quad n=\min(|\lambda|,|\mu|)$$

for integral λ,μ and analogous terms for λ,μ half odd integers. Finally the background integral behaves like $|\cos\theta_t|^{-1/2}$ as $\cos\theta_t \to \infty$.

Since we are interested in the asymptotic behaviour in $\cos\theta_t$, which is equivalent to the behaviour as $s \to \infty$ except for special cases, where the linear relation between s and $\cos\theta_t$ is no longer valid, we can neglect the background integral and are left with

$$\hat{f}^{\pm}_{\{\lambda\}}(t,x) = \mp\ (-1)^{\lambda-\mu}\ \pi\ \frac{1\pm e^{i\pi(\alpha-M)}}{\sin\pi(\alpha-M)} \times$$

$$\times \frac{2^{-\alpha-1}\ \Gamma(2\alpha+2)}{[\Gamma(\alpha+\lambda+1)\Gamma(\alpha-\lambda+1)\Gamma(\alpha+\mu+1)\Gamma(\alpha-\mu+1)]^{1/2}}\ \tilde{\beta}^{\pm}_{\{\lambda\}}(t)s^{\alpha(t)-M}; \qquad (22)$$

in order to simplify notation we omitted the ± index on $\alpha(t)$ and absorbed factors which relate s and $\cos\theta_t$ in $\tilde{\beta}(t)$. In order to get the amplitude $f^{(t)}_{\{\lambda\}}$ from the amplitude $\hat{f}^{(t)}_{\{\lambda\}}$ we have to multiply the latter by

$$(1 - \cos\theta_t)^{1/2|\lambda-\mu|}(1 + \cos\theta_t)^{1/2|\lambda+\mu|}$$

which behaves like s^M for $s \to \infty$ and therefore all helicity amplitudes $f^{(t)}_{\{\lambda\}}$ behave as $s^{\alpha(t)}$ for fixed t and $s \to \infty$.

We now want to clarify our remark above concerning the linear relationship between s and $\cos\theta_t$. In analogy to (6)

$$\cos\theta_t = \frac{2\ st-t^2-t\Sigma m_i^2 - (m_1^2-m_3^2)(m_2^2-m_4^2)}{T_{13}\ T_{24}} \qquad (23)$$

with

$$T_{ij} = T^+_{ij}\ T^-_{ij} = [t - (m_i+m_j)^2]^{1/2}\ [t-(m_i-m_j)^2]^{1/2}$$

For $t \neq o$ and also different from the thresholds and pseudo-thresholds $(m_i \pm m_j)^2$ obviously $s \to \infty$ gives $\cos\theta_t \to \infty$ linearly in s. The point t = o however is slightly problematic: if $m_1 \neq m_3$ and $m_2 \neq m_4$ then for $t \to o$, $\cos\theta_t \to 1$ independent of s. But for equal masses in the crossed channel we again have the normal case $\cos\theta_t \sim s$ even for t = o. Stated mathematically the point t = o in the general mass case is a singular point of the transformation $(s,t) \to (\cos\theta_t,t)$ and exactly at this point the concept of daughters comes in, in order to establish the asymptotic s behaviour. This will be discussed later.

Let us now consider the contribution of a single Regge pole in the t-channel to the high energy cross section in the s-channel. For that purpose we have to calculate $f^{(s)}_{\{\sigma\}}$ in its physical region. This is done by applying the crossing matrices [4] to the analytically continued amplitude $f^{(t)}_{\{\tau\}}$:

$$f^{(s)}_{\{\sigma\}} = \sum_{\{\tau\}} M_{\{\sigma\},\{\tau\}} f^{(t)}_{\{\tau\}} \quad . \tag{24}$$

For our purposes it is sufficient to know that M is an orthogonal matrix in the helicity space and therefore

$$\sum_{\{\sigma\}} |f^{(s)}_{\{\sigma\}}|^2 = \sum_{\{\tau\}} \left|f^{(t)}_{\{\tau\}}\right|^2 \sim s^{2\mathrm{Re}\alpha(t)} \quad .$$

According to (5) the cross section behaves for fixed t as

$$\frac{d\sigma}{d\Omega} \underset{s\to\infty}{\sim} s^{2\mathrm{Re}\alpha(t)-1} \quad . \tag{25}$$

The situation is much more complicated if one wants to consider the variation of $d\sigma/d\Omega$ with t at high values of s, since our information on the function $\beta(t)$ is rather meager. We know some general properties of $\beta(t)$ as a function of t at the points

$$t = (m_1 \pm m_3)^2 \quad , \qquad t = (m_2 \pm m_4)^2$$

and t = o. For instance suppose $(m_1+m_3)^2$ is the threshold for the t-channel reaction, then we know from general quantum mechanics how the partial waves behave near threshold. The problem we are dealing with is that of kinematical singularities and zeros [5] . A very simple way to handle them was given by Jackson and Hite [6]. The partial wave decomposition in the spinless case in the t-channel is given by

$$\Sigma(2\ell+1)\, a_\ell(t)\, P_\ell(\cos\theta_t),$$

where

$$\cos\theta_t = \frac{\vec{p}^{\,t}_i \cdot \vec{p}^{\,t}_f}{|\vec{p}^{\,t}_i|\,|\vec{p}^{\,t}_f|} \quad .$$

Threshold means $|\vec{p}_i^{\,t}| \to 0$ and there $\cos\theta_t$ is not well defined. This can be overcome by writing

$$a_\ell(t) = |\vec{p}_i^{\,t}|^\ell |\vec{p}_f^{\,t}|^\ell \tilde{a}_\ell(t) \quad ,$$

where $\tilde{a}_\ell(t)$ is a regular function at threshold. Then the combination $a_\ell(t)\, P_\ell(\cos\theta_t)$ behaves like

$$a_\ell(t) P_\ell(\cos\theta_t) =$$

$$= \tilde{a}_\ell(t) c_\ell (\vec{p}_i^{\,t}\cdot\vec{p}_f^{\,t})^\ell + O((\vec{p}_i^{\,t}\cdot\vec{p}_f^{\,t})^{\ell-2} |\vec{p}_i^{\,t}||\vec{p}_f^{\,t}|)$$

and therefore the sum is well defined near threshold. We can now apply similar reasoning to $\hat{f}^t_{\{\lambda\}}$. From (23) we see that $\cos\theta_t$ has in its denominator the combination $T_{13}\, T_{24}$, and since the Jacobi polynomials are the analogues of the Legendre polynomials we have to take out a threshold factor $(T_{13}\, T_{24})^{J-M}$ from $F^J_{\{\lambda\}}$ in order to get rid of the complications at threshold and pseudothresholds. Actually the situation is more complicated since in reality parity is conserved in strong and electromagnetic interactions and the various partial wave amplitudes $F^J_{\{\lambda\}}$ are linearly dependent in pairs. Taking the parity conserving amplitudes one gets one combination in which the leading power of the Jacobi polynomials exactly cancels and therefore the amplitudes will exhibit a power smaller than J-M. However one can carry through the analysis, as Jackson and Hite do.

Another aspect of the same problem comes about in connection with the crossing relation (24). Since it connects the full amplitudes $f_{\{\lambda\}}$, the factors

$$(1 - \cos\theta)^{1/2|\lambda-\mu|}(1 + \cos\theta)^{1/2|\lambda+\mu|}$$

are included, and their singularity structure has to be taken into account. As an example we consider the equal mass case in the t-channel $m_1=m_3, m_2=m_4$, where in the s-channel in the forward direction the amplitude $f^{(s)}$ behaves as

$$f^{(s)}_{\lambda_3,\lambda_4;\lambda_1,\lambda_2} \underset{t\to o}{\sim} t^{1/2|\lambda_1-\lambda_2-\lambda_3+\lambda_4|}$$

and therefore goes to zero in a prescribed way unless there is no helicity flip. This is just angular momentum conservation. Then also by means of (24)

$$\sum_{\{\mu\}} M_{\{\lambda\}\{\mu\}} f^{(t)}_{\{\mu\}} \underset{t\to o}{\sim} t^{1/2|\lambda_1-\lambda_2-\lambda_3+\lambda_4|} , \qquad (26)$$

this relation should hold in general. If we now assume our Regge pole model, i.e. we represent the amplitudes by a sum over Regge poles, the question arises whether the condition (26) is satisfied. It is seen that (26) is not fulfilled automatically and therefore (26) establishes nontrivial conditions, which have to be imposed on the parameters of the model from outside.

The analysis has been applied to the case of N-N scattering [7]. In the t channel (N-$\bar{N}$ scattering) one defines [8] five independent amplitudes taking into account parity conservation :

$$f^J_o = F^J_{1/2,1/2;1/2,1/2} - F^J_{1/2,1/2;1/2,-1/2} ,$$

$$f^J_1 = F^J_{1/2,-1/2;1/2,-1/2} - F^J_{1/2,-1/2;-1/2,1/2}$$

$$f^J_{11} = F^J_{1/2,1/2;1/2,1/2} + F^J_{1/2,1/2;-1/2,-1/2}$$

$$f^J_{22} = F^J_{1/2,-1/2;1/2,-1/2} + F^J_{1/2,-1/2;-1/2,1/2} ,$$

$$f_{12}^{J} = 2F^{J}_{1/2,1/2;1/2,-1/2} \quad .$$

These helicity amplitudes have to satisfy the condition (26) which leads to the relation [7]

$$\frac{J+2}{J+1} f_{22}^{J+1} - \frac{J-1}{J} f_{22}^{J-1} - f_{o}^{J+1} + f_{o}^{J-1} - \frac{2J+1}{J(J+1)} f_{1}^{J} \underset{t\to o}{\sim} t^{1/2} .$$

If we represent these amplitudes by the projections of the Regge pole amplitude we get connections between trajectories with different quantum numbers. A general classification of the various solutions for constraint equations (26) has been given by Leader [8]:

1. Conspiracy: Relations between different $\alpha(t)$ with different quantum numbers at the point of constraint.
2. Evasion: Relations between residue functions at the point of constraint, but no relations between trajectory functions.
3. Daughters: Existence of a sequence of different Regge poles with the same quantum numbers, which are related at the point of constraint.

2. Van Hove Model

One knows that an amplitude corresponding to a Feynman diagram has all the desirable properties of crossing and Poincaré invariance built into it. One is thus led to look for another formulation which will fulfill the constraints automatically. To see how such a thing might happen we consider the Van Hove model [9]. We consider the following Feynman diagram with spinless external particles where the scattering takes place by the exchange of particles with integer spin J in the t-channel (fig. 2)

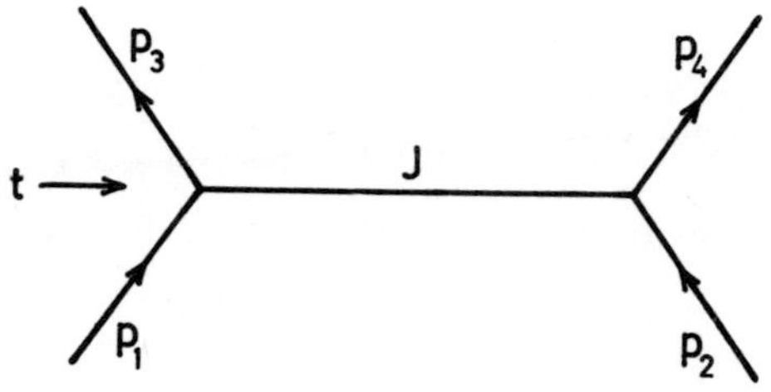

Fig. 2

We define the following kinematical quantities:

$$K_\alpha = \frac{1}{2}(p_1 + p_3)_\alpha \quad ,$$

$$K'_\alpha = \frac{1}{2}(p_4 + p_2)_\alpha \quad ,$$

$$q_\alpha = (p_3 - p_1)_\alpha = (p_2 - p_4)_\alpha \tag{27}$$

and the invariant variable $t = q^2$. The contribution of this diagram to the scattering amplitude is given by

$$f(J) = -g(J)\, g'(J)\, \frac{2J+1}{2^J (J!)^2}\, \frac{(-1)^J}{t-m^2(J)} \times \tag{28}$$

$$\times\, K^{\alpha_1} \dots K^{\alpha_J}\; \Gamma^J_{\alpha_1 \dots \alpha_J;\beta_1 \dots \beta_J}\; K'^{\beta_1} \dots K'^{\beta_J} \quad ,$$

where $g(J)$, $g'(J)$ represent the coupling constants, Γ^J is the numerator of the propagator of a spin J particle One important remark about Γ^J is to be made. If we look at the case J=1, then

$$\Gamma^1_{\alpha\beta} = g_{\alpha\beta} - \frac{q_\alpha q_\beta}{m^2(1)} \tag{29}$$

Γ^J is essentially the direct product of J such expressions with the difference that we have to take $m^2(J)$ instead of $m^2(1)$. On the mass shell we have $q^2 = m^2(J)$ and thus propagate exactly spin J particles. However in going away from the mass shell of the propagated particle not only spin J but also the whole series J-1, J-2,...,0 will be exchanged. If one wanted to get rid of these redundant components one has to replace m^2 by q^2, then Γ^J is exactly a projection operator for every value of q^2. But this introduces another difficulty since (29) would diverge at q^2=o. We therefore will carry with us all these redundant components to ensure regularity at $t=q^2=o$. In a Regge picture the image of these redundant components are the daughters.

Carrying out the products in (28) we arrive at the following form

$$f(J) = (2J+1)\ g(J)\ g'(J)(\bar{K}_J)^J(\bar{K}'_J)^J P_J(\bar{z}_J)\frac{1}{m^2(J)-t} , \tag{30}$$

where we introduced

$$(\bar{K}_J)^2 = -\left[K^2 - \frac{(K.q)^2}{m^2(J)}\right], \quad (\bar{K}'_J)^2 = -\left[K'^2 - \frac{(K'.q)^2}{m^2(J)}\right] \tag{31}$$

$$\bar{z}_J = \frac{-1}{\bar{K}_J\bar{K}'_J}\left[K.K' - \frac{(K.q)(K'.q)}{m^2(J)}\right] , \quad \bar{K}_J = \sqrt{\bar{K}^2_J} .$$

It should be noted that the products (Kq), (K'q) are differences of squares of mass as

$$(K.q) = \frac{1}{2}(m_3^2 - m_1^2) \quad , (K'.q) = \frac{1}{2}(m_2^2 - m_4^2) . \tag{32}$$

It is obvious that the equal mass case (in the t-channel) will be considerably simpler than the general one.

Now in the Van Hove model the scattering amplitude

is represented by a sum over an infinite number of J values with corresponding masses m(J) and coupling constants g(J), g'(J)

$$f = \sum_J (2J+1)g(J)g'(J)(\bar{K}_J)^J(\bar{K}'_J)^J P_J(\bar{z}_J) \frac{1}{m^2(J)-t} \quad (33)$$

In this model it can be shown that the exchange of a single Regge trajectory is equivalent to the exchange of infinitely many spins in the Feynman language. In order to see how this comes about we first consider the equal mass case $m_1=m_3$, $m_2=m_4$. From (31), (32) one sees that $\bar{K}_J$, $\bar{K}'_J$ and $\bar{z}_J$ are independent of J:

$$\bar{K}_J = \sqrt{-K^2} = \frac{1}{2}\sqrt{t-4m_1^2} = K \ ,$$

$$\bar{K}'_J = K' \ ,$$

$$-\bar{z}_J = -z = \frac{s-u}{4} \frac{1}{K.K'} \ ,$$

and we can drop in (33) the bars and the subscript J. We have made the hidden assumption that the masses and coupling constants can be made analytic functions of J. We may now perform the Sommerfeld-Watson transformation to obtain

$$f = \frac{i}{2} \int_c dJ \frac{2J+1}{\sin\pi J} (K)^J (K')^J P_J(-z) \frac{g(J)g'(J)}{m^2(J)-t} \ . \quad (34)$$

Opening up the contour we pick up a pole in the J-plane at $J = \alpha(t)$ which is a solution of $m^2(J) - t = 0$. In general there may be many solutions, for the sake of definiteness we choose the leading one. Having thus defined $\alpha(t)$ we get the leading contribution as

$$f = \frac{g(\alpha)\ g'(\alpha)(2\alpha+1)}{\sin\pi\alpha} (K)^{\alpha} (K')^{\alpha} P_{\alpha}(-z) \frac{d\alpha}{dt} + \text{con-}$$

tributions of other poles + background integral

(35)

For $|z| \to \infty$ we see from (35)that

$$(K)^{\alpha}(K')^{\alpha}P_{\alpha}(-z) \xrightarrow[|z|\to\infty]{} \left(\frac{s-u}{4}\right)^{\alpha} \frac{\Gamma(2\alpha+1)}{2^{\alpha}[\Gamma(\alpha+1)]^2} \quad .$$

and therefore in the equal mass case under consideration the dominant term as $s \to \infty$ is $s^{\alpha(t)}$.

In the unequal mass case, however, the quantities defined in (31) depend on J through $m^2(J)$; therefore, due to the factors $(\bar{K}_J)^J$ $(\bar{K}_J')^J$ in the analogue to (34), we pick up poles at $J = \alpha(o)$ (where $m^2(J) = 0$),in addition to the ones ($J = \alpha(t)$) coming from $m^2(J) - t = 0$. The contribution of these new poles has a singularity at $t = 0$, since in this case when $m^2(J) = 0$

$$\frac{1}{m^2(J) - t} \to -\frac{1}{t} \quad .$$

In order to do the calculation in a more systematic way we use the property of the Legendre polynomial

$$P_J(x) = \frac{(2J)!}{2^J(J!)^2} x^J + O(x^{J-2}) \quad .$$

With this we evaluate the leading contribution of

$$(\bar{K}_J)^J(\bar{K}_J')^J\ P_J(\bar{z}_J) = \frac{(2J)!}{2^J(J!)^2}\{\cdot\ [(K.K') - \frac{(K.q)(K'.q)}{m^2(J)}]\}^J +$$

$$+ \ldots \qquad (36)$$

To bring out the singularity at $t = o$ we rewrite the

r. h. s. of (36) as

$$\frac{(2J)!}{2^J(J!)}\left\{-\left[(K.K') - \frac{(K.q)(K'.q)}{t}\right] - \frac{m^2(J)-t}{m^2(J).t}(Kq)(K'.q)\right\}^J + \\ + \;.... \qquad (37)$$

Defining the quantities K_t, K'_t, z_t by the same expressions as (31) except that we replace $m^2(J)$ by t everywhere and developing (37) in a binomial series we obtain

$$(\bar{K}_J)^J(\bar{K}'_J)^J P_J(\bar{z}_J) = (K_t)^J(K'_t)^J P_J(z_t) - \\ - (2J-1)(K.q)(K'.q)\frac{m^2(J)-t}{m^2(J).t}(K_t)^{J-1}(K'_t)^{J-1}P_{J-1}(z_t) + \;... \qquad (38)$$

The extra terms of order J-1 and so on occurring in (38) vanish in the equal mass case because of the factors (K.q)(K'.q) which are zero. If we carry out this procedure for each term in the sum (33) we arrive at the form

$$f = \sum_{J=0}^{\infty}\frac{(2J+1)g(J)g'(J)}{m^2(J)-t}(K_t)^J(K'_t)^J P_J(z_t) + \\ + \sum_{J=1}^{\infty}\frac{(2J+1)g(J)g'(J)}{m^2(J).t}(2J-1)(Kq)(K'q)(K_t)^{J-1}\times \\ \times (K'_t)^{J-1}P_{J-1}(z_t) + \;... \qquad (39)$$

Again we perform the Sommerfeld-Watson transformation and we end up with the Regge expression

$$f = \pi \frac{g(\alpha)g'(\alpha)(2\alpha+1)}{\sin\pi\alpha} (K_t)^{\alpha}(K'_t)^{\alpha}P_{\alpha}(-z_t) \frac{d\alpha}{dt} +$$

$$+ \pi \frac{g(\alpha(o)) \; g'(\alpha(o))(2\alpha(o)+1)(2\alpha(o)-1)}{\sin\pi[\alpha(o)-1]} \left.\frac{d\alpha}{dt}\right|_{t=o} \times$$

$$\times \frac{(Kq)(K'q)}{t} (K_t)^{\alpha(o)-1} (K'_t)^{\alpha(o)-1} \times$$

$$\times P_{\alpha(o)-1}(-z_t) + \text{terms with } (\alpha(o)-2) \text{ and so on,}$$

$$\text{down to } -\infty \qquad (40)$$

We therefore automatically get daughters together with the parent trajectory which at $t = 0$ are spaced one unit apart. There is, however, a slight defect in the expression (40), since except for the first term all the other ones contain just $\alpha(o)$ and therefore do not represent moving trajectories. This difficulty has been overcome [10] by unitarizing the pole diagrams, which doesn't change the behaviour at $t = 0$.

Again the leading term as $|z_t| \to \infty$ is of the form

$$\left\{\frac{(s-u)}{4} - \frac{(Kq)(K'q)}{t}\right\}^{\alpha(t)} \qquad (41)$$

and for $t \neq o$ fixed we get the usual $s^{\alpha(t)}$ behaviour for s large enough. If we consider the binomial series of (41) we get

$$\left(\frac{s-u}{4}\right)^{\alpha(t)} - \alpha(t)\left(\frac{s-u}{4}\right)^{\alpha(t)-1} \frac{(K.q)(K'.q)}{t} + \dots .$$

Now the second term here is exactly cancelled by the leading term of the first daughter trajectory (second term in (40)) in the limit $t \to 0$ and so on. Thus we are left simply with $\left(\frac{s-u}{4}\right)^{\alpha(t)}$ even for $t = 0$ and we can safely let s go to infinity and again obtain the familiar $s^{\alpha(t)}$ behaviour for the scattering amplitude. It should be kept in mind that in order to achieve this we had to take into account the infinite number of daughter trajectories by which we get rid of the singularity at t = 0 of the contribution of the parent trajectory.

3. O(4) Symmetry and Regge Pole Theory

The work of Freedman and Wang [11] represents another point of view of the appearance of daughter trajectories. This treatment works for spinning particles, however, only for the case of equal masses. The idea behind this analysis is the following: One knows how to make a partial wave expansion in the c.m. system in the physical region. If we make the transformation to the c. m. system, then the sum of the two incoming momenta has only a time component and obviously the rotation group on the three space components is the only Lorentz transformation which conserves this sum. We therefore have an expansion in the representations of the rotation group. Now the idea of Freedman and Wang was to continue analytically the vector sum of the incoming momenta in the c.m. system to zero, then the full complex Lorentz group is the group of transformations which conserves this zero vector. Due to the analytic continuation used the space components of the various momenta become pure imaginary and if we do not want to alter the property in the Lorentz transformation we are left with a subgroup which is the rotation group in four dimensions O(4).

The general transformation properties of the scattering amplitude in the t-channel $f^{(t)}_{\{\lambda\}}$ under Lorentz transformations are given by (in contrast to our previous notation we consider as the t-channel reaction $1+2\rightarrow 3+4$)

$$f^{(t)}_{\lambda_3,\lambda_4;\lambda_1,\lambda_2}(p_3,p_4;p_1,p_2) =$$

$$= \sum_{\{\mu\}} D^{s_3}_{\lambda_3,\mu_3}[R_w^{-1}(\Lambda,p_3)] \; D^{s_4}_{-\lambda_4,-\mu_4}[R_w^{-1}(\Lambda,p_4)] \times$$

$$\times \; f^{(t)}_{\mu_3,\mu_4;\mu_1,\mu_2}(\Lambda p_3,\Lambda p_4;\Lambda \dot{p}_1,\Lambda p_2) \; D^{s_1}_{\mu_1,\lambda_1}[R_w(\Lambda,p_1)] \times$$

$$\times \; D^{s_2}_{-\mu_2,-\lambda_2}[R_w(\Lambda,p_2)].$$

$$(42)$$

The arguments of the rotation matrices $D^s_{\lambda,\mu}$ are the Wigner rotations

$$R_w(\Lambda,p) = L^{-1}(\Lambda p)\Lambda L(p) \quad,$$

where $L(p)$ is the Lorentz transformation which takes a particle of mass m from rest to momentum p_i

$$L(p) = R(\phi,\theta,o) \; e^{-i\bar{\delta}N_3}$$

$$p_o = m \; \cosh \bar{\delta}, \qquad \vec{p} = m \sinh \bar{\delta} \; \vec{n} \; (\theta,\phi)$$

N_3 is the generator of pure Lorentz transformation along the z direction. In our case we adopt the convention [12] $\bar{\delta}_1, \bar{\delta}_3 \geq 0, \bar{\delta}_2, \bar{\delta}_4 \leq 0$, corresponding to the opposite directions of the respective particles. In the c.m. system of the physical region of the t-channel

$$\phi_1 = \phi_2 \;,\; \theta_1 = \theta_2 \;,\; \bar{\delta}_1 = -\bar{\delta}_2 = \sinh^{-1}\{[\frac{t}{4m_1^2} -1]^{1/2}\}$$

for $m_1 = m_2$ and $t = (p_1+p_2)^2 = q_\mu q^\mu$, $s = (p_3-p_1)^2$. (43)

If we consider the Mandelstam plot we can say that the

usual partial wave expansion which is made for fixed t above threshold in the t c.m. system involves all amplitudes corresponding to (s,t) values on the segment l which is bounded by the physical region boundary (cf. fig. 3).

In order to get the O(4) symmetry we continue the amplitudes from the physical region (line l) down to t = 0 as shown by the line L in fig. 3. By this $\bar{\delta}_{1,2}$ becomes purely imaginary

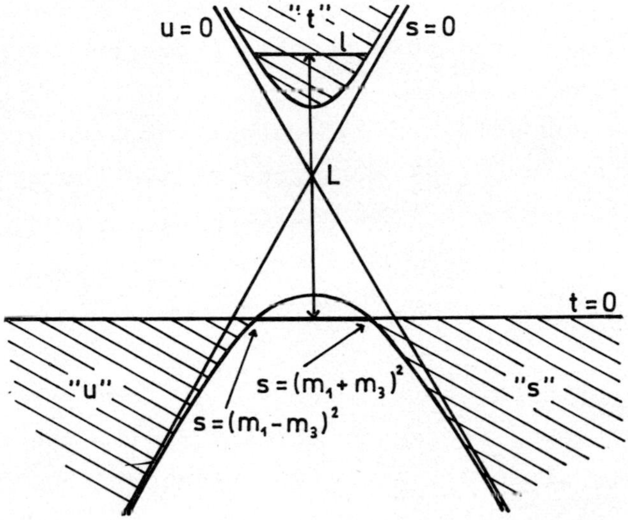

Fig. 3

Mandelstam plot

$$t = 0 \quad \rightarrow \bar{\delta}_1 = -\bar{\delta}_2 = i\delta \quad , \quad \delta = \pi/_2$$

and

$$q_\mu = (\sqrt{t},0,0,0) \rightarrow (0,0,0,0). \qquad (44)$$

From the last line we see that $p_1 = -p_2$, $p_3 = -p_4$.

The scattering amplitude in (42) could be written as a function of the sums and differences of incoming and outgoing momenta respectively; by (44) it then depends on the differences only which we denote by

$$P = \frac{1}{2}(p_1 - p_2) \quad , \; P' = \frac{1}{2}(p_3 - p_4) \quad ,$$

and the components of, e.g., P are

$$P_0 = 0 \quad , \qquad \vec{P} = i\, m_1\, \vec{n}\,(\theta,\phi)$$

It should be noted that by this procedure we remain on the mass-shell.

We are now going to make a partial wave analysis for $f^t_{\{\lambda\}}(p',-p',p,-p)$ analogous to the one in the physical t-region in the region t=0,

$$(m_1 - m_3)^2 \le s \le (m_1 + m_3)^2 \quad ,$$

where we have SO(4) symmetry.

Before doing so let us briefly review the essential properties of SO(4). Any given element $g \varepsilon SO(4)$ can be represented by

$$g = R(\phi,\theta,o)\; L_3(\delta)\; R(\alpha,\beta,\gamma) \; ,$$

$L_3(\delta)$ is a rotation by an angle δ in the 34 plane; with the invariant volume element on the group space

$$dg = d\phi d(\cos\theta)\, \sin^2\delta d\delta d\alpha d(\cos\beta) d\gamma \quad .$$

The commutation relations of the six generators J_i, K_i (i = 1,2,3) read

$$[J_i, J_j] = i\,\varepsilon_{ijk}\,J_k,\quad [J_i, K_j] = i\varepsilon_{ijk}\,K_k,$$

$$[K_i, K_j] = i\,\varepsilon_{ijk}\,J_k\ .$$

By forming sums and differences of these

$$A_i = \frac{1}{2}(J_i+K_i)\quad ,\quad B_i = \frac{1}{2}(J_i-K_i)$$

we can "diagonalize" this set of commutators so that

$$[A_i,A_j] = i\varepsilon_{ijk}\,A_k,\quad [B_i,B_j] = i\varepsilon_{ijk}\,B_k,$$

$$[A_i,B_j] = 0\ .$$

The generators A_i and B_i have the commutation relations of two independent angular momenta , and therefore there is a homomorphism between SO(4) and SU(2) ⊗ SU(2). In our case we have a four-vector whose space components are pure imaginary numbers with which we can in a one-to-one way associate a 2 × 2 matrix

$$(t,i\vec{x}) \leftrightarrow t\,I + i\,\vec{x}\,.\,\vec{\sigma}\ .$$

The transformation

$$t'I + i\,\vec{x}'\,.\,\vec{\sigma} = U(t\,I + i\,\vec{x}\,.\,\vec{\sigma})\,V^{\dagger}$$

with U, V ε SU_2 is then our SU(2) ⊗ SU(2) transformation.

Since we have decomposed the commutation relations by the introduction of two commuting sets of angular momenta the irreducible representations of SO(4) are given by a set of numbers (a,b), where a,b = 0,1/2,1, 3/2,... independently. The state vectors can be labelled

$$\vec{A}^2|(a,b),a_3,b_3> = a(a+1)|(a,b),a_3,b_3> ,$$

$$\vec{B}^2|(a,b),a_3,b_3> = b(b+1)|(a,b),a_3,b_3> ,$$

$$A_3|(a,b),a_3,b_3> = a_3|(a,b),a_3,b_3> ,$$

$$B_3|(a,b),a_3,b_3> = b_3|(a,b),a_3,b_3> .$$

A set of state vectors which is more convenient is the set

$$|(a,b),J,J_3> ,$$

where $\vec{J} = \vec{A} + \vec{B}$ and $J_3 = A_3 + B_3$ with eigenvalues $j(j+1)$ and m respectively. The transformation between these two sets is given by the usual Clebsch-Gordan coefficients

$$<(a,b),a_3,b_3|(a,b),J,J_3> .$$

In the latter basis the representation matrices of $SU(2) \otimes SU(2)$ can be written as

$$D^{(a,b)}_{jm,j'm'} = \sum_{m''} D^{j}_{mm''}(\phi,\theta,o)d^{(a,b)}_{jj'm''}(\delta)D^{j'}_{m''m'}(\alpha,\beta,\gamma) ,$$

where $D^{j}_{mm'}$ are the representation matrices of $SU(2)$ and the remaining matrix $d^{(a,b)}_{jj'm}(\delta)$ is given by a trigonometric polynomial

$$d^{(a,b)}_{jj'm}(\delta) = \sum_{\mu} <(a,b),\mu,m-\mu|j,m><(a,b),\mu,m-\mu|j'm> \times$$
$$\times \; e^{-i(2\mu-m)\delta}$$

It will be convenient in the following to use the sum and difference of a and b, denoted by

$$n = a + b \quad , \; M = a - b$$

For the partial wave expansion we write the scattering amplitude, continued to t = 0, in the form

$$f_{\{\lambda\}}(P',-P';P,-P) = \langle P',\lambda_3,\lambda_4|F|P,\lambda_1,\lambda_2\rangle \quad . \qquad (45)$$

In complete analogy to the rotation group [1] the state vectors $|P,\lambda_1,\lambda_2\rangle$are now related to a set of standard vectors ("north-pole" states)

$$|P,\lambda_1,\lambda_2\rangle = R(\phi,\theta,0)e^{-i\delta K_3}|\bar{P},\lambda_1,\lambda_2\rangle \quad ,$$

where

$$\bar{P}_\mu = (0;0,0,i\ m) \ .$$

In continuing the analysis it is convenient to add together the angular momenta carried by the incoming (outgoing) particles and form states

$$|P,s,\lambda\rangle = \sum_{\lambda_1,\lambda_2} (-1)^{s_2-\lambda_2} \langle s_1,\lambda_2,s_2,-\lambda_2|s,\lambda = \lambda_1-\lambda_2\rangle \times$$
$$\times \ |P,\lambda_1,\lambda_2\rangle$$

with

$$|s_1 - s_2| \leq s \leq s_1 + s_2 \quad .$$

It should be commented that s is not the total spin of the two incoming particles, although it has the properties of a rotational quantum number. Only the "north-pole" state $|\bar{P},s,\lambda\rangle$ transforms according to the irreducible representations of SU(2), but going to the state $|P,s,\lambda\rangle$

we need the boost $\exp\{-i\delta K_3\}$ which doesn't commute with the rotations J_1 and J_2.

Finally, we construct states which are labelled by the irreducible representations of SO(4)

$$|(a,b),j,m,s\rangle =$$

$$= N_s^{(a,b)} \sum_\lambda \int d\Omega\, D^{j*}_{m,\lambda}(\phi,\theta,o)\, d^{(a,b)*}_{j,s,\lambda}(\delta)\,|P,s,\lambda\rangle, \quad (46)$$

$$|a-b| \le j,\ s \le a+b\ , \qquad \text{and } d\Omega = \sin^2\delta\, d\delta\, d(\cos\theta)\, d\phi\ .$$

With this we have the tools for making the partial wave expansion. We insert in (45) a complete set of states $|P,s,\lambda\rangle$ to the left and right of F and then the set $|(a,b),j,m,s\rangle$. Finally we are left with the partial waves of the O(4) group

$$f^{n,M}_{s's} = \langle (a,b),j,m,s'|F|(a,b),j,m,s\rangle \quad , \qquad (47)$$

which are diagonal in the labels of the representation (a,b) and independent of j and m because of Schur's lemma. With these we get a lengthy expression for the partial wave expansion for $f_{\{\lambda\}}$. We will now consider an application of this formalism to the process $N\bar{N}\to N\bar{N}$. We get the following set of independent partial waves (since s = 0,1 and because of parity conservation $f^{n,M}_{s's} = \delta_{s's} f^{n,M}_s$)

$$f^{n,o}_o \quad , \qquad f^{n,1}_o \quad , \qquad f^{n,1}_1 \quad .$$

The five helicity amplitudes defined at the end of section 1 can be expressed as an infinite sum over these O(4) partial waves, where we have taken into account already the signature:

$$f_o^{J\pm} = \frac{1}{3\pi^2(2J+1)}\Big\{\sum_{\substack{K=1\\ \text{odd}}}^{\infty} (J+K+1)^2 \,|d^{(J+K,0)}_{J,1,0}(\pi/2)|^2 f_1^{J+K,0\mp} +$$

$$+ \sum_{\substack{K=0\\ \text{even}}}^{\infty} 2(J+K)(J+K+2)\,|d^{(J+K,1)}_{J,1,0}(\tfrac{\pi}{2})|^2 f_1^{J+K,1\pm}\Big\},$$

$$f_1^{J\pm} = \frac{2}{3\pi^2(2J+1)}\Big\{\sum_{\substack{K=0\\ \text{even}}}^{\infty} (J+K+1)^2\,|d^{(J+K,0)}_{J,1,1}(\tfrac{\pi}{2})|^2 f_1^{J+K,0\pm} +$$

$$+ \sum_{\substack{K=1\\ \text{odd}}}^{\infty} 2(J+K)(J+K+2)\,|d^{(J+K,1)}_{J,1,1}(\tfrac{\pi}{2})|^2 f_1^{J+K,1\mp}\Big\},$$

$$f_{11}^{J\pm} = \frac{1}{\pi^2(2J+1)} \sum_{\substack{K=0\\ \text{even}}}^{\infty} (J+K+1)^2\,|d^{(J+K,0)}_{J,0,0}(\tfrac{\pi}{2})|^2 f_o^{J+K,0\pm},$$

$$f_{22}^{J\pm} = \frac{1}{3\pi^2(2J+1)} \sum_{\substack{K=0\\ \text{even}}}^{\infty} 2(J+K)(J+K+2)\,|d^{(J+K,1)}_{J,1,1}(\tfrac{\pi}{2})|^2 f_1^{J+K,1\pm};$$

$$f_{12}^{J} = 0 \quad . \tag{48}$$

Under the usual assumptions we can perform a Sommerfeld-Watson transformation and in opening up the contour we again assume that only simple poles in the $f_s^{n,M\pm}$ contribute, which are called Lorentz poles. According to the three independent amplitudes we had we now have three classes of Lorentz poles:

Class I : $M = 0,\ s = 0.$

Near such a pole the amplitude behaves like

$$f_o^{J+K,0\pm} \simeq \frac{\gamma^\pm}{J+K-\alpha^\pm}$$

From (48) we see that this contributes only to $f_{11}^{J\pm}$ and therefore this simple Lorentz pole corresponds to infinitely many Regge poles of $f_{11}^{J\pm}$ at

$$J = \alpha^\pm\ ,\quad \alpha^\pm-2,\quad \alpha^\pm-4\quad ,\ldots\ldots$$

with the quantum numbers $P = C = (-1)^J$.

Class II : $M = 0,\ s = 1$.

Here the Lorentz poles contribute to $f_1^{J\pm}$ leading to Regge poles at

$$J = \alpha^\pm\ ,\quad \alpha^\pm-2,\ldots \qquad P = C = -(-1)^J$$

and to $f_o^{J\pm}$ with

$$J = \alpha^\pm - 1,\quad \alpha^\pm-3,\ldots. \qquad P = -C = -(-1)^J\ .$$

Class III : $M = 1,\quad s = 1$.

This Lorentzpole causes Regge poles of the amplitudes $f_o^{J\pm}$ at

$$J = \alpha^\pm\ ,\quad \alpha^\pm-2,\ldots\ ;\qquad P = -C = -(-1)^J\ ,$$

and $f_{22}^{J\pm}$ at

$$J = \alpha^{\pm}, \alpha^{\pm} -2, \ldots; \qquad P = C = (-1)^J$$

and also f_1^J at

$$J = \alpha^{\pm} - 1, \quad \alpha^{\pm} -3, \ldots; \quad P = C = -(-1)^J$$

In this approach daughters automatically come in and are therefore connected with the higher symmetry of the problem in the case of equal masses at $t = 0$, which corresponds to an identically vanishing q_μ . This however is not true for unequal masses where the point $t = 0$ corresponds to a light-like q_μ . Thus it is not clear how to generalize the above treatment to unequal masses, except possibly by an analytic continuation in the masses.

The author is indebted to Dr. E. Gotsman and Dr. D. Horn for their critical reading of the manuscript. He would also like to thank Dr. R. Baier and Dr. H. Latal who helped him in preparing these notes.

References

1. M. Jacob , G. C. Wick, Ann. Phys. 7, 404 (1959).
2. H. Lehmann, Nuovo Cim. 10, 579 (1958)
3. M. Gell-Mann, M. L. Goldberger, F. E. Low, E. Marx, F. Zachariasen, Phys. Rev. 133, B145 (1964); V. N. Gribov, JETP 14, 478 (1962); M. Froissart, La Jolla Conference 1961; F. Calogero, J. M. Charap, E. J. Squires, Ann. Phys. 25, 325 (1963).
4. T. L. Trueman, G. C. Wick, Ann. Phys. 46,239 (1968).

F. J. Muzinich, J. Math. Phys. 5, 1481 (1964).

5. Y. Hara, Phys. Rev. 136, B507 (1964);, L. L. Chan Wang, Phys. Rev. 142, 1187 (1965); G. Cohen-Tannoudji, A. Morel, H. Navelet, Ann. Phys. 46, 239 (1968).
6. J. D. Jackson, G. E. Hite, Phys. Rev.
7. D. V. Volkov, V. N. Gribov, JETP, 17, 720 (1963).
8. E. Leader, Phys. Rev. 166, 1599 (1968).
9. L. Van Hove, Phys. Letters 24B, 183 (1967).
10. R. L. Sugar, J. D. Sullivan, Phys. Rev. 166,1515(1968); R. Blankenbecler, R. L. Sugar, J. D. Sullivan, Phys.Rev to be published; L. Durand, to be published. J. C. Taylor, to be published
11. D. Z. Freedman, J. M. Wang, Phys. Rev. 160, 1560 (1967).
12. G. C. Wick, Ann. Phys. 18, 65 (1962).

ON FEYNMAN DIAGRAMS WITH INFINITE DIMENSIONAL REPRESENTATIONS OF THE LORENTZ GROUP O(3,1) †

By

K. KOLLER

Imperial College, London

Conventional field theory hitherto has been exclusively based on finite dimensional representations of the Lorentz group O(3,1). However there exists no convincing argument why one should restrict oneself to finite dimensional representations only. We will therefore consider besides finite dimensional representations also infinite dimensional (not necessarily unitary) representations of the Lorentz group O(3,1). One may argue that these infinite dimensional representations are needed for a complete dynamical theory. Recently infinite dimensional representations have been used in infinite multiplet field theories [1], in generalized Regge theories [2] and in current algebras [3].

We will in this lecture consider a field theory with infinite dimensional representations of the Lorentz group. In general such theories may be very complicated. We will talk mainly about the mathematical structure

† Seminar given at the VII.Internationale Universitätswochen f.Kernphysik, Schladming, February 26-March 9,1968.

of simple models. In the first lecture we will give a derivation [4] of the matrix elements of infinite dimensional representations in an SU(2), SU(1,1) or Euclidean group U(1) ∧ T(2) basis. The generalized "Dirac spinor" wave functions will be given by columns of these infinite matrices. We will show explicitly the relation between finite and infinite dimensional representations and emphasize that in many respects there should be a continuous transition between a field theory based on finite dimensional representations to theories involving infinite dimensional representations. In the second lecture we will set up the free local fields and free propagators for infinite dimensional unitary fields and construct simple invariant 3-point couplings among these. These correspond to the Feynman rules of local quantized field theory. The aim will be then to discuss simple (supposedly dominant) pole diagrams of the scattering amplitude [5]. In general we will use only fields of the Lorentz group, where all states in one representation have the same mass.

Usually one realizes the states of a unitary representation $[\mu s]$ of the Poincaré group (μ = mass, s = spin) as the manifold of solutions of a relativistic invariant equation. Such field equations serve only as a relation between different representations. We will not use field equations but work instead with the matrix elements of infinite dimensional representations directly. To derive these we define the six generators of the Lorentz group O(3,1) or its covering group SL(2,C) as

$$\underline{J} = (J_{23}, J_{31}, J_{12}) \quad , \quad \underline{K} = (J_{01}, J_{02}, J_{03}) \qquad (1)$$

with the commutation relations

$$[J_i, J_j] = -[K_i, K_j] = i\varepsilon_{ijk} J_k$$

$$[J_i, K_j] = [K_i, J_j] = i\varepsilon_{ijk} K_k \qquad (2)$$

In the 2×2 dimensional fundamental representation, $\underline{J}$ and $-i\underline{K}$ are simply the Pauli matrices $1/2\ \underline{\tau}$. The two Casimir operators are then given by

$$\underline{J}^2 - \underline{K}^2 = \sigma^2 + \kappa^2 - 1$$

$$\underline{J} \cdot \underline{K} = i\sigma \,.\, \kappa \qquad (3)$$

All irreducible representations are then specified by two labels (κ,σ); κ is an integer or half integer, σ is an arbitrary complex number (where $\kappa = k_0$, $\sigma = c$ in Naimarks notation) [6].

Specific examples of representations are the following:

$(0,\sigma)$ $o<\sigma<1$, $\mathrm{Im}\ \sigma = 0$	unitary representations of the complementary series.
(κ,σ) $o\leq \mathrm{Im}\sigma<\infty$, $\mathrm{Re}\sigma = 0$	unitary representations of the principal series.
$\mathrm{Im}\sigma = o$, $\sigma = \lvert\kappa\rvert + n$, where $n = 1,2,\ldots\infty$	finite dimensional representations
$(0,1)$	scalar representation
$(\frac{1}{2}, \frac{3}{2}) \oplus (-\frac{1}{2}, \frac{3}{2})$	Dirac representation.

It is not necessary to write the matrix elements of the Lorentz group in a SU(2) basis, one can equally well diagonalize the SU(1,1) or Euclidean group $U(1) \wedge T(2)$ contained within the Lorentz group. ($\wedge$ denotes the semi-direct product). Therefore let us denote the three kinds of basis states of SL(2,C) as

$\|\kappa\sigma, jm>$	when in a SU(2) basis
$\|\kappa\sigma, km>$	when in a SU(1,1) basis
$\|\kappa\sigma, \ell m>$	when in a $U(1)\wedge T(2)$ basis ,

(4)

where SU(2) will be generated by J_3, J_1, J_2

SU(1,1)	by J_3, K_1, K_2
$U(1)\wedge T(2)$	by J_3; $L_1=K_1+J_2$; $L_2=J_1-K_2$

(5)

with the following Casimir operator equations

$$(J_1^2+J_2^2+J_3^2)|\kappa\sigma,jm> = j(j+1)|\kappa\sigma,jm>$$

$$(J_3^2-K_1^2-K_2^2)|\kappa\sigma,km> = k(k+1)|\kappa\sigma,km>$$

$$(L_1^2+L_2^2)|\kappa\sigma,\ell m> = \ell^2 |\kappa\sigma,\ell m> \qquad (6)$$

Note that all three kinds of states are diagonalized in J_3 with eigenvalue m, the third component of spin.
We will now determine the matrix elements of the boost in the third direction

$$d^{\kappa\sigma}_{AmB}(\xi) = <\kappa\sigma Am|\exp\{-i\xi K_3\}|\kappa\sigma Bm> \qquad (7)$$

where A, B refer to any of the subgroup Casimir labels (A,B = k,j,ℓ). The choice of A may be different from B. In this case one just obtains in the limit $\xi \to 0$ the transformation functions $<\kappa\sigma Am|\kappa\sigma Bm> \equiv <Am|Bm>$

which provide us with transformation functions between the representations in a SU(2), SU(1,1) or U(1)∧T(2) basis.

The derivation of the representations $d^{\kappa\sigma}_{AmB}(\xi)$ will be done for the unitary representations of the principal series, but other representations may be obtained by analytic continuation of the final results (18). To derive (7) one has to employ the following two tricks.

The first trick is to construct the identities

$$\exp\{-i\xi K_3\}\exp\{-i\chi_B J_B\}=\exp\{-i\chi_A J_A\}\exp\{-i\eta M_1\}\times$$

$$\times\exp\{-i\alpha K_3\} \qquad M_1 = K_1 - J_2 \qquad (8a)$$

where χ_A stands for the little group parameter and J_A is the infinitesimal generator corresponding to Casimir operator A (and similarly for χ_B, J_B). If A or B represents the SU(1,1) group one has to supplement (8a) by a second identity

$$\exp\{-i\xi K_3\}\exp\{-i\chi_B J_B\}\exp\{-i\pi J_2\} =$$

$$= \exp\{-i\chi_A J_A\}\exp\{-i\eta M_1\}\exp\{-i\alpha K_3\} \qquad (8b)$$

in order to cover the full range of the involved group parameters. Similar identities may be written down for all semi-simple Lie groups. In the mathematical literature this is known as the Iwasawa decomposition of Lie groups [7].

The second trick is to construct a special set of states $|\kappa\sigma\}$ such that the operation with the factor $\exp\{-i\eta M_1\}\exp\{-i\alpha K_3\}$ from (8) on these states always gives

$$\exp\{-i\eta M_1\}\exp\{-i\alpha K_3\}|\kappa\sigma\} = \exp\{\alpha(\sigma-1)\}|\kappa\sigma\} \qquad (9)$$

The $|\kappa\sigma\}$ states are defined by considering the $|j_1 j_2, m_1 m_2\}$ states of the SU(2) ⊗ SU(2) group with the Casimir operators

$$\underline{J}^{(1)2} = \frac{1}{4} (\underline{J} - i \underline{K})^2 = j_1(j_1+1) \cdot 1$$

$$\underline{J}^{(2)2} = \frac{1}{4} (\underline{J} + i \underline{K})^2 = j_2(j_2+1) \cdot 1 \qquad (10)$$

By comparing (10) with (3) one obtains the relations

$$\sigma - 1 = j_1 + j_2$$

$$\kappa = j_1 - j_2 \qquad (11)$$

The $|\kappa\sigma\}$ states are now defined to be $|j_1 j_2, m_1 m_2\}$ states with the extreme values:

$$m_1 = j_1 = \frac{1}{2} (\sigma-1+\kappa)$$

$$m_2 = -j_2 = \cdot\cdot \frac{1}{2}(\sigma-1-\kappa) \qquad (12)$$

Since $J_+^{(1)}|\kappa\sigma\} = 0$, $J_-^{(2)}|\kappa\sigma\} = 0$ one has the relations

$$J_3|\kappa\sigma\} = \kappa|\kappa\sigma\} \quad , \quad K_3|\kappa\sigma\} = i(\sigma-1)|\kappa\sigma\}$$

$$M_1|\kappa\sigma\} = M_2|\kappa\sigma\} = 0 \quad \text{where } M_1 = K_1 - J_2 \, , \quad M_2 = K_2 + J_1 \qquad (13)$$

and equation (9) follows. Now one sandwiches the group parameter identities (8a) between $\langle Am|\ldots|\kappa\sigma\}$ states and obtains

$$\langle Am|\exp\{-i\xi K_3\}|Bm\rangle\langle Bm|\exp\{-i\chi_B J_B\}|B\kappa\rangle\langle B\kappa|\kappa\sigma\} =$$

$$= \langle Am|\exp\{-i\chi_A J_A|A\kappa\rangle\langle A\kappa|\kappa\sigma\}\exp\{\dot{\alpha}(\sigma-1)\} \qquad (14)$$

by using symbolic completeness relations of the little group states

$$|Bm\rangle\langle Bm| = 1 \quad , \quad |B\kappa\rangle\langle B\kappa| = 1 \quad \text{etc.}$$

Observing that $\langle Bm|\exp\{-i\chi_B J_B\}|B\kappa\rangle \equiv d^{\dot{B}}_{m\kappa}(\chi_B)$ are the little group representations and using orthogonality of $d^{B}_{m\kappa}(\chi_B)$ one obtains the final result:

$$d^{\kappa\sigma}_{AmB}(\xi) = N_{AB}\int d\chi_B \; d^{B}_{m\kappa}(\chi_B) \; d^{A}_{m\kappa}(\chi_A) \; \exp\{\alpha(\sigma-1)\} \qquad (15)$$

where $N_{AB} \equiv \frac{\langle A\kappa|\kappa\sigma\}}{\langle B\kappa|\kappa\sigma\}}$ is a normalization factor.

In (15) χ_A and α are always definite functions of χ_B and ξ. These relations may be determined from (8) by using the 2×2 dimensional spinor representation. If (8b) is needed there will be a second expression, analogous to (15).

In the case of representations SL(2,C) in a SU(2) basis eq. (8a) reads (taking $A = j'$, $B = j$, $\chi_A = \Theta'$, $\chi_B = \Theta$).

$$\exp\{-i\xi K_3\}\exp\{-i\Theta J_2\} = \exp\{-i\Theta' J_2\}\exp\{-i\eta M_1\} \times$$

$$\times \; \exp\{-i\alpha K_3\} \qquad (16)$$

i.e.

$$\begin{pmatrix} \exp\{\frac{\xi}{2}\} & 0 \\ 0 & \exp\{-\frac{\xi}{2}\} \end{pmatrix} \begin{pmatrix} \cos\frac{\Theta}{2} & -\sin\frac{\Theta}{2} \\ \sin\frac{\Theta}{2} & \cos\frac{\Theta}{2} \end{pmatrix} =$$

$$= \begin{pmatrix} \cos\frac{\Theta'}{2} & -\sin\frac{\Theta'}{2} \\ \sin\frac{\Theta'}{2} & \cos\frac{\Theta}{2} \end{pmatrix} \begin{pmatrix} 1 & \eta \\ 0 & 1 \end{pmatrix} \begin{pmatrix} \exp\{\frac{\alpha}{2}\} & 0 \\ 0 & \exp\{-\frac{\alpha}{2}\} \end{pmatrix} \tag{17}$$

from which we find

$$\tan\frac{\Theta'}{2} = \exp\{-\xi\}\tan\frac{\Theta}{2}$$

$$e^{\alpha} = \frac{\sin\Theta}{\sin\Theta'}$$

and the integral (15) becomes

$$d^{\kappa\sigma}_{j'mj}(\xi) =$$

$$= N_{j'j}\,\frac{1}{2}\int_{-1}^{+1} d(\cos\Theta)\; d^{j}_{m\kappa}(\Theta)\,d^{j'}_{m\kappa}(\Theta')\left[\frac{\sin\Theta}{\sin\Theta'}\right]^{\sigma-1} =$$

$$= N_{j'j}\,\frac{1}{2}\int_{-1}^{+1} d(\cos\Theta)\frac{d^{j}_{m\kappa}(\Theta)\,d^{j'}_{m\kappa}(\Theta')}{\{\cosh\xi+\cos\Theta\sinh\xi\}^{1-\sigma}} \tag{18}$$

with

$$N_{j'j} = \left[(2j+1)(2j'+1)\,\frac{\sin\pi(\sigma-j')\Gamma(1+j'-\sigma)\Gamma(1+j+\sigma)}{\sin\pi(\sigma-j)\Gamma(1+j'+\sigma)\Gamma(1+j-\sigma)}\right]^{\frac{1}{2}}$$

where as usual $\cosh\xi = \frac{E}{\mu}$, (E = energy, μ= mass)

$$\sinh\xi = \frac{|\vec{p}|}{\mu} \qquad (\vec{p} = \text{three momentum})$$

and $d^j_{m\kappa}(\Theta)$, $d^{j'}_{m\kappa}(\Theta')$ are the familiar SU(2) representations (8). The spin j and third component m take the values $j = |\kappa|, |\kappa| + 1, |\kappa| + 2 \ldots \infty$; $-j \leq m \leq +j$.

The integral expressions where SU(1,1) and $U(1) \wedge T(2)$ are diagonalized may be found in ref. 4.

A general integral representation $D_{j'm'jm}(L_p)$ of a Lorentz transformation L_p parametrized in the form

$$U(L_p) =$$

$$= \exp\{-i\phi J_3\}\exp\{-i\Theta J_2\}\exp\{-i\psi J_3\}\exp\{-i\xi K_3\}\exp\{-i\Theta' J_2\}\times$$

$$\times \exp\{-i\phi' J_3\}$$

is obtained by multiplying (18) with two SU(2) representations D^j

$$D^{\kappa\sigma}_{j'm'jm}(L_p) = \sum_{m''} D^{j'}_{m'm''}(\phi,\Theta,\psi)\, d^{\kappa\sigma}_{j'm''j}(\xi)\, D^j_{m''m}(0,\Theta',\phi') \tag{19}$$

Expression (18) can be evaluated in terms of a finite series of hypergeometric functions ${}_2F_1$. In the special case $j' = m = \kappa = o$ this sum reduces to a single term

$$d^{o\sigma}_{ooj}(\xi) = \tag{20}$$

$$= (2j+1)^{\frac{1}{2}}\,(j+1)\left[\frac{\Gamma(\sigma)\Gamma(\sigma-j)}{\Gamma(\sigma+1)\Gamma(\sigma+j+1)}\right]^{1/2} (\sinh\xi)^j C^{j+1}_{\sigma-j-1}(\cosh\xi),$$

in particular for j=o

$$d^{o\sigma}_{ooo}(\xi) = \frac{1}{\sigma}\,\frac{\sinh\sigma\xi}{\sinh\xi} \quad ; \quad C^{\nu}_{n} \text{ are the Gegenbauer functions.} \tag{21}$$

Eq. (20) is derived from (18) by inserting Rodrigues formula for the Legendre function and integrating by parts. The derivation of (18) was strictly valid only for the principal series of unitary representations, but by analytic continuation in σ one obtains a much larger class of representations. In fact one gets all continuous linear representations in a reflexive Banach space [6], i.e. all representations given by κ = integer or half integer and σ an arbitrary number of the complex plane $\mathcal{C}$. Among those linear infinite dimensional representation of the Lorentz group one also recovers from the integral expression (18) the familiar finite dimensional representations as a special case, given by σ = integer or half integer, $\mathrm{Im}\sigma = 0$, $\sigma = |\kappa| + n$, $n = 1,2,\ldots\infty$. Hence (15), (18) now provide us with a much larger exploitation of Lorentz invariance. Let us clarify now the special role of the finite dimensional representations. For the scalar representation $(\kappa,\sigma) = (0,1)$ one gets from (21)

$$d^{o1}_{ooo}(\xi) = 1$$

and for the Dirac representation $(\frac{1}{2}\,,\,\frac{3}{2}) \oplus (-\frac{1}{2}\,,\,\frac{3}{2})$ from (18)

$$u = \begin{pmatrix} d^{\frac{1}{2}\,\frac{3}{2}}_{\frac{1}{2}m',\,\frac{1}{2}m}(\xi) & 0 \\ 0 & d^{-\frac{1}{2}\,\frac{3}{2}}_{\frac{1}{2}m',\,\frac{1}{2}m}(\xi) \end{pmatrix} = \begin{pmatrix} e^{+\frac{1}{2}} & & & 0 \\ & e^{-\frac{\xi}{2}} & & \\ & & e^{-\frac{\xi}{2}} & \\ 0 & & & e^{+\frac{\xi}{2}} \end{pmatrix} \tag{22}$$

$$m',m = \pm \frac{1}{2}$$

where

$$d^{\kappa\sigma}_{j'm'jm}(\xi) = \delta_{m'm}\; d^{\kappa\sigma}_{j'mj}(\xi)$$

The usual Dirac spinor matrix is obtained by a similarity transformation of u with

$$X = \begin{pmatrix} 1 & -1 \\ 1 & 1 \end{pmatrix}$$

where 1 is the 2×2 dimensional unit matrix. Hence

$$XuX^{-1} = \cosh\frac{\xi}{2}\begin{pmatrix} 1 & 0 \\ 0 & 1 \end{pmatrix} + \sinh\frac{\xi}{2}\begin{pmatrix} 0 & \tau_3 \\ \tau_3 & 0 \end{pmatrix}$$

$$\cosh\frac{\xi}{2} = \frac{1}{\sqrt{2}}\left[\frac{E}{\mu} + 1\right]^{1/2}, \quad \sinh\frac{\xi}{2} = \frac{1}{\sqrt{2}}\left[\frac{E}{\mu} - 1\right]^{1/2} \tag{23}$$

This corresponds to writing the 4×2 dimensional Dirac particle spinor u(p) in the following general form

$$u(p) = \begin{pmatrix} D^{\frac{1}{2}\,\frac{3}{3}}_{j'm'\frac{1}{2}m}(L_p) + D^{-\frac{1}{2}\,\frac{3}{2}}_{j'm'\frac{1}{2}m}(L_p) \\ \\ D^{\frac{1}{2}\,\frac{3}{2}}_{j'm'\frac{1}{2}m}(L_p) - D^{-\frac{1}{2}\,\frac{3}{2}}_{j'm'\frac{1}{2}m}(L_p) \end{pmatrix} \tag{24}$$

$$j' = \frac{1}{2}\;; \qquad m,m' = \pm\frac{1}{2}$$

which gives the right symmetric and antisymmetric parity eigenstates, because the parity operator always maps a representation (κ,σ) into $(-\kappa,\sigma)$. To obtain generalized infinite dimensional Dirac spinor analogues we will only have to complexify the number $\frac{3}{2}$ and use instead of $(\pm\frac{1}{2}\,,\,\frac{3}{2})$ infinite dimensional representation $(\pm\frac{1}{2},\sigma)$, $\sigma \in \mathbb{C}$. To see clearly the relation between finite and infinite dimensional representations one observes from the integral expression (18), that the finite dimensional representations have the following matrix structure:

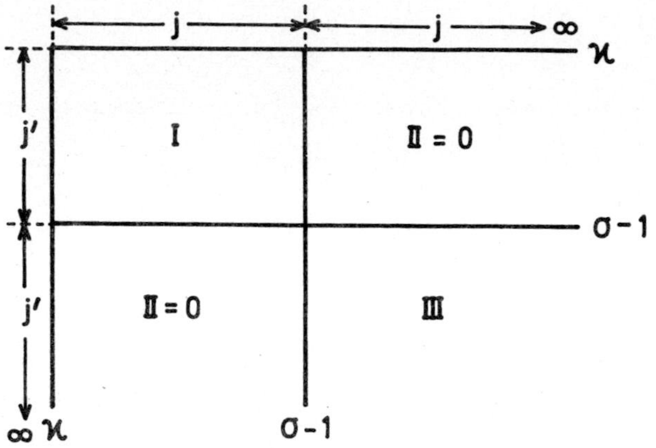

I. usual matrix elements of finite dimensional representation.

II. subspaces with only zeros as matrix elements.

III. non-zero matrix elements.

Fig. 1

The matrix elements of finite dimensional representations lie in domain I, $\kappa \leq j, j' \leq \sigma-1$, but there exists a second infinite dimensional subspace III with both $j, j' > \sigma-1$ which has non-vanishing transition elements for all spin values higher than $\sigma-1$. If either j or j' does not lie in subspace I or III, one obtains from (18) for all finite dimensional (κ,σ)

$$d^{\kappa\sigma}_{j'mj}(\xi) = 0 \qquad \text{when} \qquad \begin{array}{l} \kappa \leq j \leq \sigma-1 \;\; j' > \sigma - 1 \\ \kappa \leq j' < \sigma-1 \;\; j > \sigma - 1 \end{array} \tag{25}$$

Hence the finite dimensional representations are characterized in such a way that there exist sub-spaces II which contain only zeros as matrix elements, thus leaving two invariant subspaces I and III.

If we now deform a finite dimensional representation by an infinitesimal amount in σ , there exist $d^{\kappa\sigma}_{j'mj}(\xi) \neq 0$ in II, hence there will be transitions between subspace I and III and we will have an infinite dimensional representation which contains all the spin states $j = |\kappa|, |\kappa|+1 \ldots \infty$. In this way one obtains a continuous deformation from finite dimensional to infinite dimensional representations. Finite dimensional representations will be precisely those representations which have subspaces II with only zeros, and infinite dimensional representations which are infinitesimal neighbours to finite dimensional representations will have in region II small but non-zero matrix elements. One even may combine an irreducible finite dimensional representation given by an invariant subspace I with the supplementary infinite dimensional invariant subspace III and II to form operator irreducible representation [9]. This leads us to the discussion of irreducibility. For finite dimensional representations one usually defines irreducibility as 1. nonexistence of invariant subspaces or 2. by Schur's Lemma which says that a representation is irreducible if and only if all operators which commute with all representation operators are a multiple of the unit matrix. For unitary representations of non-compact groups one can generalize 1. and 2. to infinite dimensional representations and, what is more important, show that both definitions are equivalent. But in general for infinite dimensional (non-unitary) representations of non-compact groups there does not exist a natural and unique notion of irreducibility. If one generalizes Schur's Lemma to operator irreducibility: A representation on a space $H^{\kappa\sigma}$ is called operator irreducible if every closed* operator on $H^{\kappa\sigma}$ that commutes with all representation operators is a multiple of the unit operator. One can then show that all representations (κ,σ) are operator irreducible representations [9]. Thus operator irreducibility is not equivalent to non-existence of invariant subspaces, and finite dimensional representations occur as invariant subspaces of operator irreducible infinite dimensional representations. Therefore one may consider the Dirac spinor as an infinite dimensional object, which has only 4 nonvanishing components and obtain in this way a continuous transition to neighbouring infinite dimensional representations which have transitions between all half integer spin states $j = \frac{1}{2}, \frac{3}{2}, \ldots\infty$.

Hence finite dimensional representations are surrounded by a sea of infinite dimensional representations, they are only finite dimensional invariant subspaces of operator irreducible infinite dimensional representations; even, when one wishes to maintain a conservative view of conventional field theory, one can never ignore the fact that the slightest deformation of the representation space brings into appearance infinite towers of spin states.

Lecture II

In this lecture we will discuss pole diagrams with infinite-dimensional representations. Considering only irreducible representations of the Lorentz group and using the expression (18), (19) let us write down free local infinite dimensional fields [5].

---

* An operator O in $H^{\kappa\sigma}$ is closed when its graph $(x, Ox) \in H^{\kappa\sigma} \oplus H^{\kappa\sigma}$ is a closed set where x runs through the domain of definition of O.

$$\psi^{\kappa\sigma}_{jm}(x) = \int d\Omega_p \sum_{s,\lambda} \{D^{\kappa\sigma}_{jm,s\lambda}(L_p)a(p,s\lambda)\, e^{-ipx} +$$

$$+ [D^{\kappa\sigma}(L_p)C^{-1}]_{jm,s\lambda}\; a^*(p,s\lambda)\, e^{+ipx}\} \quad (26)$$

with

$$d\Omega_p = \Theta(p_o)\ \delta(p^2-\mu^2)\ \frac{d^4p}{(2\pi)^3}$$

which transforms under the Poincaré group (Λ,a) as

$$U(\Lambda,a)\psi^{\kappa\sigma}(x)\ U^{-1}(\Lambda,a) = \psi^{\kappa\sigma}(\Lambda x + a)\ ; \qquad (27)$$

$a^*(p,s\lambda)$ and $a(p,s\lambda)$ are the creation and annihilation operators, respectively . κ,σ are the representation labels; j,m the field labels and s,λ the Poincaré group labels. We have considered only fields with irreducible representations. In the more general case of taking fields based on reducible representations there will appear projection operators between the representation functions $D^{\kappa\sigma}$ and the particle operators $a(p,s\lambda)$. In our case these projection operators are

$$\delta_{j's}\ \delta_{m'\lambda}$$

which change $D^{\kappa\sigma}_{jm,j'm'}$ into $D^{\kappa\sigma}_{jm,s\lambda}$. j,s take all values $j,s = |\kappa|,\ |\kappa|+1,\ldots \infty$. The spin flip matrix C is also an infinite-dimensional matrix consisting of a direct sum $\bigoplus_s C^s$ of finite dimensional spin-flip matrices C^s where

$$C^s_{\lambda\lambda'} = (\exp\{-i\pi J_2\})_{\lambda\lambda'} = (-1)^{s-\lambda}\delta_{\lambda,-\lambda'}$$

In an S-matrix diagram the external lines will be represented by one of the four infinite dimensional wave-functions:

$D^{\kappa\sigma}_{jm,s\lambda}(L_p)\ .\ e^{-ipx}$	if a particle is destroyed
$D^{*\kappa\sigma}_{jm,s\lambda}(L_p)\ .\ e^{+ipx}$	if a particle is created
$[D^{\kappa\sigma}(L_p)\ C^{-1}]_{jm,s\lambda}\ e^{+ipx}$	if an antiparticle is created
$[D^{\kappa\sigma}(L_p)\ C^{-1}]^{*}_{jm,s\lambda}\ e^{-ipx}$	if an antiparticle is destroyed.

(28)

In diagrams one has to sum j,m over all values $j = |\kappa|, |\kappa|+1, \ldots \infty;\ -j \leq m \leq +j$. The representation (κ,σ) and the spin s and λ take definite fixed values, according to the physical process one is dealing with. Clearly the external line wave-functions for finite dimensional fields are a special case of (28) [10].

Propagators, 3-point functions and pole diagrams

The aim here is to write down a pole diagram with infinite dimensional representations. For general infinite dimensional representations this would require a treatment in terms of distributions with built in regularization of divergent integrals. Therefore we use only a very special case where all our fields belong to irreducible infinite dimensional unitary representations of the principal series. In this case one does not need a more general formalism and the propagator in momentum space, after quantizing with a commutator, is simply given by

$$\Delta^{\kappa\sigma}_{jm,j'm'}(q) = i \sum_{s,\lambda} \frac{D^{\kappa\sigma}_{jm,s\lambda}(L_q)\; D^{*\kappa\sigma}_{j'm',s\lambda}(L_q)}{q^2 - \mu^2} =$$

$$= i \, \frac{\delta_{jj'}\, \delta_{mm'}}{q^2-\mu^2} \quad . \tag{29}$$

In order to couple three infinite dimensional wave functions to form an invariant, one needs to know the coupling coefficients (generalized Clebsch-Gordan coefficients) for these unitary representations of SL(2,C). The coupling coefficients which couple three fields to form an invariant 3-point function will be denoted formally by

$$\begin{pmatrix} \kappa_1\sigma_1 & \kappa_2\sigma_2 & \kappa_3\sigma_3 \\ j_1 m_1 & j_2 m_2 & j_3 m_3 \end{pmatrix} \tag{30}$$

and taking the Poincaré invariant

$$\mathcal{H}(x) = G \sum_{\substack{j_i, m_i \\ i=1,2,3}} \begin{pmatrix} \kappa_1\sigma_1 & \kappa_2\sigma_2 & \kappa_3\sigma_3 \\ j_1 m_1 & j_2 m_2 & j_3 m_3 \end{pmatrix} \psi^{\kappa_1\sigma_1}_{j_1 m_1}(x)\; \psi^{\kappa_2\sigma_2}_{j_2 m_2}(x) \times \psi^{\kappa_3\sigma_3}_{j_3 m_3}(x) \tag{31}$$

as our interaction Hamiltonian density, the Feynman rule for representing a vertex gives

$$- i\, G \begin{pmatrix} \kappa_1\sigma_1 & \kappa_2\sigma_2 & \kappa_3\sigma_3 \\ j_1 m_1 & j_2 m_2 & j_3 m_3 \end{pmatrix} \tag{32}$$

where G is a coupling constant.

To determine the coupling coefficients (30) for the principal series of SL(2,C) one can generalize Wigner's derivation of the Clebsch-Gordan coefficients for SU(2). The SU(2) 3-j symbols are given by the well known formula [8]

$$\int_{SU(2)} dg\, D^{*j_3}_{m_3 m'_3}(g)\, D^{j_2}_{m_2 m'_2}(g)\, D^{j_1}_{m_1 m'_1}(g) =$$

$$= \begin{pmatrix} j_1 & j_2 & j_3 \\ m_1 & m_2 & m_3 \end{pmatrix} \begin{pmatrix} j_1 & j_2 & j_3 \\ m'_1 & m'_2 & m'_3 \end{pmatrix} \qquad g \in SU(2) \qquad (33)$$

which is obtained by expanding $D^{j_1} \otimes D^{j_2}$ in a Clebsch-Gordan series and using orthogonality relations for D^{j_3}. The generalization of the Clebsch-Gordan sum in SU(2)

$$D^{j_1} \otimes D^{j_2} = \sum_j D^j \quad \text{with} \quad |j_1 - j_2| \leq j \leq j_1 + j_2$$

to a discrete and continuous sum in SL(2,C)

$$D^{\kappa_1\sigma_1} \otimes D^{\kappa_2\sigma_2} = \sum_\kappa \int_0^{i\infty} d\sigma\, D^{\kappa\sigma}$$

with $\kappa_1 + \kappa_2 + \kappa$ = integer (see Naimark [11] for a proof of this expansion) gives straight away [12]

$$\int_{SL(2,C)} dg\, D^{*\kappa_3\sigma_3}_{j_3 m_3 s_3 \lambda_3}(g)\, D^{\kappa_2\sigma_2}_{j_2 m_2 s_2 \lambda_2}(g)\, D^{\kappa_1\sigma_1}_{j_1 m_1 s_1 \lambda_1}(g) =$$

$$= \begin{pmatrix} \kappa_1\sigma_1 & \kappa_2\sigma_2 & \kappa_3\sigma_3 \\ j_1 m_1 & j_2 m_2 & j_3 m_3 \end{pmatrix} \begin{pmatrix} \kappa_1\sigma_1 & \kappa_2\sigma_2 & \kappa_3\sigma_3 \\ s_1\lambda_1 & s_2\lambda_2 & s_3\lambda_3 \end{pmatrix} \qquad g \in SL(2,C) \qquad (34)$$

where the direct integral in σ and the sum in κ is knocked out again by the orthogonality property of the D functions. We will need later the Clebsch-Gordan theorem for SL(2,C) in the following different form

$$\sum_{j_1 m_1} D^{\kappa_1\sigma_1}_{j_1 m_1 s_1 \lambda_1} D^{\kappa_2\sigma_2}_{j_2 m_2 s_2 \lambda_2} \begin{pmatrix} \kappa_1\sigma_1 & \kappa_2\sigma_2 & \kappa\sigma \\ j_1 m_1 & j_2 m_2 & jm \end{pmatrix} =$$

$$= \sum_{s\lambda} D^{\kappa\sigma}_{j\,m\,s\,\lambda} \begin{pmatrix} \kappa_1\sigma_1 & \kappa_2\sigma_2 & \kappa\sigma \\ s_1\lambda_1 & s_2\lambda_2 & s\lambda \end{pmatrix} \qquad (35)$$

A second method to determine the form of the coupling coefficients (30) is given by analytic continuation from the coupling coefficients of SU(2) $\otimes$ SU(2). The coupling of two SL(2,C) representations corresponds then to a coupling of four SU(2) representations which is a 9-j symbol. The coupling coefficients of SL(2,C) are therefore analytic continuations of 9-j symbols and the continuation is given by Casimir variable relations (11) between the group SL(2,C) and SU(2) $\otimes$ SU(2), that is $j_1 = \frac{1}{2}(\sigma+\kappa-1)$, $j_2 = \frac{1}{2}(\sigma-\kappa-1)$ and with this one has

$$\begin{pmatrix} \kappa_1\sigma_1 & \kappa_2\sigma_2 & \kappa\sigma \\ j_1 m_1 & j_2 m_2 & jm \end{pmatrix} = \qquad (36)$$

$$= N \begin{pmatrix} \frac{1}{2}(\sigma_1-\kappa_1-1) & \frac{1}{2}(\sigma_2-\kappa_2-1) & \frac{1}{2}(\sigma-\kappa-1) \\ \frac{1}{2}(\sigma_1+\kappa_1-1) & \frac{1}{2}(\sigma_2+\kappa_2-1) & \frac{1}{2}(\sigma+\kappa-1) \\ j_1 & j_2 & j \end{pmatrix} \begin{pmatrix} j_1 & j_2 & j \\ m_1 & m_2 & m \end{pmatrix}$$

where N is a normalization factor. In eq. (36) the first factor denotes the complex 9-j symbol, the second factor is a usual 3-j symbol of SU(2), which occurs because one has to transform the $|j_1 m_1, j_2 m_2>$ basis into a $|jm>$ basis.

This concludes the derivation of the vertex functions. Let us then finally apply the obtained Feynman rules to write down the scattering amplitude T corresponding to the following pole diagram,

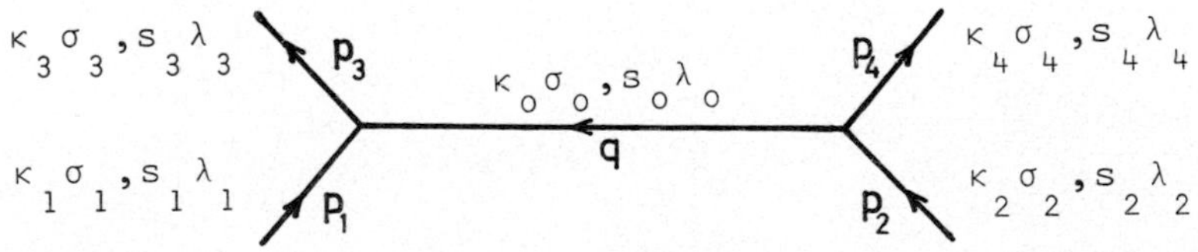

$q=p_3-p_1=p_2-p_4$; κ_i,σ_i are infinite dimensional (unitary) representations, and $s_i,\lambda_i (i=1,2,3,4)$ the (fixed) external physical spins of ingoing and outgoing particles.

An infinite set of Poincaré states $s_o\lambda_o$ is exchanged. Because (for simplicity) we take only a unitary representation with degenerate mass states as the exchanged tower this infinite sum over $s_o\lambda_o$ gives just δ-functions in the numerator of the propagator (see eq. (29)), and one obtains

$$T(q^2) = g(q^2)g'(q^2) \sum_{\substack{j_i,m_i \\ i=o,1,2,3,4}} \overset{*}{D}{}^{\kappa_4\sigma_4}_{j_4m_4s_4\lambda_4}(L_{p_4}) \times$$

$$\times \overset{*}{D}{}^{\kappa_3\sigma_3}_{j_3m_3s_3\lambda_3}(L_{p_3}) \begin{pmatrix} \kappa_4 - \sigma_4 & \kappa_2\sigma_2 & \kappa_o(-\sigma_o) \\ j_4 \; m_4 & j_2m_2 & j_om_o \end{pmatrix} \times$$

$$\times \frac{(+i)}{q^2-\mu^2} D^{\kappa_2\sigma_2}_{j_2m_2s_2\lambda_2}(L_{p_2}) D^{\kappa_1\sigma_1}_{j_1m_1s_1\lambda_1}(L_{p_1}) \begin{pmatrix} \kappa_3(-\sigma_3) & \kappa_1\sigma_1 & \kappa_o\sigma_o \\ j_3m_3 & j_1m_1 & j_om_o \end{pmatrix} \tag{37}$$

with $L_{p_3} - L_{p_1} = L_{p_2} - L_{p_4}$.

In eq. (37) one has to sum over all auxiliary labels

$$j_i = |\kappa_i|, |\kappa_i| + 1, \ldots \infty \; ; \; -j_i \leq m_i \leq j_i \quad .$$

Evaluation of (37)for the zero momentum transfer limit $L_{p_1} = L_{p_3} = L_p$, $L_{p_2} = L_{p_4} = L_{p'}$ (having equal masses for particles 1 and 3, and for particles 2 and 4) gives with the aid of (35) the following expression

$$T(q=0) = \frac{(-i)}{\mu^2} gg' \sum_{\substack{j_om_o \\ j,j'}} D^{\kappa_o\sigma_o}_{j_om_ojm}(L_p) \overset{*}{D}{}^{\kappa_o\sigma_o}_{j_om_oj'm'}(L_{p'}) \times$$

$$\times \begin{pmatrix} \kappa_4(-\sigma_4) & \kappa_2\sigma_2 & \kappa_0(-\sigma_0) \\ s_4\lambda_4 & s_2\lambda_2 & j'm' \end{pmatrix} \begin{pmatrix} \kappa_3(-\sigma_3) & \kappa_1\sigma_1 & \kappa_0\sigma_0 \\ s_3 \lambda_3 & s_1\lambda_1 & jm \end{pmatrix} \qquad (38)$$

Or putting, in addition, the second particle at rest $\vec{p}_2 = \vec{p}_4 = 0$ and taking L_p in the third direction, one obtains

$$T(q=0) = \frac{(-i)}{\mu^2} g\, g' \sum_{j,j'} d^{\kappa_0\sigma_0}_{j'mj}(\xi) \times$$

$$\times \begin{pmatrix} \kappa_3(-\sigma_3) & \kappa_1\sigma_1 & \kappa_0\sigma_0 \\ s_3\lambda_3 & s_1\lambda_I & jm \end{pmatrix} \begin{pmatrix} \kappa_4(-\sigma_4) & \kappa_2\sigma_2 & \kappa_0(-\sigma_0) \\ s_4\lambda_4 & s_2\lambda_2 & j'm \end{pmatrix} \qquad (39)$$

because

$$D^{\kappa_0\sigma_0}_{j_om_oj'm'}(0) = \delta_{j_oj'}\,\delta_{m_om'}$$

Let us discuss finally a model π-N scattering. Assigning the π-meson to $(0,\sigma_1)$ and the nucleon to $(\frac{1}{2},\sigma_2)$, exchanging a $(0,\sigma)$ representation and putting the nucleon at rest, only one term remains in (39) because of the triangular conditions among the s_i in the coupling coefficients.

$$T(q=0) = g_N g_\pi \frac{(-i)}{\mu^2} d^{o\sigma}_{ooo}(\xi) \times$$

$$\times \begin{pmatrix} o(-\sigma_1) & 0\sigma_1 & o\sigma \\ oo & oo & oo \end{pmatrix} \begin{pmatrix} \frac{1}{2}(-\sigma_2) & \frac{1}{2}\sigma_2 & 0(-\sigma) \\ \frac{1}{2}(-\frac{1}{2}) & \frac{1}{2}\,\frac{1}{2} & o\;o \end{pmatrix} = \text{const}\; d^{0\sigma}_{000}(\xi) =$$

$$= \text{const}\;\frac{1}{\sigma}\,\frac{\sinh\sigma\xi}{\sinh\xi} \qquad (40)$$

Taking the asymptotic limit $\xi \to \infty$ one gets

$$T(q=o) \xrightarrow{\xi\to\infty} \text{const } e^{-\xi} \sinh\sigma\xi = \text{const } s^{-1} \sinh\sigma\xi$$

$$\text{where } s = (p_\pi + p_N)^2 \sim e^{\xi} \tag{41}$$

For unitary representations σ is purely imaginary, hence

$$|T| \xrightarrow{\xi\to\infty} \text{const. B.} s^{-1} \tag{42}$$

where B is a function oscillating between 0 and 1. If one continues the exchanged representation away from a unitary representation to an arbitrary linear representation , then the asymptotic behaviour will be

$$|T| \overset{\xi\to\infty}{\sim} \text{const. B.} s^{\mathrm{Re}\,\sigma - 1} \tag{43}$$

Note that eq. (43) gives the familiar high energy behaviour of pole diagrams with finite dimensional representations, that is Im $\sigma = 0$, $j = \sigma - 1$, $|T| \sim s^j$, where j is the highest spin value in the exchanged finite dimensional representation. For non-forward direction $t \neq 0$ one can postulate the exchange of a Toller trajectory [2] $\sigma(t)$, i.e. we consider the exchanged representation σ to be a function of t.

Eq.(39) gives the forward amplitude for a pole diagram in which one representation (κ_o,σ_o) is exchanged. One could think of exchanging a continuous infinity of infinite dimensional representations. This would require one to integrate over σ_o and sum over κ_o with μ^2 , g, g', dependent on (κ_o,σ_o) .

To apply this to our π-N scattering model (with the notation used in equation (40)), let us choose, for simplicity, $\sigma_1 = \sigma_2 = 0$ for the external particles, and take g_π, g_N, μ^2 to be independent of σ . Integrating over unitary representations $(0,\sigma) \equiv (0, \pm i\frac{\rho}{2})$ with the Plancheral measure $\rho^2 d\rho$ (see ref.6.) one obtains

$$T(q=0) = g_N g_\pi \frac{(-i)}{\mu^2} \int_0^\infty \rho^2 d\rho \; d^{o\rho}_{ooo}(\xi) \times$$

$$\times \begin{pmatrix} 00 & 00 & 0\rho \\ 00 & 00 & 00 \end{pmatrix} \begin{pmatrix} \frac{1}{2} & 0 & \frac{1}{2} & 0 & 0\rho \\ \frac{1}{2} & -\frac{1}{2} & \frac{1}{2} & \frac{1}{2} & 00 \end{pmatrix} =$$

$$= g_N g_\pi \frac{(-i)}{\mu^2} \frac{32\pi^5}{\sinh\xi} \int_0^\infty \rho d\rho \frac{\sin\rho \frac{\xi}{2}}{\cosh^2 \rho\frac{\pi}{4}} =$$

$$= \frac{i(4\pi)^4}{\mu^2} g_N g_\pi \frac{1}{\sinh\xi} \frac{d}{d\xi} \left(\frac{\xi}{\sinh\xi}\right) \qquad (44)$$

The Clebsch-Gordan coefficient of SL(2,C) in eq. (44) are evaluated by using eq. (34), see for instance [12]. As pointed out above, in general one would have a weighting factor which depends on the exchanged representations. The generalized Regge pole theory corresponds then to the special case where the weighting function ist just a δ-function.

References

1. C. Fronsdal, Feynman Rules for Reggeons, Preprint Trieste IC/70(1967); G. Feldman, P. T. Matthews, Phys. Rev. 151, 1176 (1966), Phys. Rev. 154, 1241 (1967).
2. M. Toller, CERN Preprints 770, 780 (1967); G. Domokos and G. L. Tindle, "On the algebraic classification of Regge-Poles" (Preprint, Berkeley 1967).
3. M. Gell-Mann, D. Horn, J. Weyers, amplified version of the report delivered at the Heidelberg Conference on High Energy Physics and Elementary Particles 1967. (Preprint, Inst. for Advanced Studies, Prince-

ton, October 1967); H. Leutwyler, Models of Local Current Algebra (Schladming, Suppl.Acta Phys.Austr. 1968).

4. R. Delbourgo, K. Koller, P. Mahanta, Nuovo Cim. LIIA, 1254 (1967).
5. K. Koller, Feynman Diagrams for Unitary Representation of SL(2,C). Nuovo Cim. LIVA, 79 (1967).
6. M. A. Naimark, Linear Representation of the Lorentz Group (Pergamon Press).
7. R. Hermann, Lie Groups (Benjamin, 1966).
8. M. E. Rose, Theory of Angular Momentum (Wiley & Sons).
9. I. M. Gel'fand, M. I. Graev and N. Ya. Vilenkin, Generalized Functions Vol. V.
10. S. Weinberg, Phys. Rev. 133B, 1318 (1964).
11. M. A. Naimark, Am. Math. Soc. Translations Ser. 2, Vol. 36.
12. R. L. Anderson, R. Raczka, M. A. Rashid and P. Winternitz, Clebsch-Gordan Coefficients for the Lorentz Group, Trieste preprint IC/50 (1967).

EFFECTIVE LAGRANGIANS IN PARTICLE PHYSICS†

By

F. GÜRSEY

Middle East Technical University, Ankara
Turkey

I. Introduction

Chiral transformations connect fermion field components with opposite chirality, or, alternatively, states which differ by a soft pion. Hence, they are associated with symmetries of an unconventional type, since massive particle states are unstable with respect to chiral transformations. The introduction of chiral groups was mainly motivated by the symmetrical V-A form of the leptonic weak interactions [1],[2],[3] . This structure will persist in the presence of strong interactions only if hadronic systems exhibit some sort of chiral symmetry [1]. Preliminary examples of such Lagrangians for massive nucleons coupled to massless mesons were constructed [4],[5],[6],[7]. This led to a class of (π-N) Lagrangians exactly invariant under the chiral SU(2) ⊗ SU(2) group in the zero pion mass

† Lecture given at the VII. Internationale Universitätswochen f.Kernphysik,Schladming,February 26 -March 9,1968.

limit [8],[9],[10]. These models turned out to be unrenormalizable and nonlinear. The first renormalizable chiral invariant theory is the σ-model proposed by Gell-Mann and Lévy [11]. It involves an auxiliary scalar field σ that linearizes the previous models. In all these theories chiral invariance is broken by the pion mass term. In the same way that an isovector current is associated with the isospin group we have an axial isovector current associated with chiral transformations. This axial vector is, however, conserved only in the limit of vanishing pion mass. One way of breaking chiral symmetry is to introduce pion terms in such a way that the divergence of the axial current is proportional to the pion field. This is the exact PCAC hypothesis and it was also introduced by Gell-Mann and Lévy [11]. It is a special case of the principle of perfect chiral symmetry in the limit of zero pion mass [8],[12].

The trouble that plagued these chiral Lagrangians was the difficulty of using them for a reliable quantitative analysis of (π-N) processes. Thus, in spite of the initial successes of Nambu and his collaborators in this direction [13], the approach was abandoned.

Recently Weinberg [14] and others [15] calculated (π-N) and (π-π) S-wave scattering lengths from the SU(2) ⊗ SU(2) current algebra, by going to the soft pion limit, that is, by extrapolating T-matrix elements to zero pion mass. Weinberg then remarked [16] that the same expressions can be obtained from an effective Lagrangian which is meant to be used only to lowest order in the pseudovector π-N coupling constant. The effective Lagrangian is obtained from the σ-model in two steps.

1. The mass of the auxiliary σ-field is made to tend to infinity. The elimination of σ by this me-

thod results in a nonlinear partially chiral invariant Lagrangian that is identical with the old models.

2. By a unitary transformation all nonderivative π-N couplings are changed to a derivative coupling form. This is like the Foldy-Dyson transformation. It had previously been applied to chiral π-N systems by the author [9]. The new nucleon field has a complicated transformation property under the chiral group, but the π-N coupling is now explicitly P-wave coupling. The new Lagrangian can be more readily used as an effective Lagrangian and it leads exactly to the same predictions as current algebra for scattering lengths.

The last form, which is a chiral generalization of the pseudovector coupling of pions with nucleons, can also be directly derived by postulating directly the new transformation law for the transformed nucleon field [17]. This law involves the pion field $\vec{\pi}$ instead of the Dirac matrix γ_5. The two methods are completely equivalent in spite of remarks to the contrary in a recent paper by Weinberg [18]. This is what we propose to show in the following. Chiral systems were recently enlarged by Schwinger [17],[19] through the introduction of S = 1 mesons of either parity (ρ and A_1) into the scheme, following current algebraic results of Weinberg [20]. Wess and Zumino [21], among others, proposed similar schemes, basing their argument on local isotopic and chiral invariance. In fact we shall see that the direct extension of the Yang-Mills theory [22] to a theory invariant under the local chiral gauge group SU(2) $\otimes$ SU(2) involves an interacting system consisting of the particles (N, π, ρ, A_1). Furthermore, the partially invariant Yang-Mills theory allows finite bare masses for the vector and pseudovector mesons, shifting the stigma of masslessness to the pion. The chiral invariant Lagrangian thus provides a much better approximation to the real world, since pions, with their mass μ = 140 MeV, are the lightest hadrons and are separated from the next non-strange hadron (ρ) by a mass gap of more than 4μ. Finally, used as an effective Lagrangian, the chiral theory makes quantitative predictions in agreement with current algebra and gives insight into the origin of the ρ-dominance picture [23].

If SU(6) relations are superimposed to the chiral scheme, new results in striking agreement with experiment can be obtained, as shown by Schwinger [19], Wess and Zumino [21], and Freund [24]. However, in that case there is no field-theoretical rationale since SU(6) invariant Lagrangian models have not been discovered. It may be argued that this successful numerology gives us enough motivation to look for new invariance principles that encompass and generalize the chiral group, SU(6) and the Yang-Mills concept of local invariance.

Speculations in this direction will be indicated after we deal with SU(2) ⊗ SU(2) systems.

Part of the present work was done jointly with Mr. Philip Chang [25], [26] and I have benefited greatly from conversations and correspondence with Dr. V. Ogievetsky [27].

2. π-N Lagrangians with Exact Chiral Invariance

Let us start from the renormalizable π-N Lagrangian of Kemmer,

$$\mathcal{L} = -\bar{\psi}(\gamma_\mu \partial_\mu + m)\psi - \frac{1}{4}\,\mathrm{Tr}(\partial_\mu \Phi \partial_\mu \Phi) - \frac{\mu^2}{4}\,\mathrm{Tr}\Phi\Phi + \mathcal{L}_{int}$$

where

$$\Phi = \vec{\tau} \cdot \vec{\phi}$$

$\vec{\phi}$ being the pion field and $\mathcal{L}_{int}$ is

$$\mathcal{L}^{ps}_{int} = -i\, g\, \bar{\psi}\, \gamma_5\, \Phi\, \psi \;.$$

In the case of unrenormalizable derivative coupling we would have

$$\mathcal{L}^{pv}_{int} = -\frac{F}{\mu}\, \bar{\psi}\, i\, \gamma_5 \gamma_\mu (\partial_\mu\, \Phi)\psi$$

where μ is the pion mass and F is the π-N pseudovector coupling constant. According to the equivalence theorem, the same T matrix is obtained in first order of g for both types of coupling if we have

$$F = \frac{g\mu}{2m} \;.$$

The above Lagrangian is invariant under the isospin group

$$\psi \to \psi' = \exp\{\frac{i}{2}\, \vec{\tau} \cdot \vec{\omega}\}\psi$$

$$\Phi \to \Phi' = \exp\{\frac{i}{2}\, \vec{\tau}.\vec{\omega}\}\Phi \cdot \exp\{-\frac{i}{2}\, \vec{\tau}.\vec{\omega}\}$$

but not under chiral transformations

$$\psi \to \psi'' = \exp\{\frac{i}{2}\, \gamma_5\, \vec{\tau} \cdot \vec{\omega}\}\psi$$

which, on the other hand, leave invariant the free, massless part of $\mathcal{L}$, namely $\mathcal{L}_o = -\bar{\psi}\, \gamma_\mu\, \partial_\mu\, \psi$.

The two transformations form a group isomorphic to SU(2) $\otimes$ SU(2) with generators$((1\pm\gamma_5)/2)\cdot(\vec{\tau}/2)$. Since invariance is obstructed by the nucleon mass term we make the replacement

$$\bar{\psi}\,\psi \rightarrow \bar{\psi}\,U\,\psi$$

where U is a function of the dimensionless operator $i\,f\,\gamma_5\,\Phi$ where $[\Phi]$ is a length. We choose U so that it satisfies the conditions

a) $U = U(i\,f\,\gamma_5\,\Phi)$, with $U(o) = 1$.

b) U is unitary , $U\,U^+ = U^+\,U = 1$

c) $U^+ = U(-i\,f\,\gamma_5\,\Phi)$, so that f is real and $\mathcal{L}$ is hermitian

The general form of U is

$$U = \sigma(f^2\Phi^2) + 2\,i\,f\,\gamma_5\,\rho(f^2\Phi^2)\,\Phi$$

where

$$\sigma(o) = 1 \quad \text{and} \quad \sigma^2 + 4\,f^2.\Phi^2.\rho^2 = 1.$$

Let

$$\alpha = f^2\,\Phi^2 = f^2\,\vec{\phi}\cdot\vec{\phi}\ .$$

Then, if we choose a real function $\sigma(\alpha)$, with $\sigma(o) = 1$, the function ρ is completely determined and given by

$$\rho(\alpha) = \frac{1}{2}\sqrt{[1-\sigma^2(\alpha)]/\alpha}\ .$$

We shall choose three examples

1. $\sigma(\alpha) = \cos 2\sqrt{\alpha}$

then

$$\rho(\alpha) = \frac{\sin 2\sqrt{\alpha}}{2\sqrt{\alpha}}$$

and

$$U = U_1 = \exp\{2i\, f\gamma_5 \vec{\tau}\cdot\vec{\phi}\} = \exp\{2i\, f\gamma_5 \Phi\} .$$

This is the exponential model [28] .

2. $$\sigma(\alpha) = \frac{1-\alpha}{1+\alpha} ,$$

then

$$\rho = \frac{1}{1+\alpha}$$

and

$$U = U_2 = \frac{1+if\gamma_5\vec{\tau}\cdot\vec{\phi}}{1-if\gamma_5\vec{\tau}\cdot\vec{\phi}} = \frac{1+if\gamma_5\Phi}{1-if\gamma_5\Phi} .$$

This is the rational model used by Kramer et al.[7], Schwinger [17], Wess and Zumino [21] .

3. $$\sigma(\alpha) = \sqrt{1 - 4\alpha}$$

then

$$\rho(\alpha) = 1$$

and

$$U = U_3 = \sqrt{1-4f^2\phi^2} + 2\, if\gamma_5\vec{\tau}\cdot\vec{\phi} .$$

This is the square root model used by Gell-Mann and Lévy [11], Weinberg [16] and L. Brown [29].

We can also add a pion kinetic term that gives the usual expression in the limit $f \to 0$. We now consider the Lagrangian

$$\mathcal{L}' = -\bar{\psi}(\gamma_\mu \partial_\mu + mU)\psi - \frac{1}{16f^2} \mathrm{Tr}(\partial_\mu U \partial_\mu U^+)$$

that reduces to $\mathcal{L}$ with non-derivative coupling for $f \to 0$ and is invariant under the chiral transformation

$$\psi \to \psi' = \exp\{\frac{i}{2}\gamma_5 \vec{\tau}\cdot\vec{a}\}$$

$$\bar{\psi} \to \bar{\psi}' = \bar{\psi} \exp\{\frac{i}{2}\gamma_5 \vec{\tau}\cdot\vec{a}\}\psi$$

$$U \to U' = \exp\{-\frac{i}{2}\gamma_5 \vec{\tau}.\vec{a}\}\, U \exp\{-\frac{i}{2}\gamma_5 \vec{\tau}.\vec{a}\}.$$

Since $\mathcal{L}$ is also invariant under the isospin transformations

$$\psi \to \exp\{\frac{i}{2}\vec{\tau}\cdot\vec{\omega}\}\psi \quad ,$$

$$U \to \exp\{\frac{i}{2}\vec{\tau}\cdot\vec{\omega}\}\, U \exp\{-\frac{i}{2}\vec{\tau}\cdot\vec{\omega}\}$$

we obtain invariance under the chiral SU(2) $\otimes$ SU(2).

The above Lagrangian [8],[9], [10],[25] has no pion mass term and generalizes the pseudoscalar interaction. We can also find a chiral invariant Lagrangian which has an additional pseudovector interaction [27]. We write

$$\mathcal{L}'' = -\bar{\psi}\gamma_\mu\partial_\mu\psi + \frac{1}{2} g'\bar{\psi}U\gamma_\mu(\partial_\mu U)\psi - m\bar{\psi}U\psi - \\ - \frac{1}{16f^2}\mathrm{Tr}(\partial_\mu U \partial_\mu U^+)$$

where g' is dimensionless. To first order in f we obtain

$$\mathcal{L}'' \cong -\bar{\psi}\gamma_\mu\partial_\mu\psi - m\bar{\psi}\psi - \frac{1}{2}\partial_\mu\vec{\phi}\cdot\partial_\mu\vec{\phi} - 2mf\bar{\psi}i\gamma_5\vec{\tau}\cdot\vec{\phi}\psi - \\ - g'f\bar{\psi}i\gamma_5\gamma_\mu\vec{\tau}\psi\partial_\mu\vec{\phi} \quad .$$

Hence $g'f\mu$ is the pseudo-vector coupling constant, while $g = 2mf$ is the pseudoscalar coupling constant.

3. Axial Vector Current and Partial Chiral Invariance

To obtain the axial current, following Gell-Mann and Lévy we make an infinitesimal coordinate dependent chiral transformation, namely,

$$\psi \to \psi + \delta\psi \qquad \text{with} \quad \delta\psi = \frac{i}{2}\gamma_5\, \vec{\tau}\cdot\vec{a}\,\psi$$

$$U \to U + \delta U \qquad \text{with} \quad \delta U = -\frac{i}{2}[\gamma_5\, \vec{\tau}\cdot\vec{a}, U]_+$$

on neglecting a^2. The Lagrangian undergoes the change

$$\mathcal{L} \to \mathcal{L} + \delta\mathcal{L}$$

with

$$\delta\mathcal{L} = (\partial_\nu \vec{J}_{5\nu}) \cdot \vec{a} + \vec{J}_{5\nu}(\partial_\nu \vec{a})$$

so that

$$J^i_{5\nu} = \frac{\partial\mathcal{L}}{\partial(\partial_\nu a^i)} .$$

For a chiral invariant Lagrangian the axial vector is conserved :

$$\partial_\nu \vec{J}_{5\nu} = 0 .$$

If we want exact PCAC we must have a symmetry breaking term such that

$$\partial_\nu \vec{J}_{5\nu} = C\vec{\phi} = c\mu^2\vec{\phi}$$

or

$$\frac{\partial \mathcal{L}}{\partial a^i} = c\,\mu^2\,\vec{\phi} \quad .$$

To obtain the vector current we make the parameters $\vec{\omega}$ coordinate dependent and replace $\vec{a}$ by $\vec{\omega}$ and $\vec{J}_{5\nu}$ by $\vec{J}_\nu$ in the above. For simplicity consider the non-derivative case. For the vector current we find

$$\vec{J}_\mu = -\bar{\psi}\, i\gamma_\mu \frac{\vec{\tau}}{2}\psi + \rho^2(\alpha)\vec{\phi} \times \partial_\mu \vec{\phi} \quad .$$

The axial vector current is

$$\vec{J}_{5\mu} = i\,\bar{\psi}\,\gamma_5\gamma_\mu \frac{\vec{\tau}}{2}\psi + \frac{\rho\sigma}{2f}\partial_\mu\vec{\phi} + \frac{f}{2}\vec{\phi}(\partial_\mu\phi^2)(\sigma\rho' - \rho\sigma') \ .$$

Let us take the symmetry breaking term to be a function of α. Then, to ensure exact PCAC we must have

$$\mathcal{L}^{S.B.}(\alpha) = \frac{\mu^2}{2f^2}\int_0^{f^2\phi^2} \frac{\sqrt{\alpha}\,\sigma'(\alpha)}{\sqrt{1-\sigma^2(\alpha)}}\,d\alpha$$

which gives

$$\delta\mathcal{L}^{S.B.} = \frac{\mu^2}{2f}\ \vec{\phi}\cdot\vec{a}$$

so that

$$\delta(\mathcal{L}^{S.} + \mathcal{L}^{S.B.}) = \frac{\mu^2}{2f}\vec{\phi}\cdot\vec{a} + \vec{J}_{5\nu}(\partial_\nu\vec{a}) \quad .$$

Thus $c = \frac{1}{2f}$, so that $\partial_\lambda\vec{J}_{5\lambda} = \frac{\mu^2}{2f}\vec{\phi}$.

For the various special cases we have

1. Exponential model: $\sigma = \cos 2\sqrt{\alpha}$

$$\mathcal{L}^{S.B.} = -\frac{\mu^2}{2f^2}\alpha = -\frac{1}{2}\mu^2\vec{\phi}\cdot\vec{\phi} \quad .$$

This is the ordinary mass term.

2. Rational model : $\sigma = (1-\alpha)/(1+\alpha)$

$$\mathcal{L}^{S.B.} = - \frac{\mu^2}{2f^2} \log (1+\alpha) = - \frac{\mu^2}{2f^2} \log (1+ f^- \vec{\phi}\cdot\vec{\phi})$$

3. Square root model: $\sigma = \sqrt{1-4\alpha}$

$$\mathcal{L}^{S.B.} = \frac{\mu^2}{4f^2} (\sqrt{1-4\alpha} -1) = - \frac{\mu^2}{4f^2}(1-\sqrt{1-4f^2\vec{\phi}\cdot\vec{\phi}})$$

In case both non-derivative and derivative couplings are present we get an additional term to $\delta\mathcal{L}$ from the derivative coupling. To lowest order in f and for any form of $\sigma(\alpha)$ we have

$$j^i_{5\nu} = \frac{\partial \mathcal{L}}{\partial(\partial_\nu a^i)} = i\bar{\psi}\gamma_5\gamma_\nu\frac{\tau^i}{2}\psi + ig'\bar{\psi}\gamma_5\gamma_\nu\frac{\tau^i}{2}\psi + \frac{1}{2f}\partial_\nu\vec{\phi} =$$

$$= (1+g')i\bar{\psi}\gamma_5\gamma_\nu\frac{\tau^i}{2}\psi + \frac{1}{2f}\partial_\nu\vec{\phi} + O(f) ,$$

since

$$\vec{\phi} \to \vec{\phi} - \frac{1}{2f}\vec{a} + O(f)$$

under the transformation, to lowest order in f.

On the other hand, the vector current is, to lowest order in f,

$$\vec{j}_\nu = -i\bar{\psi}\gamma_\nu \frac{\vec{\tau}}{2}\psi + \vec{\phi} \times \partial_\nu\vec{\phi} + O(f^2)$$

If now the lepton current ℓ_ν is coupled to the vector and axial vector currents in the (V-A) form, we have

$$\mathcal{L}_W = \frac{G}{\sqrt{2}} \ell_\nu (j^1_\nu + i\, j^2_\nu - j^1_{5\nu} - i j^2_{5\nu}) + \text{H.C.} =$$

$$= \frac{1}{\sqrt{2}} G\, \ell_\nu (\bar\psi i \gamma_\nu \tau^+ \psi - g_A \bar\psi i \gamma_5 \gamma_\nu \tau^+ \psi + \sqrt{2}\, \phi^o \overleftrightarrow{\partial}_\nu \phi^{(+)} -$$

$$- \frac{1}{\sqrt{2} f} \partial_\nu \phi^{(+)})$$

where

$$g_A = 1 + g' = - G_A/G_V \quad .$$

We see that this chiral invariant theory does not imply $g_A = 1$ contrary to a statement by Weinberg in a recent review paper [18]. Only in the absence of derivative couplings do we have $g_A = 1$.

Let us now check the commutation relations in the case with derivative coupling.

If f^2 is neglected then the axial vector current is

$$\vec{J}_{5\mu} = \frac{\partial \mathcal{L}}{\partial(\partial_\mu \vec{a})} = (1+g') i \bar\psi \gamma_5 \gamma_\mu \frac{\vec\tau}{2} \psi + \frac{1}{2f} \partial_\mu \vec\phi + O(f)$$

$$\vec{J}_\mu = \frac{\partial \mathcal{L}}{\partial(\partial_\mu \vec\omega)} = -i \bar\psi \gamma_\mu \frac{\vec\tau}{2} \psi + \vec\phi \times \partial_\mu \vec\phi + O'(f^2) \ .$$

We have to show that the current algebra is satisfied, so that

$$[J^i_{5o}(t,\vec{x}), J^j_{5o}(t,\vec{y})] = i \varepsilon^{ijk} \delta(\vec{x}-\vec{y}) J^k_o(t,\vec{x}) \quad .$$

Let us find the canonical variables. We have

$$\frac{\partial \mathcal{L}}{\partial(\partial_o \psi)} = \psi^+ \quad .$$

Hence ψ and ψ^+ are canonical conjugates.

$$\frac{\partial \mathcal{L}}{\partial(\partial_o \vec\phi)} = \partial_o \vec\phi + g' f\, \psi^+ \gamma_5 \vec\tau\, \psi = \vec\Pi_o \quad .$$

Hence $\partial_0\vec{\phi}$ does not commute with ψ, but

$$[\vec{\phi}\,,\psi] = 0 \quad , \qquad [\vec{\phi},\ \psi^+] = 0$$

$$[\vec{\Pi}_o,\psi] = 0 \quad , \qquad [\vec{\Pi}_o,\psi^+] = 0 \quad .$$

Now, we have

$$\vec{J}_{50} = (1+g')\psi^+\gamma_5\,\frac{\vec{\tau}}{2}\psi + \frac{1}{2f}\,\partial_0\vec{\phi} + O(f) =$$

$$= \psi^+\gamma_5\frac{\vec{\tau}}{2}\psi + \frac{1}{2f}\,\vec{\Pi}_0 + O(f)$$

where $O(f)$ only includes $\vec{\phi}$ and its canonical conjugate $\vec{\Pi}_0$.

Since $\vec{\Pi}_0$ commutes with $\psi^+\gamma_5\,\frac{\vec{\tau}}{2}\,\psi$, we have

$$[J^i_{50}(t,\vec{x}),\ J^j_{50}(t,\vec{y})] =$$

$$= [\psi^+(x)\gamma_5\,\frac{\tau^i}{2}\,\psi(x),\ \psi^+(y)\gamma_5\frac{\tau^j}{2}\,\psi(y)] + \text{pion terms} \quad .$$

It can be shown that the pion terms give $\vec{\phi}\times\partial_0\,\vec{\phi}$. On the other hand, because ψ and ψ^+ are canonical conjugate variables we get

$$[J^i_{50}(t,\vec{x}),J^j_{50}(t,\vec{y})] =$$

$$= i\varepsilon^{ijk}\delta(\vec{x}-\vec{y})[\psi^+(t,\vec{x})\frac{\tau^k}{2}\,\psi(t,\vec{x}) + (\vec{\phi}\times\partial_0\vec{\phi})_k] =$$

$$= i\varepsilon^{ijk}\ \delta(\vec{x}-\vec{y})\ J^k_0(t,\vec{x}) \quad .$$

4. Removal of the Non-Derivative Coupling by a Unitary Transformation

It was pointed out many years ago [9] that non-derivative terms in the chiral Lagrangian can be removed by a unitary transformation. We do that in order to have a manifestly P-wave π-N interaction for direct comparison with experiment.

Let W a unitary operator in Hilbert space such that

$$N = W \psi W^{+} = U^{1/2} \psi$$

$$\vec{\pi} = W \vec{\phi} W^{+} = \vec{\phi} .$$

As an example, for the exponential model and non-derivative coupling we have

$$W = \exp\{if\int \psi^{+}(x)\gamma_5 \vec{\tau}\psi(x)\vec{\phi}(x)\, d^3x\} .$$

In general the form of W depends on the function $\sigma(\alpha)$. Since $U = \sigma + if\gamma_5 \Phi\rho$, we find

$$U^{1/2} = \sqrt{(\sigma+1)/2} + if\gamma_5 \Phi \frac{\sqrt{2}}{\sqrt{\sigma+1}} \rho = \frac{1+U}{\sqrt{2(1+\sigma)}} .$$

Another solution is obtained by replacing $\sigma + 1$ by $\sigma - 1$. $U^{1/2}$ is also unitary as can be verified from the last form. This gives for the exp., Schwinger and square root models

$$U_1^{1/2} = \exp\{if \gamma_5 \Phi\}$$

$$U_2^{1/2} = \frac{1+if\gamma_5\Phi}{\sqrt{1+f^2\phi^2}}$$

$$U_3^{1/2} = \frac{(\sqrt{1-4f^2\phi^2}+1)^{1/2}}{\sqrt{2}} + i\,\frac{\sqrt{2}\ f\gamma_5\Phi}{\sqrt{(1-4f^2\phi^2+1)^{1/2}}} \,.$$

We also have

$$\psi = U^{-1/2}\,N\,, \qquad \bar{\psi} = \bar{N}\,U^{-1/2}\,,$$

so that

$$\bar{\psi}\,U\,\psi = W\,\bar{\psi}\,W^+\,W\,\psi\,W^+ = \bar{N}N\,.$$

The Lagrangian $\mathcal{L}$ can now be reexpressed in terms of the new fields N and π . It takes the form

$$\mathcal{L} = -\,\bar{\psi}\gamma_\nu\partial_\nu\psi + \frac{g'}{2}\,\bar{\psi}\,U\gamma_\nu(\partial_\nu U)\psi + m\bar{\psi}U\psi + \mathcal{L}'(\phi) + + \mathcal{L}^{(\mu)}(\phi) =$$

$$= -\,\bar{N}\gamma_\nu\partial_\nu N - \bar{N}\gamma_\nu(U^{1/2}\partial_\nu U^{-1/2})N - - \frac{g'}{2}\,\bar{N}\gamma_\nu U^{1/2}(\partial_\nu U^{-1})U^{1/2}N - - m\,\bar{N}N + \mathcal{L}'(\phi) + \mathcal{L}^{(\mu)}(\phi) =$$

$$\cong -\,\bar{N}\gamma_\nu\partial_\nu N + f^2(\pi\times\partial_\nu\vec{\pi})\bar{N}\,i\gamma_\nu\vec{\tau}N - - (1+g')f\bar{N}i\gamma_5\gamma_\nu\vec{\tau}N\partial_\nu\vec{\pi} - m\,\bar{N}N + + \mathcal{L}'(\pi) + O(f^3)$$

We note that, although $\vec{\phi}$ has not changed, its canonical conjugate is now different. The pseudovector coupling constant F is then given by

$$\frac{F}{\mu} = (1 + g')\,f = g_A f$$

or

$$F = \mu f g_A = g\frac{\mu}{2m} .$$

This gives

$$f = \frac{g}{2m\, g_A} = \frac{F}{\mu g_A} .$$

We have

$$\frac{F^2}{4\pi} = 0.08 \quad , \qquad F = \sqrt{4\pi \times 0.08} \sim 1$$

so that

$$f \simeq (\mu g_A)^{-1} = 0.81\, \mu^{-1} \text{ , if } g_A = 1.23 .$$

The PCAC relation reads

$$\partial_\lambda \vec{J}_{5\lambda} = \frac{\mu^2}{2f}\vec{\phi} = \frac{m g_A}{g}\mu^2\vec{\phi} \quad \text{or } C_\pi = \frac{1}{2f} = \frac{m}{g} g_A = \frac{\mu}{2F} g_A .$$

The pion decay constant can be found from the effective leptonic interaction

$$\frac{G}{\sqrt{2}}\frac{1}{2f}(\partial_\mu \pi^1 + \partial_\mu \pi^2)\ \ell_\mu$$

which is the pion part of

$$\frac{G}{\sqrt{2}}(J^1_{5\mu} + J^2_{5\mu})\ell_\mu = \frac{G}{\sqrt{2}} F_\pi \partial_\mu \pi^+ \ell_\mu = \frac{G}{\sqrt{2}}(\frac{1}{f\sqrt{2}}\partial_\mu \pi^+ \ell_\mu) .$$

Hence

$$F_\pi = C_\pi\sqrt{2} = \frac{1}{f\sqrt{2}} = \frac{2m g_A}{g\sqrt{2}} \sim \frac{\mu g_A}{\sqrt{2}} = 0.87\mu .$$

We have

$$(F_\pi)_{\text{exp.}} = 0.95\mu .$$

In general, including also off mass shell corrections when the pion mass is changed from zero to its physical value we have

$$F_{\pi} = \frac{1}{f\sqrt{2}} \frac{1}{K(o)} = \frac{\sqrt{2}}{K(o)} \frac{m}{g} g_A = \frac{\sqrt{2}}{K(o)} C_{\pi}$$

which is the Goldberger-Treiman relation. We see that K(o) differs from 1 by about 10%.

Now study the transformation law for N under chiral transformations. Let

$$V(\vec{a}) = \exp\{i\gamma_5 \vec{\tau}\cdot\vec{a}\} .$$

Under the chiral transformations we have

$$\psi' = V\psi \quad , U' = V^{-1} U V^{-1} .$$

We also have

$$N = U^{1/2} \psi \quad .$$

The chiral transformation law for N reads

$$N' = \Lambda(\vec{a},\vec{\pi})N = U'^{1/2} \psi' = (V^{-1} U V^{-1})^{1/2} V\psi$$

or

$$N' = (V^{-1}(\vec{a}) U(\vec{\pi}) V^{-1}(\vec{a}))^{1/2} V(\vec{a}) U^{-1/2}(\vec{\pi}) N$$

so that

$$\Lambda(\vec{a},\vec{\pi}) = (V^{-1}(\vec{a}) U(\vec{\pi}) V^{-1}(\vec{a}))^{1/2} V(\vec{a}) U^{-1/2}(\vec{\pi}) .$$

We can verify the following properties of Λ :

$$\Lambda \Lambda^{+} = 1 \qquad \text{and} \qquad [\Lambda , \gamma_{\mu}] = 0 \quad .$$

Thus Λ must have the form

$$\Lambda(\vec{a},\vec{\pi}) = \exp\{\frac{i}{2}\ \vec{\tau}.\vec{b}(\vec{a},\vec{\pi})\}\ .$$

For infinitesimal transformations, to lowest order in f we have

$$\Lambda \overset{\sim}{=} 1 + \frac{i}{2} f\ \vec{\tau}\cdot(\vec{a}\times\vec{\pi})$$

so that

$$\vec{b}(\vec{a},\vec{\pi}) \overset{\sim}{=} f\ \vec{a} \times \vec{\pi}\ .$$

Since

$$\delta\ \vec{\pi} = -\frac{1}{2f}\ \vec{a}$$

we can also write

$$\vec{b} \overset{\sim}{=} -\ 2f^2\ \delta\vec{\pi} \times \vec{\pi} = 2f^2\ \vec{\pi} \times \delta\vec{\pi}$$

so that

$$\Lambda \overset{\sim}{=} 1 + i\ f^2\vec{\tau}\cdot(\vec{\pi} \times\delta\vec{\pi})\ .$$

Thus we have

$$\vec{\pi} \rightarrow \vec{\pi} + \delta\vec{\pi}$$

$$N \rightarrow N + i\ f^2\ \vec{\tau}\cdot(\vec{\pi}\times\delta\vec{\pi})\ N$$

under chiral transformations. This is exactly the form of the chiral transformations in Schwingers work [17]. We see that it is equivalent to

$$\psi \rightarrow \psi + \frac{i}{2}\gamma_5 \vec{\tau}\cdot\vec{a}\psi = \psi - i f \gamma_5 \vec{\tau}\cdot\delta\vec{\pi}\psi$$

for ψ . It has the same form as an isospin transformation, except that the parameters depend on $\vec{\pi}$.

If we postulate Schwinger's transformation law for N and $\vec{\pi}$, we take $\sigma(\alpha) = (1-\alpha)/(1+\alpha)$ and require exact PCAC,then we are led back to the Lagrangian

$$\mathcal{L} = - \bar{N}(\gamma_\nu \partial_\nu + m)N + \bar{N}\gamma_\nu \frac{ig_A f\gamma_5\vec{\tau}.\partial_\nu\vec{\pi} - if^2\vec{\tau}.(\vec{\pi}\times\partial_\nu\vec{\pi})}{1 + f^2\pi^2} N -$$

$$- \frac{1}{2}\partial_\nu\vec{\pi}\partial_\nu\vec{\pi}\frac{1}{(1+f^2\pi^2)^2} - \frac{\mu^2}{2f^2}\log(1+f^2\pi^2) .$$

Note that in Schwinger's notation our f is his $\frac{f_o}{\mu}$ and Schwinger's f is our F.

5. π-N and π-π S-Wave Scattering Lengths

From the phenomenological Lagrangian we can derive the following results:

The axial vector current is

$$\vec{J}_{5\nu} = + g_A \bar{N} i\gamma_5\gamma_\nu \frac{\vec{\tau}}{2} N + \frac{1}{2f}\partial_\nu\vec{\pi} + O(f) .$$

The interaction terms are

$$\mathcal{L}_{int}(N-\pi) = - \bar{N}i\gamma_\nu[-fg_A\gamma_5\vec{\tau}.\partial_\nu\vec{\pi} + f^2\vec{\tau}.(\vec{\pi}\times\partial_\nu\vec{\pi})]N + + O(f^3)$$

$$\mathcal{L}_{int}(\pi-\pi) = f^2[(\vec{\pi}.\vec{\pi})(\partial_\mu\vec{\pi}.\partial_\mu\vec{\pi}) + \frac{1}{4}\mu^2(\vec{\pi}.\vec{\pi})^2] + O(f^4)$$

To order f^2 , the π-N interaction is the same for all three forms of σ. The derivative part of the (π-π) interaction is also the same. However, the non-derivative part depends on the function σ.

-Let us write

$$\mathcal{L}_{int}(\pi-\pi) = f^2[\lambda'(\vec{\pi}.\vec{\pi})(\partial_\mu\vec{\pi}.\partial_\mu\vec{\pi}) + \lambda\mu^2(\vec{\pi}.\vec{\pi})^2].$$

We have the table

Model	λ'	λ
Exp.	1	$\frac{1}{3}$
Rational	1	$\frac{1}{4}$
Square root	1	$\frac{1}{2}$

To first order in λ' and λ we get the following transition matrix element for the process

$$\pi^a + \pi^c \rightarrow \pi^b + \pi^d \quad :$$

$$T(ac,bd) = 16f^2\mu^2[\delta_{ab}\delta_{cd}(\frac{1}{2}(\lambda-\lambda') + \frac{t}{4\mu^2}\lambda') +$$

$$+ \delta_{ad}\delta_{bc}(\frac{1}{2}(\lambda-\lambda') + \frac{u}{4\mu^2}\lambda') +$$

$$+ \delta_{ac}\delta_{bd}(\frac{1}{2}(\lambda-\lambda') + \frac{s}{4\mu^2}\lambda')]$$

where s, t, u are the familiar Mandelstam variables. The I = 0 and I = 2 scattering lengths a_o and a_2 are then given by

$$a_o = \frac{\mu}{4\pi} f^2(5\lambda+\lambda') = \frac{F^2}{4\pi}\frac{1}{g_A^2}(5\lambda+\lambda')\mu^{-1} = 0.058(5\lambda+\lambda)\mu^{-1}$$

$$a_2 = \frac{\mu}{4\pi} f^2(2\lambda-2\lambda') = 0.058\ (2\lambda-2\lambda')\mu^{-1}.$$

We also note the combination

$$a_o - \frac{5}{2} a_2 = 6 \lambda' \frac{\mu}{4\pi} f^2$$

which is insensitive to the model. Note also that

$$(a_o - \frac{5}{2} a_2)/\mu = \frac{6}{4\pi} f^2$$

is insensitive to the pion mass. Putting in the numbers we find the values [25] shown in the following table

Model	$\mu\ a_o$	$\mu\ a_2$	a_o/a_2
exp.	0.15	- 0.075	- 2
rational	0.12	- 0.08	- 3/2
square root	0.20	- 0.06	- 7/2

We note that in the zero pion mass limit we have

$$a_o/\mu = f^2/4\pi \ , \ a_2/\mu = - 2f^2/4\pi \ , \ a_o/a_2 = - \frac{1}{2} \ .$$

The square root model gives Weinbergs values [14] while Schwingers values [17] are obtained by using the rational model.

Weinberg has recently proposed pion mass terms [18] which transform as higher order tensors of the $SU(2) \otimes SU(2)$ group. One can then get higher values for a_o. So far, these predictions agree quantitatively with K_{e4} experiments which show that π-π scattering lengths are small and of the order of μ^{-1}. But so far the experiments are not precise enough to differentiate between various models.

The π-N charge exchange scattering length is simply

$f^2 = 1/(4\pi C_\pi^2)$. In the ρ-dominance model we would get the scattering length $g_\rho^2/2m_\rho^2$, neglecting the ρ-momentum. Here g_ρ is the universal coupling constant of ρ to the isospin current and can be determined from the ρ width. Hence, if the ππNN interaction is ρ-mediated so that the coupling is

$$g_\rho \vec{\rho}_\mu \left(-i\, \bar{N}\gamma_\mu \frac{\vec{\tau}}{2} N + \vec{\pi} \times \partial_\mu \vec{\pi}\right) ,$$

we should obtain

$$f^2 = g_\rho^2/2m_\rho^2 \quad .$$

This is the Kawarabayashi-Suzuki [30] relation which will be derived in a later section. It can be written in the following alternative forms

$$f^2 = g_\rho^2/2m_\rho^2 = F^2/(\mu^2 g_A^2) \qquad \text{or}$$

$$m_\rho^2/g_\rho^2 = 1/2f^2 = 2\,\frac{1}{4f^2} = 2\,C_\pi^2$$

where F is the π-N pseudovector coupling constant. We have seen that $F \sim 1$.

$$\frac{g_\rho^2}{4\pi} = \frac{2m_\rho^2}{\mu^2}\,\frac{1}{g_A^2}\,\frac{F^2}{4\pi} = 2m_\rho^2\,\frac{f^2}{4\pi} \; .$$

Numerically

$$g_A \cong 1.2 \quad , \qquad m_\rho^2/\mu^2 \cong 30$$

so that

$$g_\rho^2/4\pi \cong 3.1 \; ,$$

which is in fair agreement with experiment. The ρ

width gives $g_\rho^2/4\pi = 2.7$. On the other hand the charge exchange scattering length is proportional to $(a_1 - a_3)$. Using dispersion relations that connect $(a_1 - a_3)$ with $\int ds\, g(s)[\sigma_{\pi^+ p}(s) - \sigma_{\pi^- p}(s)]$, where g(s) is a known kinematical factor, we obtain the Adler-Weisberger [31] relation. As Weinberg remarked [16], this is the quickest way of deriving that celebrated relation.

Finally the π-N coupling in the effective Lagrangian gives $a_1 + 2a_3 = 0$, just as in the ρ-dominance model [23], and in good agreement with experiment.

6. Relation to the σ-Model and Remarks about the Use of Effective Lagrangians

Taking the non-derivative square root model we have the coupling

$$m\,\bar\psi\,U\,\psi = m\,\bar\psi(\sqrt{1-4f^2\phi^2} + 2if\gamma_5\vec\tau.\vec\phi)\psi \;=$$

$$= m\sigma\,\bar\psi\psi + 2fm\bar\psi i\gamma_5\vec\tau\psi.\vec\phi =$$

$$= 2mf\bar\psi\sqrt{(1/4f^2)-\phi^2}\psi + 2imf\bar\psi\gamma_5\vec\tau\psi.\vec\phi \;.$$

Let

$$\sigma/2f = \sqrt{(1/4f^2) - \phi^2} = (1/2f) - \chi \text{ or } \sigma = 1-2f\chi \;.$$

Then we have

$$m\,\bar\psi\,U\psi = m\,\bar\psi\psi - 2mf\,\bar\psi\psi\chi + 2mf\,\bar\psi\,i\,\gamma_5\vec\tau\psi.\vec\phi$$

and χ has the dimension of a potential, has zero vacuum expectation value and represents a scalar field.

The interaction is now linear in $\vec{\phi}$ and χ and all we have to do to obtain the σ-model [11] is to add a kinetic term for χ, and a mass term

$$- \frac{1}{2} \partial_\lambda \chi \partial_\lambda \chi - \frac{1}{2} m^2_\chi \chi^2 .$$

The model is now renormalizable. In the limit of infinite χ mass, the kinetic term becomes negligible compared with the mass term. Then, as shown in detail by Weinberg [16], we are back to the non-linear π-N model.

Let us now pause to review our procedure. We start from a renormalizable theory that satisfies chiral invariance. When we go to higher order calculations, other couplings that are consistent with chiral invariance and do not figure in the original Lagrangian, will make their appearance. Such a term is, for example the chiral invariant derivative coupling term. Thus, if we calculate transition matrix elements from an effective Lagrangian we must include all terms consistent with chiral invariance which will be induced by the original renormalizable Lagrangian. Such a Lagrangian is phenomenological, because an induced term like the term in g' can be evaluated in principle, but its unknown renormalized value is taken from experiment and inserted into the effective Lagrangian. This Lagrangian will now give correct predictions in the low energy limit when form factors are unimportant and can be replaced by renormalized original or induced coupling constants.

If we go to higher momenta, then the induced terms which involve higher derivatives of the fields will come into play together with their induced coupling constants. If we wish to have an effective Lagrangian over the whole range of momenta, then we would need an infinite number of coupling terms, each with its own phenomenological constants. Clearly, such a Lagrangian is not useful. That is why we can talk about phenomenological Lagrangians only if we restrict ourselves to low momentum region where form factors can be replaced by low degree polynomials in the invariants formed from the momenta. It may be noted that this is exactly the region where we can get useful predictions from current algebra by combining it with low energy theorems.

7. Transformation Properties of Vector and Axial Vector Current in the π-N Scheme

Schwinger [17],[19], Wess and Zumino [21] and others have enlarged the chiral scheme for the π-N system to include the I = 1 vector meson ρ and axial vector meson A_1. I shall remain closer to the spirit of the work of Wess and Zumino which is more systematic and brings out more clearly the assumptions involved. However the treatment I shall present is both simpler and more general.

Let us go back to the π-N system with pure derivative coupling

$$i \bar{N} \gamma_\nu M_\nu N ,$$

where

$$i M_\nu = - U^{1/2} \partial_\nu U^{-1/2} - \frac{1}{2} g' U^{1/2}(\partial_\nu U^{-1})U^{1/2} .$$

In the work of Wess and Zumino we have the special case g' = 0 and the special functional form

$$U = \frac{1-if\gamma_5\vec{\tau}.\vec{\phi}}{1+if\gamma_5\vec{\tau}.\vec{\phi}} \quad , \quad U^{1/2} = \frac{1-if\gamma_5\vec{\tau}.\vec{\phi}}{\sqrt{1+f^2\phi^2}} \; .$$

Under a chiral transformation, we have

$$N \to \Lambda(\vec{a},\vec{\phi})N = N' \quad , \quad \bar{N} \to \bar{N}\,\Lambda^{-1}(\vec{a}\;,\vec{\phi}) = \bar{N}'$$

where Λ has been defined in section 4. Now, since the matrix Λ depends on coordinates through $\vec{\phi}$, even when $\vec{a}$ is constant, the kinetic term $\bar{N}\gamma_\nu\partial_\nu N$ is changed to

$$\bar{N}\,\Lambda^{-1}\gamma_\nu(\partial_\nu\Lambda N) = \bar{N}\gamma_\nu\partial_\nu N + \bar{N}\gamma_\nu(\Lambda^{-1}\partial_\nu\Lambda)N \; .$$

Thus

$$\bar{N}'\gamma_\nu(\partial_\nu+iM'_\nu)N' = \bar{N}\gamma_\nu(\partial_\nu+iM_\nu)N$$

if M_ν obeys the transformation law

$$iM_\nu \to iM'_\nu \doteq i\Lambda M_\nu\Lambda^{-1} + \Lambda\partial_\nu\,\Lambda^{-1} \; .$$

Consider an infinitesimal chiral transformation, with

$$\Lambda = e^{iB} \cong 1 + iB$$

where

$$B = \frac{1}{2}\,\vec{\tau}.\,\vec{b}(\vec{a},\vec{\pi}) = f^2\vec{\tau}.\vec{\pi}\times\delta\vec{\pi} = \frac{1}{2}f\vec{\tau}.(\vec{a}\times\vec{\pi}) \; .$$

The transformation law is

$$N' = N + iBN$$

$$M'_\nu = M_\nu + i[B,M_\nu] - \partial_\nu B \; .$$

Now the covariant $\vec{M}_\nu$ defined by

$$M_\nu = \vec{\tau} \cdot \vec{M}_\nu$$

is the sum of a vector and an axial vector. In fact

$$\vec{M}_\nu = f^2 \vec{j}_\nu^{\,(\pi)} + 2f^2(1+g') \, \vec{j}_{5\nu}^{\,(\pi)} \gamma_5$$

where $\vec{j}_\nu^{\,(\pi)}$ and $\vec{j}_{5\nu}^{\,(\pi)}$ stand respectively for the pionic contribution to the vector and axial vector currents. Putting this form of M_ν in the transformation law and remarking that $\partial_\nu B$ is a vector, we get, after disentangling the vector and axial vector parts,

$$j_\nu^{(\pi)'} = \vec{\tau} . \vec{j}_\nu^{\,(\pi)'} = j_\nu^{(\pi)} + i[B, j_\nu^{(\pi)}] + \frac{1}{f^2} \partial_\nu B$$

$$j_{5\nu}^{(\pi)'} = \vec{\tau} . \vec{j}_{5\nu}^{\,(\pi)'} = j_{5\nu}^{(\pi)} + i[B, j_{5\nu}^{(\pi)}] \quad .$$

Thus, only the vector part undergoes an inhomogeneous transformation. The axial vector part transforms like a SU(2) ⊗ SU(2) vector. On the other hand, the nucleon parts of the vector and axial vector currents $\vec{J}_\nu^{\,(N)}$ and $\vec{J}_{5\nu}^{\,(N)}$ both transform without inhomogeneous terms and are SU(2) ⊗ SU(2) vectors. It follows that, for the vector and axial vector currents, we have the transformation laws

$$\delta J_\nu = i[B, J_\nu] + \frac{1}{f^2} \partial_\nu B$$

$$\delta J_{5\nu} = i[B, J_{5\nu}]$$

under chiral transformations with constant parameters $\vec{a}$. The inhomogeneity in the transformation of the vector current can easily be understood if we make an infinitesimal chiral transformation and work in the lowest order of f. We have

$$\vec{j}_\nu = -i\,\bar{N}\,\gamma_\nu \frac{\vec{\tau}}{2}\,N + \vec{\pi}\times\partial_\nu \pi + O(f^2)$$

$$\vec{j}_{5\nu} = i(1+g')\,\bar{N}\gamma_5\gamma_\nu\vec{\tau}N + \frac{1}{2f}\,\partial_\nu\vec{\pi} + O(f)\,.$$

The transformation is

$$f\vec{\pi} \rightarrow f\vec{\pi} - \frac{1}{2}\,\vec{a} + O(f^2)\,,$$

or

$$\partial_\nu\,\vec{\pi} \rightarrow \partial_\nu\,\vec{\pi}$$

$$\vec{\pi}\times\partial_\nu\vec{\pi} \rightarrow \vec{\pi}\times\partial_\nu\vec{\pi} - \frac{1}{2f}\,\partial_\nu(\vec{a}\times\vec{\pi}) + O(f)\,,$$

which shows that the only inhomogeneous contribution is to the vector current.

8. Introduction of the ρ-Field-Local Isospin Invariance

The π-N coupling can be written as

$$\mathcal{L}(\pi\text{-}N) = 2f^2(\vec{j}_\nu^{(N)}\,\vec{j}_\nu^{(\pi)} - 2\vec{j}_{5\nu}^{(N)}\,\vec{j}_{5\nu}^{(\pi)})$$

and $\vec{j}_\nu^{(\pi)}$ and $\vec{j}_{5\nu}^{(\pi)}$ have the same transformation properties as the total vector and axial vector currents under SU(2) ⊗ SU(2). Hence, we can obtain another chiral invariant scheme by replacing for instance $\vec{j}_\nu^{(\pi)}$ by $\vec{j}_\nu$ or any local vector operator that has the same transformation properties as the vector current. Let us now consider the operator $2f^2\vec{j}_\nu(x)$. It has the dimensions of a vector field operator and it has zero divergence. It then describes an object with I = 1 and J = 1^-. No chiral property is changed if we make the replacement

$$2f^2\,\vec{j}_\nu \to g_\rho\,\vec{\rho}_\nu$$

where g_ρ is a dimensionless parameter.

We have

$$\partial_\nu\,\vec{\rho}_\nu = 0\ .$$

However, this condition can only be maintained if $\vec{\rho}_\nu$ is coupled to the total isospin current, including the isospin current of the ρ, since ρ_ν is then coupled to a conserved quantity. If we postulate for ρ the transformation law

$$\delta\rho_\nu = i[B,\rho_\nu] + \frac{2}{g_\rho}\,\partial_\nu B$$

under chiral transformations, then the part of the Lagrangian that reads

$$\mathcal{L}_1 = -\bar{N}\,\gamma_\nu\partial_\nu\,N + g_\rho(\vec{J}_\nu^{(N)} + \vec{J}_\nu^{(\rho)})\vec{\rho}_\nu$$

is chiral invariant. We have denoted the ρ contribution to the isospin current by $\vec{J}_\nu(\rho)$. As we have just seen, the spin 1 character of ρ can only be maintained if there is an additional coupling of the ρ to the pionic isospin current. A kinetic term, invariant under chiral transformations should also be added to the Lagrangian. Since chiral transformations are formally identical with local isospin gauge transformations, the kinetic term will look like the term in the Yang-Mills theory. However, in the Yang-Mills theory where the only particles present are N and ρ there cannot be an invariant mass term for the ρ.

In the present theory, however, we also have chiral transformations which change a ρ state into a state with a ρ plus a soft pion, and now there is a possibility for a mass term. In fact, the combination

$$\vec{\rho}_\nu - \frac{2f^2}{g_\rho} \vec{j}_\nu^{(\pi)}$$

transforms like a vector under chiral transformations. Thus

$$\mathcal{L}_m(\rho) = - \frac{1}{2} m_\rho^2 (\vec{\rho}_\nu - \frac{2f^2}{g_\rho} \vec{j}_\nu^{(\pi)})(\vec{\rho}_\nu - \frac{2f^2}{g_\rho} \vec{j}_\nu^{(\pi)})$$

is chiral invariant. Now besides the ρ mass term, we also get from the above expression a ρ - ππ coupling term which reads

$$\frac{2m_\rho^2 \; f^2}{g_\rho} \vec{j}_\nu^{(\pi)} \cdot \vec{\rho}_\nu$$

so that the total coupling of the ρ takes the form

$$g_\rho (\vec{J}_\nu^{(N)} + \vec{J}_\nu^{(\rho)} + \frac{2m_\rho^2 f^2}{g_\rho^2} \vec{j}_\nu^{(\pi)}) . \vec{\rho}_\nu \; .$$

The imposition of the consistency of the divergence condition on ρ, or alternatively, the requirement of invariance under local isospin gauge transformations for the interaction Lagrangian forces us to couple ρ to the total isospin current. Then we must have

$$f^2 = \frac{g_\rho^2}{2m_\rho^2}$$

which is just the Kawarabayashi-Suzuki relation, that was surmised from current algebra. It has recently been derived by Brown and Goble [32].

In our new chiral scheme there is no direct $\bar{N}N\pi\pi$ coupling. There is only an indirect one mediated by the ρ. In section 5 we have seen that this indirect coupling gives back the original scattering lengths provided the Kawarabayashi-Suzuki relation is satisfied.

We can now inquire about the existence of the direct $\pi\pi$-$\pi\pi$ coupling in our new scheme.

In the zero mass pion limit we have seen that the chiral invariant kinetic pion term also gives a $\pi\pi$ interaction term

$$f^2(\vec{\pi}\cdot\vec{\pi})(\partial_\mu\vec{\pi}.\partial_\mu\vec{\pi})$$

which is equivalent to

$$f^2(\vec{\pi}\times\partial_\mu\vec{\pi})\ (\vec{\pi}\times\partial_\mu\vec{\pi}) = f^2\vec{j}_\nu^{(\pi)}\ \vec{j}_\nu^{(\pi)}\ .$$

But the invariant ρ-mass term also provides an additional π-π interaction term, namely

$$-\frac{1}{2}m_\rho^2\ \frac{4f^4}{g_\rho^2}\vec{j}_\nu^{(\pi)}\ \vec{j}_\nu^{(\pi)}$$

which, using the K-S relation just cancels the similar term coming from the kinetic pion Lagrangian. Thus, in the present scheme, low energy πN and $\pi\pi$ scattering processes are mediated by the ρ. This is essentially the content of the ρ-dominance idea, proposed some time ago by Sakurai [23]. Thus, the chiral invariant scheme with the ρ has provided a linearization of the theory through the elimination of the direct nonlinear terms in the πN and $\pi\pi$ interactions. The scattering lengths for these processes can now be obtained to second order in the Lagrangian and they agree with the previous ones.

9. Introduction of the A_1 Field, Local Chiral Invariance

We now introduce a 1^{++} field a_μ with the quantum numbers of the A_1 meson, as a field with the same transformation properties as the axial vector current. We introduce the further assumption that $SU(2) \otimes SU(2)$ transformations which transform the vector and axial vector currents among each other also transform ρ and A_1 in the same way. Denoting the representations by $(2I_1+1,\ 2I_2+1)$ so that N belongs to the $(2,1) \oplus (2,1)$ representations this assumption means that ρ and A_1 form a 6-dimensional representation $(1,3) \oplus (3,1)$ of the group. On the other hand σ and $\vec{\pi}$ together belong of the four dimensional representation $(2,2)$. The representation to which A_1 belongs fixes the scale of a_μ since the field $(\rho_\mu - a_\mu)$ must correspond to the $(3,1)$ representation and consequently must be proportional to the difference of the vector and axial vector currents. Hence, we can make the replacement

$$2f^2\ \vec{j}_{5\nu} \rightarrow g_\rho\ \vec{a}_\nu$$

without altering the group properties and the matrix $a_\nu = \vec{\tau}.\vec{a}_\nu$ will obey the same transformation laws as $2f^2/g_\rho\ .\ j_{5\nu}$. As a result of the preceding discussion, under chiral transformations, we have

$$\delta a_\nu = i[B, a_\nu]$$

with no inhomogeneous term in the transformation. We can now add an interaction term to the Lagrangian of the form

$$2\alpha \, g_\rho \, \vec{j}^{(N)}_{5\nu} \, \vec{a}_\nu$$

with an arbitrary coefficient α in addition to the linear pion interaction term

$$4f^2 \vec{j}^{(N)}_{5\nu} \, \vec{j}^{(\pi)}_{5\nu} \quad .$$

We must also add to our Lagrangian an invariant kinetic A_1 term as well as a mass term

$$\frac{1}{2} m_a^2 \, \vec{a}_\nu \, \vec{a}_\nu \; .$$

This mass term is invariant under chiral transformations with constant parameters $\vec{a}$, since there is no inhomogeneous term in the transformation law for $\vec{a}_\nu$.

At this point let us go one step further and require invariance under local chiral transformations with the parameters $\vec{a}$ depending on x. Now, due to the presence of $\partial_\nu \vec{\pi}$ in the axial vector current, the transformation law for $\vec{j}_{5\nu}$ is no longer homogeneous. We have, for infinitesimal $\vec{a}$,

$$\delta j_{5\nu} = i\left[B, j_{5\nu}\right] - \frac{1}{4f^2} \partial_\nu (\vec{\tau}.\vec{a}) \; .$$

The transformation law for $j^{(\pi)}_{5\nu}$ is similarly

$$\delta j^{(\pi)}_{5\nu} = i[B, j^{(\pi)}_{5\nu}] - \frac{1}{4f^2} \partial_\nu (\vec{\tau}.\vec{a}) \; .$$

The field a_ν transforms like $2f^2/g_\rho . j_{5\nu}$ so that

$$\delta a_\nu = i[B, a_\nu] - \frac{1}{2g_\rho} \partial_\nu (\vec{\tau}.\vec{a}) \; .$$

It follows that the N-π and N-A_1 couplings we have written above are no longer invariant under local chiral transformations. We shall obtain a vector under the local group by taking the combination

$$\left(\frac{2f^2}{g_\rho} j^{(\pi)}_{5\nu} - a_\nu\right) .$$

Thus, if we choose $\alpha = -1$ we obtain the invariant coupling

$$4f^2 \, \vec{j}^{(N)}_{5\nu} \cdot \left(\vec{j}^{(\pi)}_{5\nu} - \frac{g_\rho}{2f^2} \vec{a}_\nu\right) .$$

Consider now the A_1 mass term and the π kinetic term. The π kinetic term, after the removal of the part $f^2 \, \vec{j}^{(\pi)}_\nu \, \vec{j}^{(\pi)}_\nu$ due to the introduction of the ρ, has been reduced to the expression

$$-2 \, f^2 \, \vec{j}^{(\pi)}_{5\nu} \, \vec{j}^{(\pi)}_{5\nu} .$$

We can have an invariant combination of the A_1 mass term and the π kinetic term by introducing the new expression

$$-\frac{1}{2} m_a^2 \left(\vec{a}_\nu - \frac{2f^2}{g_\rho} \vec{j}^{(\pi)}_{5\nu}\right)\left(\vec{a}_\nu - \frac{2f^2}{g_\rho} \vec{j}^{(\pi)}_{5\nu}\right) .$$

However, this requires

$$\frac{1}{2} m_a^2 \frac{4f^4}{g_\rho^2} = 2f^2$$

in order to yield the correct coefficient for the pion kinetic term. Consequently we must have

$$f^2 = \frac{g_\rho^2}{m_a^2} .$$

This is essentially Weinberg's sum rule. By combining with the K-S relation $f^2 = g_\rho^2/2m_\rho^2$

we obtain

$$m_a^2 = 2m_\rho^2$$

or, the famous Weinberg relation

$$m_a = \sqrt{2}\, m_\rho \ ,$$

in spectacular agreement with experiment.

It may be remarked that no unphysical A_1 $-\pi$ transition term is obtained from the invariant term because the part of the term

$$\vec{a}_\nu \, \vec{j}_{5\nu}^{(\pi)}$$

that is linear in $\vec{\pi}$ can be written as a total divergence.

This follows from the requirement

$$\partial_\nu \vec{a}_\nu = 0 \quad .$$

The combination of $\vec{a}_\nu$ and $\vec{j}_\nu^{(\pi)}$ which forms a vector under the local group can now be written as

$$\vec{j}_{5\nu}^{(\pi)} - \frac{g_\rho}{2f^2} \vec{a}_\nu \cong \frac{1}{2f}(\partial_\nu \vec{\pi} - \frac{g_\rho}{2f} \vec{a}_\nu) =$$

$$= \frac{1}{2f}(\partial_\nu \vec{\pi} - m_a \vec{a}_\nu)$$

which is a combination noted in the recent literature by many authors. It can be verified that the Lagrangian we have written is invariant under local isospin gauge transformations only if a scalar counterpart to π is introduced. Otherwise the ρ mass term will violate the local isospin gauge group.

To recapitulate, we see that the extension of the

chiral scheme to cover invariance under local SU(2)⊗ SU(2) transformations gives a generalized partially invariant Yang-Mills theory with a massless pion but massive gauge fields ρ and A_1 . It yields the Kawarabayashi-Suzuki relation, the Weinberg sum rule for the masses of ρ, A_1, fixes the A_1N coupling constant relatively to the ρ-N coupling and linearizes the chiral (π-N) theory to replace it by a ρ -dominance picture.

Finally, the lepton current is assumed to transform like the charged part of the (3,1) representation. This gives immediately the coupling of the lepton current to the weak hadronik (V-A)current considered in section 4.

Electromagnetic couplings can also be introduced by assuming that the isovector part of the electromagnetic current transforms like the neutral part of the (3,1) ⊕ (1,3) representation.

The generalization of the chiral scheme to include SU(3) will not be discussed in these lectures.

10. Injection of SU(6) Results into the Chiral Scheme

As we have remarked in the introduction, there has been up to now no Lagrangian theory which, in some limit, exhibits invariance under SU(6), a group that combines spin (more appropriately the difference between fermion and antifermion spin, the so-called W-spin) with the unitary spin SU(3). However, in the static limit this phenomenological group, which puts π and ρ in the same 35-multiplet, gives a relation between the π-N and ρ-N coupling constants [33]. The same relations can be exhibited in a more covariant formalism [34],[35], although this covariance is illusory since the covariant form of the group SL(6,C) or SU(6,6) is not an invariance group for the whole Lagrangian unless one ignores the kinetic terms. Through SU(6), the pseudovector coupling constant F/μ is related to the ρ coupling constant g_ρ/m_ρ. Here F and g_ρ are dimensionless. Since the F/D ratio for pion is 2/3 and the ρ coupling is pure F, the total π-N coupling constant is 1+2/3=5/3 times the ρ-N coupling constant. Furthermore, these coupling constant relations are assumed to survive mass splitting within a SU(6) multiplet. We therefore have [34]

$$g/m = 2F/\mu = \frac{5}{3}\, g_\rho/m_\rho \quad .$$

We now use Freund's argument [24]. SU(6) invariance is broken by making pions massless and requiring SU(2) $\otimes$ SU(2) chiral invariance. As we have already seen, this requires the K-S relation which we write in the form

$$\frac{g_A}{\sqrt{2}} = \frac{m_\rho}{g_\rho}\frac{F}{\mu}$$

since we have

$$F = g\mu/2m \; ,$$

m denoting the nucleon mass. Comparing with the SU(6) relation we obtain

$$g_A = \frac{5}{3}\frac{1}{\sqrt{2}} \simeq 1.18$$

which is just Schwinger's relation [19]. A new experimental value for g_A is [36]

$$G_A/G_V = - g_A = 1.23 \pm 0.02$$

a value in excellent agreement with Adler's calculation [31]. We should not expect a better agreement from the superposition of the chiral and SU(6) schemes because our pions are massless. The above value should rather be compared with the massless pion calculation of Weisberger [31] . Then the agreement is embarrassingly good.

Another remark is in order regarding this value for g_A. If the only fermions present in the scheme are the nucleon and the N^* and if we treat N^* like a stable particle, then the saturation of the Adler-Weisberger integral by N^* gives 5/3 in agreement, with the perfect SU(6) limit. However, when we break SU(6) by imposing chiral symmetry, then the one particle states N and N^* are no longer stable. They are transformed by chiral transformations into states $|N+\text{soft}\,\pi\rangle$ and $|N^* + \text{soft}\,\pi\rangle$ which have the quantum numbers of higher πN resonances with opposite parity. That is the reason why we can get a value for g_A different from 5/3. In fact it is known that these states with opposite parity are essential in the sum rule for lowering the value 5/3 .

Other remarkable results which spring up from this not quite lawful marriage of SU(6) and the chiral SU(2) ⊗ SU(2) are Schwinger's magnetic moment formulae

$$\mu_p = \frac{5}{3}\frac{e}{2m_\rho} + \frac{1}{3}\frac{e}{2m_\omega} , \quad \mu_n = - \frac{5}{3}\frac{e}{2m_\rho} + \frac{1}{3}\frac{e}{2m_\omega}$$

which in the limit of ρ-ω degeneracy give the famous SU(6) ratio

$$\mu_n/\mu_p = - 2/3 .$$

To derive this result one would need a scheme where

the original SU(6) symmetry is broken by taking the zero mass limit for pions, but keeping the original ρ and ω mesons degenerate.

These remarks should be sufficient to provide us with enough motivation to look for a synthesis of SU(6) type groups with chiral groups.

This leads us directly to the new model [26] which shows the existence of a nonlinear Lagrangian theory invariant under covariant generalizations of SU(6), such as chiral SL(6,C) $\otimes$ SL(6,C) or SL(12,C) in the limit of all mesons having zero mass. Hence the new model is also invariant under the chiral symmetry group SU(3) $\otimes$ $\otimes$ SU(3) and synthetizes the two approaches used by physicists who believe that group theory has something to contribute to particle physics. However, the discussion of the covariant extension of SU(6) is beyond the scope of these lectures.

References

1. R. P. Feynman and M. Gell-Mann, Phys. Rev. 109, 193 (1958).
2. E. C. G. Sudarshan and R. E. Marshak, Phys. Rev. 109, 1860 (1958).
3. J. J. Sakurai, Nuovo Cim. 7, 649 (1958).
4. J. Schwinger, Ann. Phys. 2, 407 (1957). The SU(2) $\otimes$SU(2) group was also introduced for weak interactions by S. Bludman, Nuovo Cim. 9, 433 (1958).
5. J. C. Polkinghorne, Nuovo Cim. 8, 179 (1958). The model presented in this paper is referred to J. C. Taylor, in a footnote.
6. K. Nishijima, Nuovo Cim. 11, 910 (1958). This is an exponential model for isoscalar mesons. It is equivalent to the Polkinghorne-Taylor model after

a unitary transformation is performed. The group admitted by both models is an Euclidean group.
7. G. Kramer, H. Rollnik and B. Śtech, Zeits. f. Phys. 159, 564 (1959). In this paper the πN coupling is non-linear and chiral invariant, but such invariance does not hold for the pion part of the Lagrangian.
8. F. Gürsey, Nuovo Cim. 16, 230 (1960).
9. F. Gürsey, Proc. of the 1960 Rochester Conference, p. 572. Also F. Gürsey and B. Zumino, "Remarks on Asymptotic Fields in γ_5 Invariant Theories" (1960), unpublished.
10. F. Gürsey, Annals of Physics 12, 91 (1961).
11. M. Gell-Mann and M. Lévy, Nuovo Cim. 16, 705 (1960).
12. Y. Nambu, Phys. Rev. Lett. 4, 380 (1960).
13. Y. Nambu, and G. Jona-Lasinio, Phys. Rev. 122, 345 (1961); Y. Nambu, and E. Schrauner, Phys. Rev. 128, 862 (1962).
Y. Nambu and D. Lurié, Phys. Rev. 125, 1429 (1962). The modern interpretation of chiral transformations as connecting states that differ by a soft pion was first developed in these important papers.
14. S. Weinberg, Phys. Rev. Lett. 17,616 (1966).
15. Y. Tomozawa, Nuovo Cim. 46A, 704 (1966).
16. S. Weinberg, Phys. Rev. Lett. 18, 188 (1967).
17. J. Schwinger, Phys. Lett. 24B, 473 (1967).
18. S. Weinberg, Non-Linear Realizations of Chiral Symmetry, Phys. Rev. 166, 1568 (1968).
19. J. Schwinger, Phys. Rev. Lett. 18, 923 (1967).
20. S. Weinberg, Phys. Rev. Lett. 18, 507 (1967).
21. J. Wess and B. Zumino, Phys. Rev. 163, 1727 (1967).
22. C. N. Yang and R. L. Mills, Phys. Rev.95, 191 (1954).
23. J. J. Sakurai, Ann. of Phys. 11, 1 (1960).
24. P. G. O. Freund,Phys. Rev. Lett. 19, 189 (1967).
25. P. Chang and F. Gürsey, Phys. Rev. 164, 1752 (1967).

26. F. Gürsey and P. Chang, Phys.Lett. 26B, 520 (1968).
27. V. Ogievetsky, private communication.
28. An exponential pion-nucleon coupling was first introduced by R. J. Glauber in Phys. Rev. 84, 395 (1951) as a model for multiple pion production, independently of chiral invariance considerations. In Glaubers model, like in the model of ref. 7, the pion Lagrangian is not chiral invariant. The chiral invariant exponential model first appears in ref. 8.
29. L. S. Brown, Phys. Rev. 163, 1802 (1967).
30. K. Kawarabayashi and M. Suzuki, Phys. Rev. Lett. 16, 255 (1966); Riazuddin and Fayazuddin, Phys. Rev. 147, 1071 (1966); F. J. Gilman and H. J. Schnitzer, Phys. Rev. 150, 1362 (1966);. J. J. Sakurai, Phys. Rev. Lett. 17, 552 (1966); M. Ademollo, Nuovo Cimento 46, 156 (1966).
31. W. J. Weisberger, Phys. Rev. Lett. 14, 1047 (1965); S. L. Adler, Phys. Rev. Lett. 14, 1051 (1965).
32. L. Brown and R. L. Goble, Phys. Rev. Lett. 20, 346 (1968).
33. F. Gürsey, A.Pais and L. A. Radicati, Phys. Rev. Lett. 13, 299 (1964).
34. B. Sakita and K. C. Wali, Phys. Rev. 139 B, 1355 (1965).
35. A. Salam, R. Delbourgo and J. Strathdee, Proc. Roy. Soc. A, 284, 146 (1965).
36. I am indebted to Prof. A. Sirlin for this information.

ZERO MASS THEOREMS AND ANALYTIC CONTINUATION†

By

W. KUMMER

Institut für Hochenergiephysik der Österreichischen Akademie der Wissenschaften
Wien

Introduction

The extensive application of current algebra (CA) in the last years has led in many cases to remarkable successes. On some occasions it became clear recently that the assumption of smoothness, used for the required continuation (mostly in the pion - mass), must be modified. This showed that - at least sometimes - more of the dynamics than pole terms at best and simple CA must be used to explain physical phenomena. One can then think of two approaches. The first possibility consists in assuming a simple interpolation by polynomials. Sometimes, as in $K \to 3\pi$ [1] or $\eta \to 3\pi$ [2] decays, CA alone gives enough restrictions to produce moderately successful [3] predictions. In $\pi\pi$ - scattering unitarity gives further conditions and an improvement [4] of the "pure" result of CA [5]. The aim of

† Lecture given at the VII. Internationale Universitätswochen für Kernphysik, Schladming, February 26 - March 9, 1968.

this talk is the alternative, more fundamental way. We want to advertise the investigation of dynamical models by which we may be able to explain variations caused by the mass-variable in terms of other experimentally known quantities. Just in the K and η-decays mentioned above, the phenomenological method works much better for K than for η [3], in contrast with the generally accepted picture of the theoreticians who see much similarity in these decays. We shall be modest enough, not to attack at once these difficult continuations in three masses [6] . As a first step the continuation in one mass is enough. Such a problem occurs not only in the wellknown derivation of the Goldberger-Treiman (G.T.) relation from the hypothesis of pion pole dominance in the appropriate matrix element of the divergence of the axial vector current (PDDAC) [8], but also in the decay - amplitudes of $\pi^o, \eta \to 2\gamma$. It is usually stated that these amplitudes vanish in the limit $m^2_{\pi,\eta} \to 0$ from CA [9]. It seems to us that CA is not needed for this result (see below). On the other hand clearly in this case the "smoothness assumption" fails badly, the variations with the mass are necessarily not small compared to the vanishing value at $m^2_{\pi,\eta} = 0$. Our calculations are based on unsubtracted dispersion relations for the invariant functions in the divergences of the currents, an assumption which has been successful in GT-relations. We emphasize the - at least pedagogical - advantages of the reduction technique with functional derivatives [10] in this context, because also the gauge method [11], which is clearer than CA (and produces the same results), is readily formulated in this way. This is contained in the first section, whereas the second one is devoted to a treatment of the dispersion relation appearing in GT - like relations and especially to the problem of $\pi^o, \eta \to 2\gamma$ decays.

The results of the dynamical theory appear in the form of integral relations, where our knowledge of the behaviour at zero mass is inserted. They do not however in general correspond to subtracted dispersion relations.

1. S-Matrix and Currents

1.a) "Classical" reduction

It is our intention to compare first the two types of formalism which are used in this connection. The by now classical [12] reduction technique uses currents defined from interpolating fields $\phi(x)$ by the action of a d'Alembert operator K_x

$$K_x \; \phi(x) = j_\phi(x) \tag{1.1}$$

The current j_ϕ carries the tensor and internal symmetry properties of ϕ. The field of the hypothetical W-meson in weak interactions, if coupled to the axial current, is an example for a field ϕ_μ with $\partial_\mu \phi_\mu \neq 0$ and a content $J^P = 1^+, 0^-$. To lowest order in a "weak" field ϕ, the matrix element of the current

$$<f; \text{out} | \phi_\mu(q), b(q'), i; \text{in}> = \tag{1.2}$$

$$= \frac{(2\pi)^4 i\delta^4(q+q'+p_i-p_f)}{(2\pi)^{3/2}\sqrt{2q_o}} <f|j_\mu^A(o)|b,i> + \delta^3(\;)\delta^3(\;)..$$

is alone important. One more reduction [13], again for the example of the axial vector current, yields

$$\langle f|j_\mu^A(o)|b(q');i\rangle = i\int d^4x\ e^{-iq'x} \times \qquad (1.3)$$

$$\times \langle f|\{[j_\mu^A(o),j_b(x)]\ \Theta(-x_o) +$$

$$+ [\dot{j}_\mu^A(o),\dot{b}(x)]\,\delta(x_o) -$$

$$- iq'_o[j_\mu^A(o),b(x)]\Theta(x_o)\}|i\rangle$$

By taking the divergence ∂_μ of (1.3), besides the equal time commutator of currents and fields in (1.3) also current-current commutators at equal time appear. We have some knowledge about them from the quark-model [14]. After partial integration an additional expression $\propto q_\mu$ is present which vanishes for $q_\mu \to 0$. The first disadvantage of (1.3) is the fact that the different commutators in an expression like the one in curly brackets are, in general, relativistically covariant only together. A related problem occurs, if "Schwinger-terms" [15] appear in the current commutators. In a model dependent way they must cancel by the interplay with current - field commutators. The analytic properties on the other hand are determined from the first, retarded, commutators; the current-field commutators produce polynomial terms which must be removed by appropriate subtractions.

1.b) Functional derivative method

Usually functional derivatives or integrations have not too much importance for practical calculations. Here the advantage lies in the more compact way causality [10] and hence analyticity are formulated. At the same time the content of CA can be represented by divergence relations between the same functional derivatives of the S-matrix. Let us recall first the reduction formalism [16]. Applying [13]

$$[b^{\pm}(q),S] = \mp \int e^{\mp iqx} d^4x \frac{\delta S}{\delta b(x)} \quad (1.4)$$

to the creation operators $b^+(q')$ and $\phi^+(q)$ we obtain, now in the interaction picture,

$$\langle f|S|\phi(q)b(q');i\rangle = (2\pi)^4 i\delta^4(q+q'+p_i-p_f) \times$$

$$\times \langle f|(-i \frac{\delta S}{\delta\phi(o)})|b;i\rangle \quad (1.5)$$

$$\langle f|(-i \frac{\delta S}{\delta\phi(o)} S^+)S|b;i\rangle = \langle f|(j_\phi S)|b,i\rangle =$$

$$= i\int e^{-iqx} d^4x \langle f|(- \frac{\delta^2 S}{\delta b(x)\delta\phi(o)} S^+)S|i\rangle \quad (1.6)$$

As usual, translation invariance has been used. We note that the bracket in eq. (1.6) contains all current-current and current-field terms of eq. (1.3). It is actually equivalent to a T-product instead of the R product used in (1.3). Clearly this reduction can be continued until all particles in i and f are replaced by functional derivatives. In no case explicitly separated equal time commutators as in eq. (1.3) appear.

The formalism allows also a simple description of causality [10], if the dependence of S with respect to external fields is considered, i.e. the fields (e.g. b and ϕ in (1.6)) are "weak", so that higher orders can be neglected and

$$j_b(o) = -i \frac{\delta S}{\delta b(o)} S^+ \quad (1.7)$$

allows the interpretation as a current. Consider $\phi(x)$ and

$$\phi'(x) = \phi(x) + \delta\phi(x) \quad (1.8)$$

where the (small) variation $\delta\phi(x)$ occurs at times later than a fixed time t so that

$$S(\phi + \delta\phi) = S(\phi + \delta\phi)_{x_o > t}\; S(\phi)_{x_o < t} \tag{1.9}$$

Under this change we obtain

$$S(\phi'(x)) = S(\phi) + \int_{x_o > t} \frac{\delta S}{\delta\phi(x)}\, \delta\phi(x)\, dx \tag{1.10}$$

The quantity (cf. (1.9), (1.10))

$$S(\phi')\, S^{+}(\phi) = S(\phi')_{x_o > t} = 1 + \int_{x_o > t} \frac{\delta S}{\delta\phi(x)}\, S^{+}(\phi)\delta\phi(x)dx$$

does not depend on the "history" of ϕ before t so that

$$\frac{\delta}{\delta\phi(y)}\left[\frac{\delta S}{\delta\phi(x)}\, S^{+}\right] = 0 \tag{1.11}$$

for $y_o < t < x_o$ or y spacelike with respect to x. Eq. (1.11) also holds for two different fields $\psi \neq \phi$ if they are independent.

We mentioned already the advantages of this approach for avoiding the often ambiguous CA by a formulation based on a gauge principle [11] and divergence equations [17]. As in CA quark-fields ψ are used as a guide. Under an infinitesimal local gauge transformation the change

$$\psi(x) \to \psi(x) + i\, \delta\varepsilon^{\ell}(x)\, \lambda^{\ell}\psi(x) \tag{1.12}$$

forces the free Lagrangian of the quarks (without mass term) to be supplemented by an interaction with external gauge fields ϕ_{μ}^{ℓ} and

$$\mathcal{L}_{\text{free(quarks)}} \to \bar{\psi}\, \gamma_{\mu}(\partial_{\mu} + i\, \lambda^{\ell}\phi_{\mu}^{\ell})\psi + \text{mass term}$$

where the ϕ_μ^ℓ obey the gauge transformation

$$\phi_\mu^\ell \rightarrow \phi_\mu^\ell + \delta\varepsilon^n(x)\phi_\mu^m \, C^{\ell mn} - \partial_\mu \delta\varepsilon^\ell(x) \tag{1.13}$$

The matrices λ^ℓ act in the internal and in the spinor space. For chiral SU(3) ⊗ SU(3) they include the Gell-Mann matrices of SU(3), $\lambda^\alpha(\ell=\alpha=0,1,\ldots 8)$ and $\lambda^\alpha\gamma_5(\ell = \alpha+9)$; ϕ_μ^ℓ are vector fields ($\ell=0,\ldots 8$) or axialvector fields ($\ell=9,\ldots 17$). The S-matrix connects asymptotic hadron states made up from quarks. If $\delta\varepsilon\,(x) \rightarrow 0$ for $t = \pm\,\infty$ the elements of the S-matrix must be invariant with respect to the gauge (1.13) of the external fields alone:

$$\int \frac{\delta S}{\delta\varepsilon(x)} \,\delta\varepsilon(x)\, dx = \int d^4x \,\frac{\delta S}{\delta\phi_\mu^\ell}(\delta\varepsilon^n \phi_\mu^m \, C^{\ell mn} - \partial_\mu \delta\varepsilon^\ell) \tag{1.14}$$

The "divergence equation"

$$\frac{\delta S}{\delta\varepsilon^\ell} = \partial_\mu \frac{\delta S}{\delta\phi_\mu^\ell} + C^{\ell mn}\phi_\mu^m \frac{\delta S}{\delta\phi_\mu^n} \tag{1.15}$$

follows immediately. In a matrixelement without further external fields $\delta S/\delta\varepsilon$ (which is of course equal to zero for an exact gauge invariance) is proportional to the divergence of a current (cf. (1.7)). Other external fields in the matrix element produce through the reduction procedure further functional derivatives. Because of

$$\frac{\delta\phi_\mu^m(x)}{\delta\phi_\nu^n\,(y)} = \delta(x-y)\,\delta_{mn}\, g_{\mu\nu}$$

from the second expression on the r.h.s. of (1.15) the typical CA-terms appear.

2. Dispersion Relations for the Divergence of the Axial Vector Current

2.a) Dispersion relations

The matrix element

$$\langle f|(j(o)S)|i\rangle \qquad (2.1)$$

where we consider only one "current" j(o) (or divergence of a current) for simplicity, can be expressed in terms of invariant functions which in turn depend on the invariants formed by internal products among the fourmomenta of the particles in the initial and final state. Besides, there is a dependence on the "mass"

$$s = q^2 = (p_f - p_i)^2 \qquad (2.2)$$

we are mainly interested in, and on invariants $(q\ p_i^{(1)}),\dots,(q\ p_f^{(1)}),\dots$. The application of a low energy theorem at $q_\mu \to 0$ therefore implies not only a continuation to $s \to o$ but also - sometimes rather important - unphysical continuations in other invariant variables. A simultaneous treatment of this large number of complex variables would be much more difficult than the continuation problem for a general amplitude at fixed mass, which is far from being solved itself. In order to separate the two problems we need the assumption that there is enough analyticity enabling the continuation in the "other" invariants towards some crossed amplitude of (2.1)

$$\langle f,\bar{i}|(j(o)S)|0\rangle \quad .$$

With a matrix element of this type we define as usual besides ($S|0\rangle = S^+|0\rangle = |0\rangle$, cf. (1.6), $b(q')$ and other particles f are in the final state)

$$T_c = \langle f|(j_\phi(o)S|0\rangle = -i\int e^{iq'x}\, d^4x \langle f|\frac{\delta^2 S}{\delta b(x)\delta\phi(o)} S^+|0\rangle \tag{2.3}$$

$$q' + p_f = q \tag{2.4}$$

the similar amplitudes

$$T^{ret} = -i\int e^{iq'x}\, d^4x \langle f|\frac{\delta}{\delta b(x)}\left(\frac{\delta S}{\delta\phi(o)} S^+\right)|0\rangle \tag{2.5}$$

$$T^{(+)} = -i\int e^{iq'x}\, d^4x \langle f|j_b(x)\; j_\phi(o)|0\rangle \tag{2.6}$$

T^{adv} and $T^{(-)}$ differ only in the exchange of ϕ and b in eqs. (2.5) and (2.6) respectively. The amplitudes are connected by

$$T_c = T^{ret} - T^{(-)} = T^{adv} - T^{(+)} \tag{2.7}$$

A complete system of intermediate states $|n\rangle$ and translation invariance in (2.6) gives

$$T^{(+)} = -i(2\pi)^4 \sum\!\!\!\!\!\!\int_n \delta^4(p_n - q)\langle f|j_b(o)|n\rangle\langle n|j_\phi(o)|0\rangle \tag{2.8}$$

where the symbol $\sum\!\!\!\!\!\!\int$ includes the phase space integrations. (2.8) can give a δ-function at $p_n^2 = m^2$ and a further contribution for $p_n^2 > (\Sigma m_n)^2$. In $T^{(-)}$ we have $\delta^4(p_n + q')$ and therefore no possible intermediate state with positive energy. $T^{(+)}$ vanishes below $q^2 = m^2$, $T^{(-)}$ is zero altogether so that in this region $T_c = T^{ret} = T^{adv}$.

For $q = \text{Re } q + i \text{ Im } q$, $(\text{Im } q)^2 > 0$, $\text{Im } q_o > 0$,

$T = T^{ret}$ is regular while for Im $q_o < 0$, $T = T^{adv}$ has the same property. The analytic function T is then analytic for both signs of Im q_o. We assume that the regularity in $s = q^2$ except for Re $s > 0$, Im $s = 0$ is not limited by a contour depending on the other complex variables. This we cannot prove from causality alone. Then the Cauchy theorem in the full s-plane yields the desired unsubtracted dispersion relation

$$T(s+i\varepsilon) = \frac{1}{2\pi i}\int_{m^2}^{\infty} \frac{T^{ret}(s') - T^{adv}(s')}{s'-s-i\,\varepsilon}\,ds' =$$

$$= \frac{1}{\pi}\int_{m^2}^{\infty} \frac{\sigma(s')}{s'-s-i\varepsilon}\,ds' \qquad (2.9)$$

which clearly should have been written down for invariant functions, with (formally)

$$\sigma(s) = \frac{(2\pi)^4}{2} \sum_n \delta^4(p_n-q)\langle f|j_b(o)|n\rangle\langle n|j_\phi(o)|o\rangle \qquad (2.10)$$

The dependence on the other invariant variables is not indicated explicitly in the last two formulas.

2.b) j_ϕ is divergence of a current

We specialize first to the divergence of an axial vector current with I = 1, S = 0. f_π in

$$\langle\pi|j_\mu^A|0\rangle = p_\mu^\pi\, f_\pi \qquad (2.11)$$

is determined from πdecay ($f_{\pi^+} = 0.95m_\pi$), so that the pion term in (2.10) reads

$$\pi m_\pi^2\, f_\pi\, \delta(s - m_\pi^2)\langle f|j_b(o)|\pi\rangle \;.$$

If we take the formal eq. (2.9) at $q^2=0$ we arrive at

$$\langle f|j_b(o)|\pi\rangle = f_\pi^{-1}\, T(o) - \frac{f_\pi^{-1}}{\pi} \int_{(3m_\pi)^2}^{\infty} \frac{\sigma(s')}{s'}\, ds' \tag{2.12}$$

This implies clearly the assumption of an unsubtracted dispersion relation for the matrix element of the divergence of the axial vector current [8] (BFGT-assumption). When the integral can be neglected, we have Bell's [18] "pion-pole dominance of the divergence of the axial vector current" (PDDAC). It is equivalent to the "smoothness assumption" in the theory of the partially conserved axial vector current (PCAC) [19]. In the formulation (2.12) at least in principle the possibility exists to calculate the difference between the matrix element for $\pi \rightarrow b + f$ and the divergence of the axial vector current coupled to b and f, which is contained in T. A necessary condition for neglecting the integral is clearly that neither of the latter expressions vanish. Naive application of PDDAC (or PCAC) in cases where T(o) happens to be zero, is therefore certainly wrong.

What has been said about $\partial_\mu j_\mu^A$ with I = 1 and S = 0 and the pion can be carried over immediately to $\partial_\mu j_\mu^A$ with I = 1/2, S = ± 1 and the kaon.

Although one should not be too generous with the BFGT-assumption, as we shall see immediately, we could apply it to the divergence of the (nonconserved) strangeness carrying vector current . The "κ-particle" which would replace the pion in this case is much less conspicuous experimentally, to say the least, so that such considerations are of small value at present.

2.c) Special cases

The simplest situation occurs for $\langle f|=\langle a|$ = one particle state. In this case also the analytic continuation in the "other" variables is trivial, as long as f and b are on their respective mass shells. Let a, $\bar{b}$ represent a nucleon-antinucleon pair. Then the divergence of the axial vector current ($q = p^a + p^b$)

$$\langle N\bar{N}|j_\mu^A|0\rangle = \bar{u}_N[F_A\gamma_\mu + F_A^{(1)}q_\mu + F_A^{(2)}(\gamma_\mu \not{q} - \not{q}\gamma_\mu)]\gamma_5 v_{\bar{N}} \tag{2.13}$$

at $s = q^2 = 0$ is an experimentally accessible quantity:

$$\langle N\bar{N}|\partial_\mu j_\mu^A|0\rangle|_{s=0} = i(m_N+m_{\bar{N}})\, F_A(0)\, \bar{u}\gamma_5\, v \tag{2.14}$$

$F_A(0)$ is the axial vector coupling constant, known from nucleon β-decay, and

$$\langle N|j_{\bar{N}}(0)|\pi\rangle = ig_{N\bar{N}\pi}\, \bar{u}\, \gamma_5\, v \tag{2.15}$$

is expressed in terms of the pion-nucleon coupling constant. The formula

$$g_{N\bar{N}\pi}\, f_\pi = (m_N + m_{\bar{N}})\, F_A(0) - \frac{1}{\pi} \int_{(3m_\pi)^2}^{\infty} \frac{\sigma(s')}{s'}\, ds' \tag{2.16}$$

without the integral (PDDAC) is the old Goldberger-Treiman-relation (GT-relation) [8],[20], generalized to arbitrary baryons if "π" represents a general pseudoscalar meson (π,K,η). It is obvious that e.g. for $(N\bar{N}) = (\Sigma^+\bar{p})$ the meson pole at $q^2 = m_{K^0}^2$ is very near to the cut, starting in this case at $(m_K+2m_\pi)^2$. In the original GT-relation for $\bar{p}n$ the integral gives only 10% of the first term on the right in eq. (2.16). Even if SU(3) holds in the $(\Sigma^+\bar{p})$-case for the coupling in the first two terms, the expected larger influence of the integral may reflect the larger breaking in the K-mass with respect to the pion. The relation (2.16) permits therefore a rather transparent model of symmetry breaking. The corrections to the original GT-relation have been investigated recently [21] using an effective pole for the integral and a similar one for the axial current itself, where besides $J^P = 0^-$ also the quantum numbers $J^P = 1^+$ (e.g. the A_1) are allowed. We expect a success of such considerations only if many similar reactions are treated together or if knowledge from Weinberg sum rules [22] and the like is added.

The limitations for the application of the BFGT-assumption to the divergence of axial vector currents come from the fact that only few axial charges can be observed (via weak interactions).

As we mentioned already for the strangeness changing vector currents (which can be observed) the relation contains the decay constant of something like the mysterious κ-particle.

The BFGT-assumption gives pure nonsense for matrix elements of axial vector currents which are "transversal". An example is the axial vector current in $\pi \to e\nu\gamma$ which (after subtraction of the part that is gauge-invariant together with the internal bremsstrahlung diagrams) reads ($q=p+k$)

$$\langle \pi(p)\gamma(k,\varepsilon_\rho)|j_\mu^A|0\rangle = \varepsilon_\rho(g_{\rho\mu} - p_\rho k_\mu/(pk))\, h(q^2) .$$

Formally (because $\langle \pi\gamma|\partial_\mu j_\mu^A|0\rangle \equiv 0$) for all q^2 the BFGT-assumption would give

$$0 = \frac{m_\pi^2 f_\pi}{m_\pi^2 - s}\langle \gamma|j_\pi|\pi\rangle + \frac{1}{\pi}\int \frac{ds'}{s'-s}\,\sigma(s')$$

which cannot hold because the residue and $\sigma(s')$ are nonvanishing quantities.

The amplitudes f(s) which will be related to the decay $\pi^o \rightarrow 2\gamma$, defined in

$$<\gamma(k^1,\varepsilon^1)\ \gamma(k^2,\varepsilon^2)|j^A_\mu|0> = f(s)\ q_\mu \times$$

$$\times\ \varepsilon_{\rho\sigma\alpha\beta}\ k^1_\alpha k^2_\beta\ \varepsilon^1_\rho \varepsilon^2_\sigma$$

$$(q=k^1+k^2) \tag{2.17}$$

have no reason to be singular at s=o. Therefore

$$<\gamma\gamma|\partial_\mu j^A_\mu|0> = s\ f(s)\ \varepsilon\ \ldots.$$

and the function s f(s) vanishes at s=o.

In the literature the vanishing of this function at s=o is called a result of CA [9]. Two further functional differentiations of eq. (1.15) with respect to the electromagnetic fields A_ρ of the photons give (ℓ=12 for a π^o) with

$$\frac{\delta^2}{\delta A(y)\delta A(z)}\ \partial_\mu(-i\ \frac{\delta S}{\delta\phi^{12}_\mu(x)}) = i\ \frac{\delta^3\ S}{\delta A(y)\delta A(z)\delta\varepsilon^{12}(x)} +$$

$$+ \underbrace{C^{3,8,12}\ldots.}_{\emptyset} \tag{2.18}$$

a trivial result, because the "CA-terms" vanish for the components 3 and 8 appearing in the electromagnetic fields. In our case then after one partial integration on the left

$$+\ iq_\mu \int e^{ik_1 y + ik_2 z}\ d^4y\ d^4z\ <0|\text{l.h.s.}|0> =$$

$$= \int e^{ik_1 y + ik_2 z}\ d^4y\ d^4z <0|\text{r.h.s. of (2.18)}|0> .$$

The invariant function in the r.h.s. is appropriate for the BFGT-assumption. For $q \to 0$ a priori no statement is possible because then $k_1 \to - k_2$ and therefore the factor $\varepsilon_{\rho\sigma\alpha\beta} k^1_\alpha k^2_\beta \varepsilon^1_\rho \varepsilon^2_\sigma \to 0$ (kinematical zero). Only if in the integral on the left the same invariance consideration as in (2.17) is made, the vanishing of the invariant function at $q^2 = s \to 0$ follows. Hence CA is really superfluous.

We now use this result $T(o) = 0$ in the invariant function of (2.12). The decay amplitude g_π in

$$\langle\gamma\gamma|j_\pi(o)|0\rangle = g_\pi \; \varepsilon_{\rho\sigma\alpha\beta} \; k^1_\alpha k^2_\beta \varepsilon^1_\rho \varepsilon^2_\sigma \tag{2.19}$$

and $\sigma(s)$ defined similarly by projection with

$$- 2 \; s^{-2} \; \varepsilon_{\rho\sigma\alpha\beta} \quad k^1_\alpha \; k^2_\beta$$

into the tensor (2.10) with indices ρ, σ, after the polarizations ε^1_ρ and ε^2_σ have been split off, give

$$g_\pi = - \frac{f_\pi^{-1}}{\pi} \int_{(3m_\pi)^2}^{\infty} \frac{\sigma_\pi(s')ds'}{s'} \tag{2.20}$$

A similar formula holds for g_η in $\eta \to 2\gamma$. The main difficulty in any application comes from the a priori unknown matrix elements of ∂j in $\sigma(s')$. Nevertheless crude considerations of the type can be made, where the imaginary part is replaced by some "effective pole",

$$\sigma_\pi(s) = \pi\delta(s-M^2_\pi) \; r_{\pi M_\pi} r_{M_\pi 2\gamma}$$

$$r_{\pi M_\pi} = \langle M_\pi|(\partial j)_\pi|0\rangle$$

$$r_{M_\pi 2\gamma} = \langle \overset{1}{\gamma}|\overset{2}{j^\gamma}(o)|M_\pi\rangle$$

and similar expressions for σ_η , so that

$$\frac{g_\eta}{g_\pi} = \left(\frac{r_{M_\eta 2\gamma}}{r_{M_\pi 2\gamma}}\right)\left(\frac{r_{\eta M_\eta}}{r_{\pi M_\pi}}\right)\left(\frac{M_\pi}{M_\eta}\right)^2 \frac{f_\pi}{f_\eta} \quad . \tag{2.21}$$

What are "reasonable" assumptions for these ratios? If SU(3) invariance would hold for the r then

$$\frac{r_{M_\eta 2\gamma}}{r_{M_\pi 2\gamma}} \simeq \frac{1}{\sqrt{3}} \ , \quad \frac{r_{\pi M_\pi}}{r_{\eta M_\eta}} \simeq 1 \quad . \tag{2.22}$$

The effective masses M_η/M_π could be roughly in the ratio of the thresholds from strong interactions $(m_\eta+2m_\pi)/3m_\pi$, if it were not for the "electromagnetic" amplitude $\eta \to 3\pi$, which is observed to be of order one [23] and pushes the threshold in the η-reaction down to $3m_\pi$. So $M_\eta/M_\pi \sim 1.2 \div 1.5$ would be reasonable. This suppresses further g_η. The experimentally observed ratio $g_\eta/g_\pi \sim 1.4$ from the recent value for $\Gamma_{\eta\to 2\gamma} \simeq 0.9$ keV [24] would force us to believe that the enhancement comes entirely from $f_\pi/f_\eta \sim 4$. This last value seems to be in violent disagreement with considerations based on Weinberg-type sum rules in chiral SU(3) ⊗ SU(3) [25] which give $f_\pi/f_\eta \lesssim 0.3$. Two ways exist to avoid this difficulty: The "traditional" one introduces η - η'-mixing in the form

$$g_\pi = \sqrt{3}\, g_8 \ , \quad g_\eta = \cos\theta g_8 + \sin\theta g_1$$
$$g_{\eta'} = -\sin\theta g_8 + \cos\theta g_1$$

so that the enhancement in g_η/g_π could come from an additional factor $[\cos\theta+\sin\theta . g_1/g_8]$ which could be large even for the small $\theta \sim .19$ of the mass formula if $|g_1/g_8|$ is large enough. In this case however the $\eta' \to 2\gamma$ partial decay goes in the region of 50 keV and the width becomes slightly larger than the observed upper limit.

Another explanation could follow from a careful investigation of eq. (2.20). The additional "electromagnetic" amplitude with a threshold at $3m_\pi$ like in η-decay, could spoil completely the SU(3) relations (2.22). In this way the η→2γ-puzzle could be closely connected with the equally mysterious situation in η→3π [3],[23]

$$\sigma_\eta = \sigma_\eta^{el} + \sigma_\eta^{strong}$$

$$g_\eta/g_\pi = \frac{f_\pi}{f_\eta}\left[\int_{(3m_\pi)^2}^{\infty} \frac{ds'}{s'}\sigma_\eta^{el}(s') + \int_{(m_\eta+2m_\pi)^2}^{\infty} \frac{ds'}{s'}\sigma_\eta^{strong}(s')\right] / \left[\int_{(3m_\pi)^2}^{\infty} \frac{ds'}{s'} \times \sigma_\pi(s')\right]$$

It appears thus that most of the SU(3) breaking is likely to stem from the electromagnetic breaking, represented by σ_η^{el}. We shall not go further into details of this example here, because a mixed mechanism [η-η'-mixing plus electromagnetic breaking of SU(3)] cannot be excluded.

The author appreciates discussions with Drs. Flamm and Majerotto.

References and Footnotes

1. Y. Hara and Y. Nambu, Phys. Rev. Lett. 16, 875 (1966).
2. A. D. Dolgov, A. I. Vainshtein and V. I. Zakhorov, Phys. Lett. 24, B425 (1967); W. A. Bardeen, L. S. Brown, B. W. Lee and H. T. Nieh, Phys. Rev. Lett. 18, 1170 (1967).
3. J. S. Bell and D. G. Sutherland, "Current algebra

and $\eta \to 3\pi$", CERN preprint TH.822, August 1967.

4. J. Iliopoulos, Nuovo Cim. 52A, 192 (1967), CERN preprint, TH.854, Nov. 1967.
5. S. Weinberg, Phys. Rev. Lett. 17, 616 (1966).
6. Another important example would be the ρ-width and the Kawarabayashi-Suzuki-relation [7].
7. K. Kawarabayashi and M. Suzuki, Phys. Rev. Lett. 16, 255 (1966); Riazuddin and Fayazuddin, Phys. Rev. 147, 1071 (1966).
8. J. Bernstein, S. Fubini, M. Gell-Mann, W. Thirring, Nuovo Cim. 17, 757 (1960); Chou Kuang-Chao,JETP12, 492 (1961).
9. G. D. Sutherland, Nuclear Phys. B2, 433 (1967).
10. N. N. Bogoliubov, Doklady Akad. Nauk 82, 217 (1952); 99, 225 (1954).
11. J. Bell, Nuovo Cim. 50A, 129 (1967) and CERN preprint, TH.810 (67).
12. For the history see e.g. G. Källén, Elementary Particle Physics, Addison-Wesley, Reading 1964.
13. For simplification from now on all factors $[(2\pi)^3 2q_o]^{-1/2}$ are omitted. Our matrix elements of the currents are therefore Feynman amplitudes. We use the metric $g_{oo} = -g_{ii} = 1$.
14. M. Gell-Mann, Physics 1, 63 (1964).
15. T. Goto and T. Imamura, Progr. Theor. Phys. 14, 396 (1955).
16. N. N. Bogoliubov and D. V. Shirkov, Introduction to the Theory of Quantized Fields, Interscience, N. Y. 1959.
17. M. Veltman, Phys. Rev. Lett. 17, 553 (1966).
18. J. S. Bell, Proc. of the 1966 CERN School of Physics, Vol. I, Geneva 1966.
19. Y. Nambu, Phys. Rev. Lett. 4, 380 (1960).
20. M. L. Goldberger and S. B. Treiman, Phys. Rev. 110, 1178 (1958); 111, 354 (1958).

21. H. T. Nieh, Phys. Rev. 164, 1780 (1967); C. W. Kim and M. Ram, John Hopkins Univ. Baltimore prepr., August 1967.

22. S. Weinberg, Phys. Rev. Lett. 18, 507 (1967).

23. See for a recent discussion and extensive literature: M. Jacob, CERN preprint, TH.846, November 1967.

24. C. Bemporad, P. L. Braccini, L. Foà, L. Lübelmeyer and D. Schmitz, Phys. Lett. 25B, 380 (1967).

25. S. L. Glashow ,H.S. Schnitzer, S. Weinberg, Phys. Rev. Lett. 19, 139 (1967).

BREAKING CHIRAL SU(3) ⊗ SU(3)†

By

S. L. GLASHOW

Department of Physics, Harvard Univ., USA

I. Introduction

We shall suppose that there are eight vector currents and eight axial vector currents which satisfy the algebra of SU(3) ⊗ SU(3) ,

$$[J_o^a(x,t), J_o^b(x',t)] = \delta^3(x-x') \, f^{abc} \, J_o^c(x,t).$$

Of these sixteen currents, four are assumed to be conserved in the real world (with the neglect of weak interactions and electromagnetism). These are the vector currents associated with isotopic spin and with hypercharge. The remaining currents are not conserved, and are often assumed to satisfy a generalized PCAC relation of the form

$$\partial_\mu \, J_\mu^a = \mu_a^2 \, F_a \, \phi^a$$

---

† Lecture given at the VII. Internationale Universitätswochen für Kernphysik, Schladming, February 26 - March 9, 1968.

where the ϕ^a are local fields with parity, hypercharge, and isospin appropriate to the current J^a and they are normalized to produce one meson states with unit amplitude

$$\langle a|\phi^a(x)|0\rangle = \frac{1}{\sqrt{(2\pi)^3}}\frac{1}{\sqrt{2E}}\, e^{ipx} \quad .$$

With such a point of view, it is evident that there must exist at least one spinless meson associated with each "broken" current: we must have an octet of pseudoscalar mesons and a scalar K meson which we call κ .

What does experiment say about the existence of such a state?

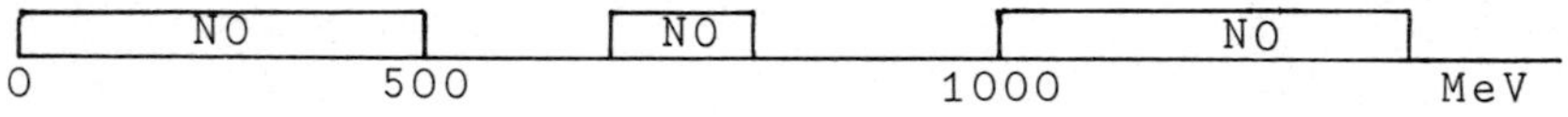

There are two "windows" in which κ may conceivably lie: in the vicinity of the K^* mass, κ would be effectively hidden; and between the K mass and $K\pi$ threshold κ would not easily have been found, for it would decay into a kaon and two photons.

Let us consider under what representation these mesons may transform, were $SU(3) \otimes SU(3)$ an exact symmetry. There are only two representations which can accomodate these particles without including particles of higher Y and T, these are the $(3,\bar{3}) + (3,\bar{3})$ and the $(8,1)+(1,8)$. (Though these are reducible representations of $SU(3) \otimes SU(3)$, they are irreducible when the discrete operation of space reflection is taken into account.) We shall concern ourselves principally with the $(3,\bar{3}) + (\bar{3},3)$, and an explicit realization of this representation is given by the nonhermitean 3×3 matrix Φ:

$$SU(3) : \quad \Phi \to U \Phi U^{-1}$$

$$\text{chiral:} \quad \Phi \to U \Phi U$$

We may write Φ in terms of its hermitean and skew-hermitean parts, and identify each as a nonet of scalar and pseudoscalar mesons, $\Phi = S + iP$. Under infinitesimal transformations, we have

$$SU(3) : \quad \delta S = i[T,S] \qquad \delta P = i[T,P]$$

$$\text{chiral} : \quad \delta S = -\{T,P\} \qquad \delta P = \{T,S\}$$

Normally, we will write these infinitesimal transformations in the form

$$\delta \phi_i = \omega_a T^a_{ij} \phi_j$$

in terms of the 16 real antisymmetric 18 × 18 matrices T^a_{ij}.

As a pure exercise, let us imagine that the effective mass Lagrangian of these mesons has the transformation properties of an invariant plus a term, itself transforming as $(3,\bar{3})$,

$$\mathcal{L}_m = \frac{m_o^2}{2}(\bar{\Phi}\Phi) + \varepsilon_{ijk}\,\varepsilon_{abc}\,(\Phi^a_i\,\Phi^b_j\,S^c_k + \text{h.c.}),$$

where S is a real matrix commuting with the isospin-hypercharge subgroup. This may be regarded as an analog of the Gell-Mann-Okubo formula for $SU(3) \otimes SU(3)$. (However, it must be stressed that the point of view now being taken is that, for the mesons, $SU(3) \otimes SU(3)$ is a good approximate symmetry in the conventional sense. This is probably absurd, for exact $SU(3)\otimes SU(3)$ invariance would predict vanishing baryon masses, and this is far from being true. We are merely playing games to develop some facility with SU(3) ⊗ SU(3).) Since the spurion matrix is determined by just two real numbers, the masses of all the scalar and pseudoscalar mesons are determined in terms of those of K,π and η . We have the pseudoscalar mass formula

$$(\eta' - \frac{4}{3} K + \frac{1}{3}\pi)(\eta - \frac{4}{3} K + \frac{1}{3}\pi) = - \frac{2}{9} (K-\pi)^2$$

which is practically identically satisfied. Moreover, all the scalar meson masses are predicted (their spectrum ist just the same as the pseudoscalar spectrum inverted about the point $\frac{1}{3}(\mu_\eta + \mu_{\eta'} + \mu_\pi)$). We find

$$\mu_{\eta_s} = 700 \text{ MeV}$$
$$\mu_\kappa = 760 \text{ MeV}$$
$$\mu_{\pi_s} = 900 \text{ MeV}$$
$$\mu_{\eta'_s} = (i)\ 300 \text{ MeV}$$

However, interesting as these numerical results may be, they are based on an unreasonable theory in which SU(3) ⊗ SU(3) is a good conventional approximate symmetry. We know that not even SU(2) ⊗ SU(2) is such a symmetry. However, many useful results may be obtained in the fake world where SU(2) ⊗ SU(2) is an exact symmetry of the Lagrangian, which is broken only by the vacuum (with the consequent appearance of massless pions). Let us make a table of some generalizations of this situation to the case of SU(3) ⊗ SU(3). Case I underlies much of the work we will consider. Case II is the familiar world of chiral SU(2) and case III is the world of the perfect eightfold-way.

Possible Idealized Worlds

	Sym. of Lagrange	Sym. of Vacuum	
	$SU(3)\otimes SU(3)$	$SU(3)\otimes SU(3)$	massless baryons 18 massive mesons
	$SU(3)\otimes SU(3)$	$SU(3)$	massless pseudo-scalar octet degenerate baryons
	$SU(3)\otimes SU(3)$	$SU(2)\otimes SU(2)$	massless K,κ and η massless nucleons
I	$SU(3)\otimes SU(3)$	$SU(2)\otimes Y$	massless K,κ,π,η
II	$SU(2)\otimes SU(2)$	$SU(2)\otimes Y$	massless π
III	$SU(3)$	$SU(3)$	degenerate K,π,η

Let $\mathcal{L}_o(\phi,\chi)$ be an $SU(3)\otimes SU(3)$ invariant Lagrangian describing the mesons ϕ as well as any other fields χ. Let us assume that the symmetry breaking takes the simple form of a ϕ tadpole, where ϕ transforms under the $(3,\bar{3})+$ $+(\bar{3},3)$ representation,

$$\mathcal{L} = \mathcal{L}_o(\phi,\chi) + \varepsilon_i \quad \phi_i \,(\chi) \quad .$$

To conserve parity, hypercharge, and isospin, ε_i can have only two nonzero components corresponding to η_s and η'_s. We define new fields according to $\phi = \hat{\phi}+\lambda$ in such a fashion that the linear terms in ϕ are removed

$$\left.\frac{\delta\mathcal{L}_o(\phi,\chi)}{\delta\phi_i}\right|_{\substack{\chi=o\\ \phi=\lambda}} = -\,\varepsilon_i \quad .$$

From the invariance of $\mathcal{L}_o$, we may write

$$\frac{\delta \mathcal{L}_o}{\delta \phi_i} T^a_{ij} \phi_j + \frac{\delta \mathcal{L}_o}{\delta \chi_\ell} T^a_{\ell m} \chi_m = 0$$

and hence, evaluated at $\chi = 0$, $\phi = \lambda$ we obtain

$$\varepsilon_i T^a_{ij} \lambda_j = 0$$

and, differentiating again, we obtain

$$\frac{\delta \mathcal{L}_o}{\delta \phi_i \delta \phi_k} T^a_{ij} \phi_j + \frac{\delta \mathcal{L}_o}{\delta \phi_i} T^a_{ik} + (\ldots) \chi_m = 0 ,$$

hence

$$\left. \frac{\delta \mathcal{L}_o}{\delta \phi_i \delta \phi_k} \right|_{\substack{\phi=\lambda \\ \chi=o}} T^a_{ij} \lambda_j - \varepsilon_i T^a_{ik} = 0$$

$$M^2_{ki} T^a_{ij} \lambda_j = T^a_{ki} \varepsilon_i \quad .$$

As we have derived this formula, it applies only to the <u>bare</u> masses, the coefficients of $\hat{\phi}_k \ \hat{\phi}_i$ in the Lagrangian. However, we will give a more general proof of this result in the next lecture.

Consider the special case of no intrinsic symmetry breaking (ε=o). Suppose, however, that there is vacuum-breaking, i.e. that $\langle \phi \rangle \neq 0$ so that $\lambda \neq 0$. Then, this becomes a proof of the Goldstone theorem and tells us which mesons are massless.

II. Meson Propagators

In the last lecture, we obtained the relation

$$M^2_{ij} T^a_{jk} \lambda^a_k = - T^a_{ik} \varepsilon_k \qquad (1)$$

where ε were parameters measuring intrinsic symmetry breakdown, and λ was the translation which was necessary to remove linear terms in ϕ from the Lagrangian, and M^2_{ij} are the coefficients of the quadratic terms (bare masses) in the translate fields.

We now give a proof of this result from current algebra. Again, let the Lagrangian be of the form $\mathcal{L}_o + \varepsilon_i \phi_i$ where $\mathcal{L}_o$ is invariant under SU(3)$\otimes$SU(3) and ϕ transforms under the $(\bar{3},3) + (3,\bar{3})$ representation,

$$\delta_a \phi_i(y,t) = i[\int J^a_o(x,t)\, d^3x,\ \phi_i(y,t)] = T^a_{ij}\phi_j(y,t) \tag{2}$$

The exact symmetry of $\mathcal{L}$ is broken by ε so that partial conservation laws are obtained

$$\partial_\mu J^a_\mu = \varepsilon_i\, T^a_{ij}\, \phi_j \ . \tag{3}$$

Let us define the vacuum expectation values

$$<\phi_i> = \lambda_i \quad . \tag{4}$$

We immediately obtain, from the partial conservation law, the result

$$\varepsilon_i\, T^a_{ij}\, \lambda_j = 0 \quad . \tag{5}$$

Moreover , we may write

$$\Delta_{jk}(p^2) = i\int d^4x\ e^{-ipx}\ <T\{\phi_j(x),\phi_k(o)\}>_o \tag{6}$$

$$\varepsilon_i T^a_{ij}\ \Delta_{jk}(p^2) = i\int d^4x\, e^{-ipx} <T\{\partial_\mu J^a_\mu,\phi_k\}>_o \tag{7}$$

$$\varepsilon_i \, T^a_{ij} \Delta_{jk}(o) = T^a_{ij} \lambda_j \quad . \tag{8}$$

This establishes our theorem once again, where M^2 is replaced by $\Delta^{-1}(o)$ and λ is replaced by the vacuum expectation value of ϕ. We may only implement this result by making an additional physical assumption:

Assumption A: Assume that those propagators appearing in the above equation may be assumed, at low values of momentum, to be simply free particle propagators

$$\Delta^{-1}_{ij}(p^2) \simeq p^2 \delta_{ij} + M^2_{ij} \qquad (0 \leq p^2 \leq m^2) \tag{9}$$

where m^2 is greater than $\sim$1 GeV. We then obtain, in terms of the two independent ε's and λ's

$$\mu^2_\pi = \varepsilon_2/\lambda_2$$

$$\mu^2_K = (\varepsilon_1+\varepsilon_2)/(\lambda_1+\lambda_2)$$

$$\mu^2_\kappa = (\varepsilon_1-\varepsilon_2)/(\lambda_1-\lambda_2) \quad , \tag{10}$$

as well as a matrix equation involving the masses of η and η'. From the known masses of π, K, η and η' we may compute the κ mass; there are two solutions lying at 720 MeV and 890 MeV.

Because the ϕ are normalized to create one meson states with unit amplitude, we may obtain the equations

$$\partial_\mu J^\pi_\mu = \mu^2_\pi F_\pi \phi_\pi = 2\,\varepsilon_2\,\phi_\pi \qquad \text{etc.} \tag{11}$$

Since the ratio of $\varepsilon_1/\varepsilon_2$ is determined by the masses, we obtain

$$F_K/F_\pi = 1.4 \ , \quad F_\kappa/F_\pi = 0.4 \quad \text{for } \mu_\kappa = 890 \text{ MeV} \tag{12}$$

$$F_K/F_\pi = 1.8 \ , \quad F_\kappa/F_\pi = 0.8 \quad \text{for } \mu_\kappa = 720 \text{ MeV.} \tag{13}$$

(The relation $F_\kappa = F_K - F_\pi$ holds in general under A). Of course, assumption A is probably too radical, so let us consider a less stringent requirement.

Assumption B: Assume that the propagators at low p^2 are given by the expression

$$\Delta^{-1}_{ij}(p^2) = Z^{-1/2}_{i\ell}(\delta_{\ell m}\ p^2 + M^2_{\ell m})Z^{-1/2}_{mj} \quad (0 \le p^2 \le m^2) \tag{14}$$

In this case, we obtain much less information from our analysis. Aside from a partial evaluation of the wave function renormalization matrix Z, the only results that are obtained from (9) are as follows:

$$(1)\quad F^2_K\ \mu^2_K + F^2_\kappa\ \mu^2_\kappa = 3/4(F^2_\eta\ \mu^2_\eta + F^2_{\eta'}\ \mu^2_{\eta'}) + \frac{1}{4}\ F^2_\pi\ \mu^2_\pi \tag{15}$$

$$(2)\quad |F_\kappa\ \mu_\kappa| \le \big||F_K\ \mu_K| - |F_\pi\ \mu_\pi|\big| \qquad (F_K\ F_\pi \ge 0), \text{ or,}$$

$$|F_\kappa\ \mu_\kappa| \ge |F_K\ \mu_K| + |F_\pi\ \mu_\pi| \qquad (F_K\ F_\pi \le 0). \tag{16}$$

We shall return to these results when we have obtained independent information about F_K, F_π and F_κ from vector meson spectral function sum rules.

Form Factors: In order to obtain information about meson form factors, like $f^+(o)$ in $K_{\ell 3}$ decay, it is desirable to replace our hypothesis (A) or (B) by hypotheses concerning 3-point functions.

Let us define the functions $f^{\pm}_{ija}$ and g_{ijk} according to

$$\Delta^{-1}_{ik}(p^2)\ \Delta^{-1}_{j\ell}(p'^2)\int d^4x\ d^4y\ e^{ipx-ip'y}\langle T\{\phi_k(x),\phi_\ell(y),\ J^\mu_a(o)\}\rangle_o$$

$$= (p+p')^\mu f^+_{ija}(p^2,p'^2,q^2) + q^\mu f^-_{ija}(p^2,p'^2,q^2) \tag{17}$$

$$\Delta^{-1}_{i\ell}(p^2)\Delta^{-1}_{jm}(p'^2)\Delta^{-1}_{kn}(q^2)\int d^4x\ d^4y\ e^{ipx-ip'y}\ \times$$

$$\times\langle T\{\phi_\ell(x),\ \phi_m(y),\phi_n(o)\}\rangle_o\ = g_{ijk}(p^2,p'^2,q^2) \tag{18}$$

where $q = p-p'$. We now may express our smoothness assumptions in terms of these functions:

Assumption A': We assume that $f^+_{ij}(p^2,p'^2,0)$ and $g_{ijk}(p^2,p'^2,0)$ are roughly constants in the region of low momenta: $0\leq p^2,\ p'^2\leq m^2$

$$f^+_{ij}\ (p^2,p'^2,0) \simeq f^+_{ij}(0,0,0)$$

$$g_{ijk}(p^2,p'^2,0) \simeq g_{ijk}(0,0,0)\ . \tag{19}$$

Assumption B' : This coincides with assumption A' except that $g_{ijk}(p^2,p'^2,0)$ is allowed to be a linear function of its argument .

By multiplying eq. (17) by q_μ and using the equal time commutation relations and partial conservation laws we obtain, at $q^2 = 0$,

$$(p^2-p'^2)f^+_{ija}(p^2,p'^2,0) = \varepsilon_n T^a_{n\ell}\Delta^{-1}_{\ell k}(o)\ g_{ijk} + \tag{20}$$

$$+ \left[T^a \Delta^{-1}(p^2) - \Delta^{-1}(p'^2)T^a\right]_{ij}$$

If we believe assumption A' or B' we see that $\Delta^{-1}(p^2)$ can be no more than a quadratic function of p^2, $\Delta^{-1}(p^2) = Z^{-1/2}(p^2-M^2)Z^{-1/2}$. For assumption A' we may write

$$f^+_{ija} = (T^a Z)_{ij} = (Z T^a)_{ij} \tag{21}$$

and from Schur's Lemma we find that Z is a multiple of the identity. Absorbing this constant into the definition of ϕ we find that assumption A' → assumption A and that

$$f^+_{ija} = (T^a)_{ij} \tag{22}$$

so that there is no renormalization of f^+ in this approximation.

For the case of B' it is again clear that assumption B' → assumption B, but the non-renormalization theorem no longer follows.

However, let us consider equation (20) for the special case of the unmixed channels K, κ and π. Counting the three permutations of the equation involving

$$(\pi\ K\ V^K_\mu),\ (\pi\ \kappa A^K_\mu),\ (K\ \kappa\ A^\pi_\mu),$$

we shall obtain nine equations in terms of 3 f^+'s, four parameters determining $g_{K\kappa\pi}$ and two ratios of renormalization constants. We obtain the results

$$f^+_\kappa(o) = \frac{F_K^2+F_\pi^2-F_\kappa^2}{2F_K F_\pi} \tag{23}$$

and cyclic results for f^+_K and f^+_π . Moreover, we obtain the relations

$$Z_\pi^{1/2} F_\pi = Z_K^{1/2} F_K - Z_\kappa^{1/2} F_\kappa$$

$$\mu_\pi^2 Z_\pi^{-1/2} F_\pi = \mu_K^2 Z_K^{-1/2} F_K - \mu_\kappa^2 Z_\kappa^{-1/2} F_\kappa \qquad (24)$$

from which the inequality (16) readily follows. It is the hypothesis B' that we shall explore when we have information about F_K, F_π, F_κ. The principal results are equations (15), (16) and (23).

At this point, we wish to introduce a stronger hypothesis about the pseudoscalar meson mass spectrum. The inverse propagator of the pseudoscalar mesons is given by the matrix expression

$$\Delta^{-1}(p^2) = Z^{-1/2}(p^2 1 + M^2)\, Z^{-1/2}$$

Assumption C is the Gell-Mann-Okubo hypothesis for this expression. We assume that Z^{-1} and $Z^{1/2}\, M^2 Z^{-1/2}$ have transformation properties of an SU(3) singlet plus the 8th component of a unitary octet.

Another Approach: Our principal result (23) depends on two hypotheses: that the symmetry breaking transforms like $(\bar{3},3) + (3,\bar{3})$ and that the smoothness assumption B' is valid. It is interesting to note that this result may be obtained in a totally independent manner. Let us define the charges

$$Q^a(t) = \int d^3x\, J_o^a(x,t) \qquad (25)$$

which have the commutation relations

$$[J_o^a(x,t),\, Q^b(t)] = f_{abc}\, J_o^c(x,t) \qquad (26)$$

where f_{abc} are the structure constants of SU(3)⊗SU(3). Consider the expression

$$\langle d|[J_o^a, Q^b]|0\rangle = f_{abc} \langle d|J_o^c|0\rangle \tag{27}$$

and assume that it is saturated by the Feynman diagrams

and

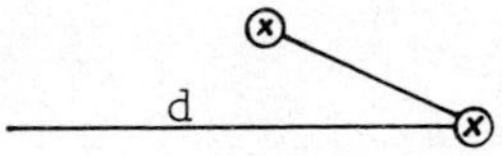

where the intermediate lines are simply scalar and pseudoscalar mesons. For simplicity let us only consider those non-zero expressions not referring to the mixed η states

$$\langle\pi|[J^K, Q^{\bar{K}}]|0\rangle = \langle\pi|J^\pi|0\rangle$$

$$\langle K|[J^\kappa, Q^\pi]|0\rangle = \langle K|J^K|0\rangle$$

$$\langle\kappa|[J^\pi, Q^K]|0\rangle = \langle\kappa|J^\kappa|0\rangle \quad . \tag{28}$$

Using the expressions

$$\langle 0|J_o^a|a\rangle = i\, F_a\, E_a \frac{1}{\sqrt{2E_a}} (2\pi)^{-3/2} e^{-ipx} \tag{29}$$

$$\langle a|J_o^b|c\rangle = (E_a + E_c)\, f_b^+((p_c - p_a)^2) + $$

$$+ (E_a - E_c)\, f_b^-((p_c - p_a)^2) \quad , \tag{30}$$

we obtain, requiring saturation, a set of coupled functional relations among the f^+'s and the f^-'s. We adopt the following Ansatz:

$$f_b^+(q^2) = \int_o^\infty \frac{\sigma_b(a^2)}{q^2-a^2}\, da^2 + f_b^+(\infty) \tag{31}$$

$$f_b^- = \int_o^\infty \frac{\sigma_b(a^2)}{a^2(q^2-a^2)} da^2(\mu_c^2 - \mu_a^2) + \frac{R_b}{q^2-\mu_b^2} \qquad (32)$$

This is, we are saying that the spin-zero form factor is given simply by a pole at the corresponding spinless meson mass, whereas the spin-one form factor is essentially arbitrary. When we put (31) and (32) into our saturation equations, we find that they may only be satisfied for

$$\sigma_a(a^2) \equiv 0 \qquad (33)$$

$$f_a^+ = f_a^+(\infty) = \frac{F_b^2+F_c^2-F_a^2}{2F_bF_c} \qquad (34)$$

and

$$R_a = \lambda F_a \qquad (35)$$

where λ is undetermined. Reasonably enough, when we put no spin-one particles into the saturation equations, we find no spin-one particle emerging in the form factors. We find constant form factors, whose magnitude agrees with eq. (23).

An interesting feature of this derivation of eq. (23) is the fact that we need make no statements about form factors off mass shell. An interesting continuation of this method would admit vector and axial mesons in the intermediate states of the saturation relations. Presumably, this would lead to momentum dependence of f^+ though we have not yet carried out this work.

III. Applications

Spectral Function Sum Rules

One way to obtain independent information about the parameters F_a is to appeal to the Weinberg spectral function sum rules as generalized to $SU(3)\otimes SU(3)$. We use only those sum rules which may be proven by assuming the algebra of fields, and we use the hypothesis of one particle dominance for the spectral functions.

In terms of the spin-one coupling constants

$$g_A^2 = (2\pi)^3\ 2E_A |\langle 0|J_\mu^a|A\rangle|^2 \tag{36}$$

where $|A\rangle$ denotes a state of one vector or axial vector meson, the first spectral function sum rule becomes

$$\begin{aligned} g_\rho^2 M_\rho^{-2} &= g_{K^*}^2 M_{K^*}^{-2} + F_\kappa^2 = g_\omega^2 M_\omega^{-2} + g_\phi^2 M_\phi^{-2} = \\ &= g_{A_1}^2 M_{A_1}^{-2} + F_\pi^2 = g_{K_A}^2 M_{K_A}^{-2} + F_K^2 = \\ &= g_D^2 M_D^{-2} + g_E^2 M_E^{-2} + F_\eta^2 + F_{\eta'}^2 = 2F_\pi^2 \end{aligned} \tag{37}$$

where $A_1(1080)$, $K_A(1250$ or $1360)$, $D(1286)$ and $E(1410)$ are the conjectured members of a $J = 1^+$ nonet. The second spectral function sum rule applies only to states of the same isospin and hypercharge, and yields

$$g_\rho^2 = g_{A_1}^2 = g^2$$

$$g_{K^*}^2 = g_{K_A}^2 = f^2 \quad , \quad g_D^2 + g_E^2 = g_\omega^2 + g_\phi^2 = h^2 \tag{38}$$

We may extract from (37) and (38) the following useful relation between F_K, F_κ and F_π

$$\frac{2F_\pi^2 - F_\kappa^2}{2F_\pi^2 - F_K^2} = \frac{M_{K_A}^2}{M_{K^*}^2} = \begin{cases} 1.98 \ (M_{K_A} = 1250 \text{ MeV}) \\ 2.30 \ (M_{K_A} = 1360 \text{ MeV}) \end{cases} \quad (39)$$

We have quoted only the relevant consequences of the sum rules - detailed discussions of their derivations may be found in the literature. Another hypothesis, first put forward by Okubo et al., is the requirement that the vector or axial vector coupling constants satisfy the Gell-Mann-Okubo formula

$$f^2 = 3/4 \ h + 1/4 \ g^2 \ . \quad (40)$$

This relation replaces the suggestion that $f^2=g^2=h^2$ which was shown to be untenable by Sakurai and others.

Weak Decay of Mesons

Let us assume the conventional Cabibbo model of weak interactions where the current is given by

$$J_\mu^W = L_\mu + \cos\Theta(V_\mu^1 + iV_\mu^2 + A_\mu^1 + iA_\mu^2) +$$
$$+ \sin\Theta(V_\mu^4 + iV_\mu^5 + A_\mu^4 + iA_\mu^5) \quad (41)$$

where L_μ is the usual lepton current and V_μ^i and A_μ^i denote the vector and axial currents of $SU(3)\otimes SU(3)$. Then, from a measurement of the rates of $K \to \mu\nu$ and $\pi \to \mu\nu$ we may determine the quantity $F_K/F_\pi \tan\Theta$. $\tan\Theta$ is not independently measurable in a manner that is free from renormalization effects. However, the comparison of $K_{\ell 3}$ rate to muon decay yields

$f^+(o)\tan\Theta$. Comparing these observations, we obtain

$$F_K/(F_\pi f^+(o)) = 1.28 \pm 0.06 \quad . \qquad (42)$$

Using equations (23), (39) and (42), we can now determine the ratios F_κ/F_π, F_K/F_π and $f^+(o)$

$$(F_K/F_\pi)^2 = 1.17$$

$$(F_\kappa/F_\pi)^2 = 0.34 \qquad \text{for } M_{K_A} = 1250 \text{ MeV}$$

$$f^+(0) = 0.85$$

or

$$(F_K/F_\pi)^2 = 1.25$$

$$(F_\kappa/F_\pi)^2 = 0.28 \qquad \text{for } M_{K_A} = 1360 \text{ MeV}$$

$$f^+(0) = 0.88$$

Note that we anticipate a departure from the non-renormalization of $f^+(o)$ of order 15%. This value of $f^+(o)$ also determines cos Θ . This result may be compared to that obtained from comparing muon decay and G_A. According to Prof. Sirlin, our value of Θ seems to be consistent with an intermediate vector meson of mass ∿ 10 GeV.

Decays of Vector Mesons into Lepton-Pairs

The vector mesons φ, ω and ρ^o may all decay into $e\bar{e}$ (or $\mu\bar{\mu}$) pairs electromagnetically. These decay modes may be computed in terms of the coupling constants g_A, if it is assumed that the electromagnetic current is a member of a unitary octet, i.e. if there is no SU(3) singlet part of the electromagnetic current. The result is

$$\Gamma_\rho \propto 3g_\rho^2 \, m_\rho^{-3}$$

$$\Gamma_\phi \propto g_\phi^2 \, m_\phi^{-3}$$

$$\Gamma_\omega \propto g_\omega^2 \, m_\omega^{-3} \tag{41}$$

as is easily verified by dimensional argument, the factor 3 being an SU(3) coefficient. From the first spectral function sum rule, eq. (37), we obtain

$$m_\phi \, \Gamma_\phi + m_\omega \, \Gamma_\omega = \frac{1}{3} m_\rho \, \Gamma_\rho \ , \tag{42}$$

a result which may be directly compared with experiment when data concerning Γ_ω and Γ_ϕ becomes available. It is interesting to speculate on the possible existence of an SU(3) singlet current. We must introduce a new conserved vector current K which is a unitary singlet, so that

$$J^Q = \frac{1}{2}(J^3 + \frac{1}{\sqrt{3}} J^8) + K \ , \tag{43}$$

and the new coupling constants

$$f_\omega \sim \langle 0|K|\omega\rangle$$

$$f_\phi \sim \langle 0|K|\phi\rangle \ . \tag{44}$$

We assume that the first spectral function sum rule is satisfied for the off-diagonal combination. $\langle 0|K(J^Q-K)|0\rangle$, which is again just the requirement of unitary singlet Schwinger terms. We obtain

$$m_\omega^{-2} f_\omega g_\omega + m_\phi^{-2} f_\phi g_\phi = 0 , \tag{45}$$

so that we may write f_ω and f_ϕ in terms of a single parameter

$$f_\omega = \lambda \frac{m_\omega^2}{g_\omega}$$

$$f_\phi = -\lambda \frac{m_\phi^2}{g_\phi}$$

to obtain

$$m_\phi \Gamma_\phi + m_\omega \Gamma_\omega = \frac{1}{3} m_\rho \Gamma_\rho + \lambda^2 \left(\frac{m_\omega^2}{g_\omega^2} + \frac{m_\phi^2}{g_\phi^2}\right) > \frac{1}{3} m_\rho \Gamma_\rho . \tag{46}$$

Thus, if the decay rates of ω and ϕ are found to be too large to satisfy (42), we would have evidence for the existence of a singlet current. Let us, for the moment, put aside this possibility and assume that there is no such singlet current.

We may obtain further information about these rates by making use of the eq. (40), the Gell-Mann-Okubo-like hypothesis

$$g_\omega^2 + g_\phi^2 = \frac{4}{3} f^2 - \frac{1}{3} g^2 = \tag{47}$$

$$= 2 F_\pi^2 \left(\frac{4}{3} M_{K^*}^2 - \frac{1}{3} M_\rho^2\right) - \frac{4}{3} F_\kappa^2 M_{K^*}^2 .$$

This result is precisely that of Das, Mathur and Okubo, with the exception of the appearance of F_κ^2 which they neglect. Combining eq. (47) with eq. (42), we obtain (with $F_\kappa^2 = 0.34$)

$$\Gamma_\omega / \Gamma_\phi = 6.4 \tag{48}$$

which, surprisingly, indicates that ω should decay into leptons more copiously than ϕ . On the other hand, with the neglect of F_κ^2 , one obtains

$$\Gamma_\omega/\Gamma_\phi = 0.9 \qquad (49)$$

so that the existence of $F_\kappa^2 \neq 0$ makes quite a lot of difference.

Since this result is so very sensitive to F_κ^2, it is worthwhile to point out that the errors in M_{K_A} and in leptonic decay rates both contribute uncertainties, so we merely know

$$0.44 \geq F_\kappa^2 \geq 0.22$$

In the following illustration (fig. 1), eq. (42) becomes a band because of uncertainty in $\Gamma(\rho \to e\bar{e})$, and the predicted values of Γ_ω and Γ_ϕ must lie in the trapezoidal area bounded by this band and the lines corresponding to extreme values of F_κ^2 .

Properties of κ

Evidently, one of the most startling predictions of our formalism is the existence of a scalar kaon. We can say something about its mass from the inequalities we have deduced, making use of our results for $F_\kappa^2 = 0.34\ F_\pi^2$ and $F_K^2 = 1.17\ F_\pi^2$. We obtain

$$\mu_\kappa \leq 670 \text{ MeV for } F_K F_\pi > 0$$

$$\mu_\kappa \leq 1100 \text{ MeV for } F_K F_\pi < 0 \quad . \qquad (50)$$

If SU(3) is exact, we must have $F_\pi = F_K$, so that the second inequality is ruled out if SU(3) breaking is

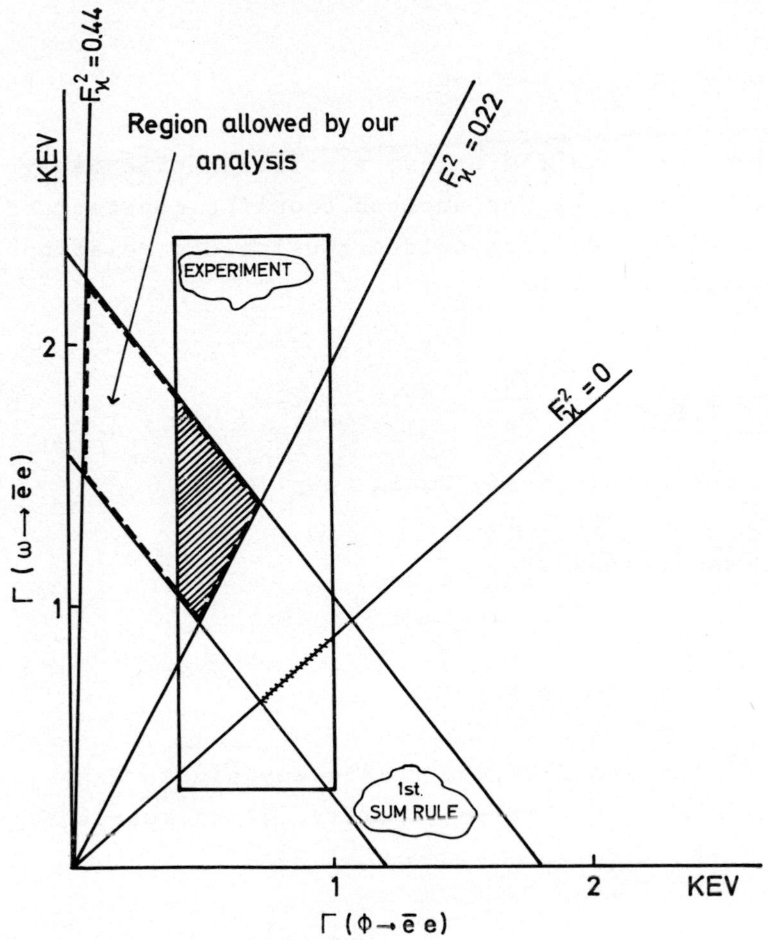

Fig. 1: Leptonic decays of ω, ϕ and ρ .

Better experiments can clearly differentiate between $F_\kappa = 0$ and $0.44 \geq F_\kappa^2 \geq 0.22$ as obtained from our analysis. (We have used

$\Gamma(\rho \to \bar{e}e) = 6.0 \pm 1.2$ KeV ,

$\Gamma(\omega \to \bar{e}e) = 1.1^{+1.3}_{-0.8}$ KeV , and

$\Gamma(\phi \to \bar{e}e) = 0.7 \pm 0.3$ KeV .)

small. Nc $K\pi$ resonance near threshold has been seen, and since κ cannot lie below the kaon mass, we may conclude

$$500 \text{ MeV} \leq \mu_\kappa \leq 630 \text{ MeV} \tag{51}$$

and that κ enjoys the unique electromagnetic decay mode $\kappa \to K + 2\gamma$. The nucleon coupling constant of κ may be computed from Goldberger-Treiman relations. We obtain

$$\frac{g_{\kappa N\Lambda}}{g_{KN\Lambda}} \simeq \frac{F_K}{F_\kappa} \; \frac{M_\Lambda - M_N}{M_\Lambda + M_N} \simeq 0.13 \tag{52}$$

indicating a relatively small coupling .

An examination of

$$K^\pm + p \to \begin{cases} K^\pm + p + \pi^o \\ \kappa^\pm + p \end{cases}$$

below K^* threshold is one relatively simple experiment to detect κ . Another possibility is the search for

$$D \text{ or } E \to \bar{K} + \kappa .$$

Finally,let us return to our formula

$$F_K^2 \, \mu_K^2 + F_\kappa^2 \mu_\kappa^2 = \frac{3}{4}(F_\eta^2 \, \mu_\eta^2 + F_{\eta'}^2 \, \mu_{\eta'}^2) + \frac{1}{4} F_\pi^2 \, \mu_\pi^2 \quad .$$

With eq. (51) we obtain the inequality

$$1.6 \leq \frac{F_\eta^2 + 3F_{\eta'}^2}{F_\pi^2} \leq 1.85 \quad . \tag{53}$$

It is interesting to see whether this inequality is consistent with one derivable from the spectral function sum rules. From (37) and (38) we obtain

$$F_{\eta}^2 + F_{\eta'}^2 = 2F_{\pi}^2 - g_D^2 M_D^{-2} - g_E^2 M_E^{-2}$$

and with (40)

$$g_D^2 + g_E^2 = g_{\omega}^2 + g_{\phi}^2 = \frac{4}{3} g_{K^*}^2 - \frac{1}{3} g_{\rho}^2 \simeq 2F_{\pi}^2(0.70 \text{ GeV}^2)$$

Thus, we again obtain an inequality

$$1.15 \leq \frac{F_{\eta}^2 + F_{\eta'}^2}{F_{\pi}^2} \leq 1.30 \tag{54}$$

which is consistent with (53) providing $F_{\eta'}^2$ is quite small and $F_{\eta}^2 \simeq F_{\pi}^2$.

We have not yet discussed the implications of assumption C, the Gell-Mann-Okubo hypothesis for pseudoscalar mesons. S. S. Shei has studied this problem and finds that it may only be satisfied for a κ- mass in the vicinity of 575 MeV.

References

T. Das, V. Mathur, and S. Okubo, Phys. Rev. Letters 18, 761 (1967).

T. Das, V. Mathur, and S. Okubo, Phys. Rev. Letters 19, 470 (1967).

S. Glashow, H. Schnitzer, and S. Weinberg, Phys. Rev. Letters 19, 139 (1967).

S. Glashow, and S. Weinberg, Phys. Rev. Letters 20, 224 (1968).

S. Glashow, Proceedings of 1967 Summer School at Erice (In Press).

R. Oakes and J. Sakurai, Phys. Rev. Letters 19, 1266 (1967).

J. Sakurai, Phys. Rev. Letters 19, 803 (1967).

GRADIENT TERMS IN COMMUTATORS OF CURRENTS AND FIELDS[†]

By

G. KÄLLEN

Department of Theoretical Physics
University of Lund, Sweden

SUMMARY

Some elementary properties of the vacuum expectation value of a product of two vector fields (or currents) are reviewed. In particular, we emphasize that the appearance or non-appearance of certain terms involving space derivatives of a δ-function in the equal time commutator between the time component and one space component of the field is not in contradiction with the canonical formalism when the coefficient in front of this gradient term is finite. When the coefficient involves a divergent integral as happens, e.g., in electrodynamics, a certain formal contradiction exists but is removed if the theory is considered as a consistent limit of a theory with a cut-off. For the particular case of electrodynamics it is further pointed out that the coefficient of the gradient term is related to the self mass of the photon which has to vanish in a consistent theory.

[†]Lectures given at the winter schools in Karpacz and Schladming, February and March 1968.

1. Introduction

These lectures are concerned with some formal properties of the vacuum expectation values of two fields which transform like vectors under Lorentz transformations. None of the material presented is actually new and some of it is about 30 years old. Therefore, our discussion is only pedagogical.

Our basic tool is a certain representation for the vacuum expectation value of a product of two fields first described in the literature more than 15 years ago [1]. Even if this representation is rather well-known today, we here start with a derivation of it adapted to our purposes.

Consider a system described by one or more quantized fields $\phi_\alpha(x)$. We make what are nowadays standard assumptions about these fields, viz. that the system described by them is Lorentz invariant and that the mass spectrum of the theory is physically reasonable. This last phrase means that we assume the existence of one unique state, the vacuum $|0\rangle$, where the total energy and momentum both vanish. Further, every other state is assumed to have a total energy momentum vector which is time like or possibly light like with positive energy. The metric in the space of the state vectors is supposed to be positive definite. However, it should be noted that we make no assumption to the effect that the fields commute for space like separations. Also, we do not assume anything specific about the singularity of the commutator between two fields at the origin. These properties of the fields are rather the subject of our investigation and among other things we want to discuss the general behaviour of the singularity at the origin of some of the commutators of our field.

Assuming that we are working in the Heisenberg representation where the field operators are time dependent and the state vectors independent of the time coordinate, the invariance under Lorentz transformations or - more specially under space and time translations tells us that the commutator of the total energy momentum operators P_μ and an arbitrary field $\phi_\alpha(x)$ gives essentially the gradient of the field operator

$$[P_\mu, \phi_\alpha(x)] = i \frac{\partial \phi_\alpha(x)}{\partial x_\mu} \quad . \tag{1.1}$$

We note, in particular, that this statement is independent of the transformation properties of the fields $\phi_\alpha(x)$. Further, the energy momentum operators P_μ all commute

$$[P_\mu, P_\nu] = 0 \quad , \tag{1.2}$$

and we find that it is possible to introduce a representation in the space of the state vectors where each state $|a\rangle$ is an eigenstate of all the energy momentum operators

$$P_\mu |a\rangle = p_\mu^{(a)} |a\rangle \quad . \tag{1.3}$$

Here, the eigenvalue $p_\mu^{(a)}$ has the physical meaning of being the total energy momentum vector of the state $|a\rangle$. Therefore, our assumption about a reasonable mass spectrum implies

$$p_\mu^{(o)} = 0 \quad , \tag{1.4a}$$

$$p_\mu^{(a)2} = \bar{p}^{(a)2} - p_o^{(a)2} \leq 0 \quad , \tag{1.4b}$$

$$p_o^{(a)} > 0 \quad \text{for} \quad |a> \neq |0> \tag{1.4c}$$

Taking the matrix element of eq. (1.1) between two states $|a>$, $|b>$ and using eq. (1.3) we get

$$<a|[P_\mu,\phi_\alpha(x)]|b> = (p_\mu^{(a)} - p_\mu^{(b)})<a|\phi_\alpha(x)|b> =$$
$$= i \frac{\partial}{\partial x_\mu} <a|\phi_\alpha(x)|b> \quad . \tag{1.5}$$

This simple differential equation determines the coordinate dependence of the matrix element of the field and we find quite generally

$$<a|\phi_\alpha(x)|b> = \exp\{i(p^{(b)}-p^{(a)})x\}<a|\phi_\alpha|b> \quad , \tag{1.6}$$

where the quantity $<a|\phi_\alpha|b>$ depends on the field and on the states $|a>$ and $|b>$ but is independent of the coordinate x . We note, in particular, that the plane wave behaviour of the matrix element in eq. (1.6) is true also for interacting fields. At the first moment this may appear to be physically unreasonable as we are used to having plane waves for non-interacting systems in elementary quantum mechanics but also to the fact that interactions usually introduce more complicated wave functions. However, here it should be remarked that the coordinate dependence in eq. (1.6) refers only to the total energy momentum vector and, therefore, to the center of mass behaviour of the system.

Next,we consider the vacuum expectation value of a product of two fields

$$F_{\alpha\beta}(x,x') = <0|\phi_\alpha(x)\phi_\beta(x')|0> \quad . \tag{1.7}$$

Note that the quantity under investigation here is

just the ordinary product of the two field operators and not, e.g., the time ordered product extensively used in perturbation theory. To analyze the structure of the expression (1.7) we introduce a complete set of intermediate states between the two operators and write, using eq. (1.6)

$$F_{\alpha\beta}(x,x') = \sum_{|n>} <0|\phi_\alpha(x)|n><n|\phi_\beta(x')|0> =$$

$$= \sum_{|n>} <0|\phi_\alpha|n><n|\phi_\beta|0> \exp\{ip^{(n)}(x-x')\} \ . \tag{1.8}$$

The summation over the intermediate states $|n>$ in eq. (1.8) goes over any complete set of states and we have chosen a particular set where each state is an eigen-state of the total energy momentum vector. However, it should be noted that, in general, a state is not completely characterized by its total energy and momentum. In physical situations where particles can be scattered, created and annihilated, there will normally be very many states which all have the same total energy and momentum. Therefore, we perform the summation in eq. (1.8) in such a way that we first sum over all states which have the same total energy momentum vector p and, afterwards, sum over all energies and momenta. Noting that eq. (1.8) also contains the information that the function $F_{\alpha\beta}(x,x')$ in reality only depends on the coordinate difference x-x' we get

$$F_{\alpha\beta}(x-x') = \frac{1}{(2\pi)^3} \int dp\ e^{ip(x-x')}\ G_{\alpha\beta}(p) \quad , \tag{1.9}$$

$$G_{\alpha\beta}(p) = V \sum_{p^{(n)}=p} <0|\phi_\alpha|n><n|\phi_\beta|0> \ . \tag{1.9a}$$

In the first of these two equations we have replaced

the summation over all energy momentum vectors p by an integral. The symbol V in eq.(1.9a) is the volume of periodicity used in the quantization of our system and the convention for replacing sums by integrals is

$$\lim_{V\to\infty} \frac{1}{V} \sum_{\bar{p}} f(\bar{p}) = \frac{1}{(2\pi)^3} \int d^3p \, f(\bar{p}) \quad . \tag{1.10}$$

In what follows we shall frequently interchange summations and integrations according to the prescription (1.10).

The function $G_{\alpha\beta}(p)$ in eq. (1.9) vanishes according to eq. (1.9a) and our assumption about the physical mass spectrum if the vector p is space like or if the energy p_o is negative.

We close this section by proving that the function $G_{\alpha\beta}(p)$ transforms like the product of the two fields ϕ_α and ϕ_β under proper Lorentz transformations. To see this we assume, as is usual in a quantized field theory, that the fields $\phi_\alpha(x)$ under a Lorentz transformation Λ transform with the aid of matrices $U(\Lambda)$ which are unitary for proper Lorentz transformations not involving time reflections

$$x'_\mu = \Lambda_{\mu\nu} \, x_\nu \quad , \tag{1.11a}$$

$$U(\Lambda)\phi_\alpha(x) \, U^{-1}(\Lambda) = S_{\alpha\beta}(\Lambda^{-1})\phi_\beta(\Lambda x) \quad . \tag{1.11b}$$

The matrices $S_{\alpha\beta}$ in eq. (1.11b) are numerical matrices characteristic of the fields considered. For instance, if we have only one scalar field in our model, the matrix S is the unit matrix. If we have only one vector field, it is the matrix Λ itself and so on. Further, the state vectors transform under these transformations as follows:

$$U(\Lambda)|0\rangle = |0\rangle \quad , \tag{1.12a}$$

$$U(\Lambda)|p\rangle = |\Lambda p\rangle \, s(\Lambda,p) \quad . \tag{1.12b}$$

The quantity $s(\Lambda,p)$ in eq. (1.12b) is another numerical matrix acting on the indices describing the total angular momentum of the state under consideration. It can be constructed explicitly but its detailed form is not of interest for our present purposes. Here, we only remark that $s(\Lambda,p)$ is always unitary. In general, this is not the case for the matrix $S(\Lambda)$.

From the formulae above we get

$$\langle 0|\phi_\alpha|n\rangle\Big|_{p^{(n)}=p} =$$

$$= \langle 0|U^{-1}(\Lambda)U(\Lambda)\phi_\alpha(0)U^{-1}(\Lambda)U(\Lambda)|n\rangle\Big|_{p^{(n)}=p} =$$

$$= S_{\alpha\alpha'}(\Lambda^{-1})\langle 0|\phi_{\alpha'}|n\rangle\Big|_{p^{(n)}=\Lambda p} \quad s(\Lambda,p) \quad , \tag{1.13a}$$

$$\langle n|\phi_\beta|0\rangle\Big|_{p^{(n)}=p} =$$

$$= S_{\beta\beta'}(\Lambda^{-1})s^{+}(\Lambda,p)\langle n|\phi_{\beta'}|0\rangle\Big|_{p^{(n)}=\Lambda p} \quad . \tag{1.13b}$$

Using the unitarity of the matrix $s(\Lambda,p)$ we find the following formula describing the transformation properties of the function $G_{\alpha\beta}(p)$ under Lorentz transformations

$$G_{\alpha\beta}(p) = S_{\alpha\alpha'}(\Lambda^{-1})\, S_{\beta\beta'}(\Lambda^{-1})\, G_{\alpha'\beta'}(\Lambda p) \quad . \tag{1.14}$$

2. The Appearance of Gradient Terms

We now specialize the formulae discussed in the previous section to the particular case of a vector field $A_\mu(x)$. In this case eqs. (1.9) read

$$F_{\mu\nu}(x-x') = \langle 0|A_\mu(x)\, A_\nu(x')|0\rangle =$$

$$= \frac{1}{(2\pi)^3} \int dp\; e^{ip(x-x')}\; G_{\mu\nu}(p) \quad , \tag{2.1}$$

$$G_{\mu\nu}(p) = V \sum_{p^{(n)}=p} \langle 0|A_\mu|n\rangle\langle n|A_\nu|0\rangle \quad . \tag{2.1a}$$

Equation (1.14) tells us that the function $G_{\mu\nu}(p)$ transforms as a second rank tensor under Lorentz transformations. As it depends only on the vector p, its most general form is

$$G_{\mu\nu}(p) = \delta_{\mu\nu} A(p) + p_\mu p_\nu\; B(p) \quad , \tag{2.2}$$

where the amplitudes A and B are functions of p which are invariant under Lorentz transformations without time reflections. To take account of the fact that these functions are different from zero only when the energy is positive, we introduce the step function $\Theta(p)$ defined by

$$\Theta(p) = \frac{1}{2}\left[1 + \frac{p_0}{|p_0|}\right] \quad . \tag{2.3}$$

Using a slight redefinition of our amplitudes in

(2.2) we can then write

$$G_{\mu\nu}(p) = [\delta_{\mu\nu}A(p^2) + p_\mu p_\nu \; B(p^2)] \; \Theta(p) \quad , \qquad (2.4)$$

$$A(p^2) = 0 \quad \text{for } p^2 = \bar{p}^2 - p_o^2 > 0 \; , \qquad (2.4a)$$

$$B(p^2) = 0 \quad \text{for } p^2 > 0 \; . \qquad (2.4b)$$

We note that the new amplitudes depend only on the invariant square of the energy momentum vector p. According to our assumptions about the physical mass spectrum of the theory, the amplitudes further vanish if p is space like. For future applications it will be convenient not to use the amplitudes $A(p^2)$ and $B(p^2)$ themselves but rather linear combinations $G_1(p^2)$ and $G_2(p^2)$ defined as follows

$$G_{\mu\nu}(p) = \{(p_\mu p_\nu - \delta_{\mu\nu}p^2) \; G_1(p^2) + \delta_{\mu\nu}G_2(p^2)\}\Theta(p) \; , \qquad (2.5)$$

$$G_i(p^2) =) \quad \text{for} \quad p^2 > 0 \; ; \quad i=1,2 \; . \qquad (2.5a)$$

This redefinition has the advantage that in the particular case when the vector field $A_\mu(x)$ is divergence-less or fulfills the condition

$$\frac{\partial \; A_\mu(x)}{\partial x_\mu} = 0 \quad , \qquad (2.6)$$

we get

$$<0|\frac{\partial A_\mu(x)}{\partial x_\mu} \; A_\nu(x')|0> =$$

$$= \frac{i}{(2\pi)^3} \int dp \; e^{ip(x-x')} \; p_\mu \; G_{\mu\nu}(p \;) = 0 \; , \qquad (2.7)$$

or

$$p_\nu G_{\mu\nu}(p) = p_\mu G_2(p^2)\,\Theta(p) = 0 \quad . \qquad (2.7a)$$

Consequently, the function $G_2(p^2)$ vanishes for all values of p^2 when eq. (2.6) holds.

We now proceed to calculate the vacuum expectation value of the commutator of the two vector fields in eq. (2.1).

$$<0|[A_\mu(x), A_\nu(x')]|0> =$$

$$= \frac{1}{(2\pi)^3}\int dp\; e^{ip(x-x')}[(p_\mu p_\nu - \delta_{\mu\nu}p^2)\,G_1(p^2) + $$
$$+ \delta_{\mu\nu}G_2(p^2)]\;\Theta(p) -$$

$$- \frac{1}{(2\pi)^3}\int dp\; e^{ip(x'-x)}[(p_\nu p_\mu - \delta_{\mu\nu}p^2)\,G_1(p^2) +$$
$$+ \delta_{\nu\mu}G_2(p^2)]\;\Theta(p) =$$

$$= - \frac{1}{(2\pi)^3}\int dp\; e^{ip(x'-x)}[(p_\mu p_\nu - \delta_{\nu\mu}p^2)\,G_1(p^2) +$$
$$+ \delta_{\mu\nu}G_2(p^2)]\;\varepsilon(p) \quad , \qquad (2.8)$$

$$\varepsilon(p) = \Theta(p) - \Theta(-p) = p_0/|p_0| \quad . \qquad (2.8a)$$

Next, we introduce an artificial δ-function in the integral (2.8) and write

$$<0|[A_\mu(x), A_\nu(x')]|0> =$$

$$= - \frac{1}{(2\pi)^3}\int_0^\infty da\int dp\; e^{ip(x'-x)}\delta(p^2+a)[(p_\mu p_\nu + a\delta_{\mu\nu})\times$$
$$\times G_1(-a) + \delta_{\mu\nu}G_2(-a)]\;\varepsilon(p) =$$

$$= - i \int_{0}^{\infty} da \Big[G_1(-a)\Big(a\delta_{\mu\nu} - \frac{\partial^2}{\partial x_\mu \partial x_\nu}\Big) +$$

$$+ G_2(-a)\, \delta_{\mu\nu} \Big] \Delta(x'-x,a) \quad , \qquad (2.9)$$

$$\Delta(x,a) = \frac{-i}{(2\pi)^3} \int dp \; e^{ipx} \, \delta(p^2+a) \, \varepsilon(p) \quad . \qquad (2.9a)$$

We note, in particular, the range of integration of the auxiliary variable a. This quantity which physically has the dimension of the square of a mass is positive or zero because of our assumption about the physical mass spectrum of the theory. The function $\Delta(x,a)$ which appears in eq. (2.9a) is the well-known commutator function which appears in elementary field quantization. It can be evaluated explicitly and is given by

$$\Delta(x,a) =$$

$$= - \frac{\varepsilon(x)}{2\pi} \Big[\delta(x^2) - \frac{1}{2}\, \theta(-x^2) \sqrt{a/-x^2} \; J_1(\sqrt{-a\, x^2}) \Big] . \qquad (2.10)$$

The function $J_1(z)$ is the conventional Bessel function of order one. Here, we are not so much interested in the explicit form of the function $\Delta(x,a)$ as in the following two properties which are easily derived either from the definition (2.9a) or from the explicit form (2.10)

$$\Delta(x,a) = 0 \quad \text{for} \quad x^2 > 0, \qquad (2.11a)$$

$$\frac{\partial}{\partial x_0} \Delta(x,a) \Big|_{x_0=0} = - \delta(\bar{x}) \quad . \qquad (2.11b)$$

We note the well-known but rather remarkable result that the right-hand side of eq. (2.9) vanishes for space like, non-zero values of x'-x because of eq. (2.11a). We remember that the derivation of eq. (2.9) did not use any assumption about commutativity of the field operators for space like separations or the canonical formalism.

Specializing eq. (2.9) to the case $\mu=4$ and $\nu=k\neq 4$ (and using the convention $A_4(x) = i\, A_o(x)$) we have

$$<0|\left[A_o(x),\ A_k(x')\right]|0> =$$

$$= -\, i \int_o^\infty da\ G_1(-a)\ \frac{\partial^2}{\partial x_o \partial x_k}\ \Delta(x'-x,a) \quad . \qquad (2.12)$$

Putting the two times x_o and x'_o equal we get

$$<0|\left[A_o(x),A_k(x')\right]|0>\Big|_{x_o=x'_o} =$$

$$= -\, i\ \frac{\partial}{\partial x_k}\ \delta(\bar{x}-\bar{x}')\int_o^\infty da\ G_1(-a)\ . \qquad (2.13)$$

The last step in this calculation is somewhat formal and justified only if the integral appearing on the right-hand side of eq. (2.13) is convergent. However, for the moment we assume this to be the case. The conclusion is that unless the integral over the weight function $G_1(-a)$ happens to vanish, a gradient term containing a space derivative of a three-dimensional δ -function must appear as singularity at the origin of the commutator between the two field components in eq. (2.13). We emphasize again that this result has been obtained using only invariance arguments and our assumption about the mass spectrum of the formalism but not using any starting assumption about the commutativity of our field operators, nor any details of the canonical formalism [2].

In the particular case when eq. (2.6) holds, the weight function $G_2(p^2)$ vanishes. Consequently, we have, e.g.,

$$(p_k p_\ell - \delta_{k\ell}\, p^2)\, G_1(p^2)\, \Theta(p) =$$

$$= V \sum_{p^{(n)}=p} \langle 0|A_k|n\rangle\langle n|A_\ell|0\rangle \quad , \; k \text{ and } \ell \neq 4 \; . \qquad (2.14)$$

Putting $k = \ell$ and summing from one to three we find

$$G_1(p^2)\, \Theta(p) =$$

$$= \frac{V}{\bar{p}^2 - 3p^2} \sum_{p^{(n)}=p} \langle 0|A_k|n\rangle\langle n|A_k|0\rangle > 0 \quad . \qquad (2.15)$$

In deriving the positive definite property of the right hand side of eq. (2.15) we use that p is time like. This tells us that the denominator is positive definite. Second, we assume that the space components of the vector field are Hermitian operators. Unless all the matrix elements between the vacuum and the state $|n\rangle$ of the space components of the vector field vanish identically, the function $G_1(p^2)$ is therefore positive definite. Consequently, the integral in eq. (2.13) which, by assumption, is convergent can never vanish. Summarizing, we have found that the gradient term in eq. (2.13) must appear with a non-vanishing coefficient if we assume that the vector field $A_\mu(x)$ fulfills the continuity equation (2.6) and corresponds to a real field. At the first moment this result may appear rather astonishing. We are used to canonical field theory where one of the basic assumptions is that dynamically independent components of the fields commute for equal times and, in particular, at the origin. The result (2.13) appears to be in contradiction with this intuitive idea. The main point which we shall try to make in the discussion to follow is that the result (2.13) is not in contradiction with the canonical quantization method in all elementary examples one can think of. We shall show this by discussing explicitly several examples starting with the elementary case of a free vector field and progressing in steps to the more complicated case of the electromagnetic field in interaction with a spin 1/2 field. Even if the argument we present contains no general proof that the result (2.13) is always consistent with the canonical formalism, we want to make it plausible that such is the case.

3. The Free Vector Field with Non-Vanishing Mass

We start by discussing in some detail the case of a vector field without any interaction but with a non-vanishing mass [3]. The Lagrangian for such a theory is given by

$$\mathcal{L} = -\frac{1}{4} F_{\mu\nu} F_{\mu\nu} - \frac{1}{2} m^2 A_\mu A_\mu \quad , \tag{3.1}$$

$$F_{\mu\nu} = \frac{\partial A_\nu}{\partial x_\mu} - \frac{\partial A_\mu}{\partial x_\nu} \quad . \tag{3.1a}$$

Considering the four components of the vector field as independent quantities and requiring the action integral to be stationary we find in the standard way the usual equations of motion

$$\frac{\partial F_{\mu\nu}}{\partial x_\mu} - m^2 A_\nu = (\Box - m^2)\, A_\nu - \frac{\partial}{\partial x_\nu} \frac{\partial A_\mu}{\partial x_\mu} = 0 \quad . \tag{3.2}$$

Differentiating this result with respect to x_μ and summing over μ we find after making use of the assumption that the mass m is different from zero

$$\frac{\partial A_\nu}{\partial x_\nu} = 0 \quad . \tag{3.3}$$

Consequently, eq. (3.2) simplifies and we find that our equations of motion become eq. (3.3) together with the relation

$$(\Box - m^2)A_\mu = 0 \quad . \tag{3.4}$$

We now turn to the canonical quantization procedure. To this purpose we first calculate the momentum canonically conjugate to the field A_μ

$$\Pi_\mu(x) = \frac{\partial \mathcal{L}}{\partial(\partial A_\mu/\partial x_o)} = i\ F_{4\mu} = \frac{\partial A_\mu}{\partial x_o} + \frac{\partial A_o}{\partial x_\mu} \quad . \tag{3.5}$$

We note, in particular, that the momentum canonically conjugate to the field $A_4(x)$ vanishes identically. Consequently, the component $A_4(x)$ cannot be considered to be an independent field when we use the canonical formalism. However, both this quantity and its time derivative can be expressed in terms of the other fields and the canonical momenta $\Pi_k(x)$ with $k \neq 4$ in eq. (3.5). Actually putting $\mu=4$ in eq. (3.2) we have

$$A_4(x) = \frac{1}{m^2}\frac{\partial F_{\mu 4}}{\partial x_\mu} = \frac{1}{m^2}\frac{\partial F_{k4}}{\partial x_k} = \frac{i}{m^2}\frac{\partial \Pi_k(x)}{\partial x_k} \quad , \tag{3.6}$$

On the other hand, we get directly from eq. (3.3)

$$\frac{\partial A_4(x)}{\partial x_o} = -\ i\ \frac{\partial A_k(x)}{\partial x_k} \quad . \tag{3.6a}$$

Here and below we use the convention that a summation index with a Latin letter like k,ℓ etc. always goes from one to three while asummation index with a Greek letter μ,ν etc. goes from one to four. Therefore, both the field $A_4(x)$ itself and its time derivative can be expressed in terms of the three-dimensional divergence of the other components of the field and their canonical momenta. Our quantization conditions now become

$$[A_k(x), A_\ell(x')]_{x_o=x'_o} = [\Pi_k(x), \Pi_\ell(x')]_{x_o=x'_o} = 0 \ , \tag{3.7a}$$

$$[\Pi_k(x), A_\ell(x')]_{x_o=x'_o} = -i\ \delta_{k\ell}\ \delta(\bar{x}-\bar{x}') \quad . \tag{3.7b}$$

Taking the divergence of eqs. (3.7b) and using eq. (3.6) we find directly

$$[A_4(x), A_\ell(x')]_{x_o=x'_o} = \frac{i}{m^2}\Big[\frac{\partial \Pi_k(x)}{\partial x_k}\ ,\ A_\ell(x')\Big]_{x_o=x'_o} =$$

$$= \frac{\delta_{k\ell}}{m^2}\frac{\partial}{\partial x_k}\ \delta(\bar{x}-\bar{x}') - \frac{1}{m^2}\ \frac{\partial\delta(\bar{x}-\bar{x}')}{\partial x_\ell} \ . \tag{3.8}$$

Replacing the field $A_4(x)$ by the real operator $A_o(x)$ we find instead of eq. (3.8)

$$[A_o(x),\ A_k(x')]_{x_o=x'_o} = \frac{-i}{m^2}\ \frac{\partial}{\partial x_k}\ \delta(\bar{x}-\bar{x}') \quad . \tag{3.9}$$

This result should be compared with eq. (2.13) and we see that the gradient term which was derived in section 2 using only general arguments also appears in this section as a consequence of the canonical formalism. Therefore, there is no immediate contradiction between the two alternative ways of calculating the equal time commutator between the two components of the vector field.

It remains to be checked that the coefficients on the right hand sides of eqs. (3.9) and (2.13) are the same. To do this we have to calculate the function $G_1(p^2)$ which appears as integrand in eq. (2.13). For this purpose we remark that the field operators $A_\mu(x)$ after the canonical quantization has been performed can be expanded in terms of annihilation and creation operators in the usual way. This expansion reads

$$A_k(x) = \frac{1}{\sqrt{V}} \sum_{\bar{p}} \frac{1}{\sqrt{2\omega}} \{ e^{ipx} [\sum_{\lambda=1}^{2} e_k^{(\lambda)}(\bar{p})\, a^{(\lambda)}(\bar{p}) + \frac{\omega}{m} e_k^{(3)}(\bar{p})\, a^{(3)}(\bar{p})] + e^{-ipx} [\sum_{\lambda=1}^{2} e_k^{(\lambda)}(\bar{p}) a^{*(\lambda)}(\bar{p}) + \frac{\omega}{m} e_k^{(3)}(\bar{p})\, a^{*(3)}(\bar{p})] \} , \qquad (3.10)$$

$$A_o(x) = \frac{1}{\sqrt{V}} \sum_{\bar{p}} \frac{1}{\sqrt{2\omega}} \frac{|\bar{p}|}{m} [e^{ipx}\, a^{(3)}(\bar{p}) + e^{-ipx}\, a^{*(3)}(\bar{p})] \quad . \qquad (3.10a)$$

As before, the symbol V stands for the volume of quantization while ω is the energy of a particle with momentum $\bar{p}$; $\omega = \sqrt{\bar{p}^2+m^2}$. The vectors $e_k^{(\lambda)}(\bar{p})$ are three-dimensional polarization vectors with the convention that the case $\lambda = 1$ or 2 corresponds to transverse polarizations while $\lambda = 3$ corresponds to longitudinal polarization. The operators $a^{(\lambda)}(\bar{p})$ and $a^{*(\lambda)}(\bar{p})$ are annihilation and creation operators fulfilling the standard commutation relations

$$[a^{(\lambda)}(\bar{p}),\ a^{*(\lambda')}(\bar{p}')] = \delta_{\lambda\lambda'}\ \delta_{\bar{p},\bar{p}'} \tag{3.11}$$

All other commutators vanish.

The somewhat complicated form of the coefficients in the expansions (3.10) and (3.10a) are caused by the algebraic relations (3.6) and (3.6a). By straightforward substitution it is easy to verify that these two equations as well as the canonical commutation relations (3.7) are fulfilled. On the other hand, it is also straight-forward and elementary to substitute arbitrary coefficients in the expansion (3.10) and (3.10a) and to show that the canonical quantization scheme together with equations (3.6) determine these coefficients uniquely to the values given in eqs. (3.10). From here we find by direct substitution the following formula

$$<0|A_k(x)\ A_\ell(x')|0> =$$

$$= \frac{1}{(2\pi)^3} \int \frac{d^3p}{2\omega}\, e^{ip(x-x')} \Big[\sum_{\lambda=1}^{2} e_k^{(\lambda)}(\bar{p})\ e_\ell^{(\lambda)}(\bar{p}) + \frac{\omega^2}{m^2}\, e_k^{(3)}(\bar{p})\ e_\ell^{(3)}(\bar{p}) \Big]$$

$$= \frac{1}{(2\pi)^3} \int dp\ e^{ip(x-x')}\delta(p^2+m^2)\ \Theta(p) \Big[\delta_{k\ell} + \Big(\frac{\omega^2}{m^2} - 1\Big) \frac{p_k p_\ell}{\bar{p}^2}\Big] =$$

$$= \frac{1}{(2\pi)^3} \int dp\ e^{ip(x-x')}\delta(p^2+m^2)\ \Theta(p) \Big[\delta_{k\ell} + \frac{p_k p_\ell}{m^2}\Big], \tag{3.12a}$$

$$<0|A_o(x)\ A_o(x')|0> =$$

$$= \frac{1}{(2\pi)^3} \int dp\ e^{ip(x-x')} \delta(p^2+m^2) \Theta(p) \frac{\bar{p}^2}{m^2} \tag{3.12b}$$

$$<0|A_o(x)\ A_k(x')|0> =$$

$$= \frac{1}{(2\pi)^3} \int dp\ e^{ip(x-x')} \delta(p^2+m^2)\ \Theta(p) \frac{\omega p_k}{m^2}\ . \tag{3.12c}$$

All these expressions can be summarized in the formula

$$<0|A_\mu(x')\ A_\nu(x')|0> =$$

$$= \frac{1}{(2\pi)^3} \int dp\ e^{ip(x-x')} \delta(p^2+m^2)\ \Theta(p) [\delta_{\mu\nu} + \frac{p_\mu p_\nu}{m^2}]\ . \tag{3.13}$$

Comparing this result with eq. (2.5) we find

$$G_1(p^2) = \frac{\delta(p^2+m^2)}{m^2}\ . \tag{3.14}$$

As expected, this expression is positive definite. The integral appearing in eq. (2.13) can be calculated and we have

$$\int_0^\infty da\ G_1(-a) = 1/m^2\ . \tag{3.15}$$

Consequently, the coefficients on the right hand sides of eqs. (2.13) and (3.9) are, indeed, identical. This establishes the consistence of the canonical quantization and the appearance of the gradient term in eq. (2.13) for the special case which we have considered here.

4. The Free Electromagnetic Field

The discussion in section 3 depends critically on two things, viz. first of all that the mass of the vector particle considered is different from zero, so that we can freely divide by it (cf., e.g., eq. (3.14) and, second, that the divergence condition (3.3) holds as an operator equation. None of these conditions are fulfilled in electrodynamics.

As is well-known, there are several ways of quantizing the electromagnetic field. One technique consists essentially of separating the field in a transverse part and a Coulomb field. Only the transverse part is quantized. Clearly, such a formalism is not explicitly Lorentz invariant and the discussion in section 2 is not immediately applicable. Alternatively, one can treat all the components of the electromagnetic potential as quantum mechanical operators but the divergence condition (2.6) does not hold as an operator equation but only as a supplementary condition. Even in this language certain formal inconsistencies remain at the first moment but can be avoided in different ways. The way we particularly want to discuss here is the technique with an indefinite metric in the space of the state vectors originally introduced by Gupta and Bleuler [4]. An alternative way would be to treat electrodynamics as a limit of a vector theory with a small photon mass. [5]. However, in this case the formalism becomes essentially identical with the discussion in the previous section [6] and we do not want to enter upon the details here.

Electrodynamics without interaction and using the Gupta-Bleuler technique can be summarized in the following way.

The Lagrangian of the system is

$$\mathcal{L} = - \frac{1}{4} F_{\mu\nu} F_{\mu\nu} - \frac{1}{2} \left(\frac{\partial A_\nu}{\partial x_\nu}\right)^2 , \qquad (4.1)$$

$$F_{\mu\nu} = \frac{\partial A_\nu}{\partial x_\mu} - \frac{\partial A_\mu}{\partial x_\nu} . \qquad (4.1a)$$

The condition that the action integral should be stationary gives

$$\Box A_\mu(x) = 0 . \qquad (4.2)$$

This equation of motion corresponds rather directly to eq. (3.4) but we have no counterpart of eq. (3.3). Further, the canonical momenta become

$$\Pi_k(x) = i F_{4k} = \frac{\partial A_k}{\partial x_o} + \frac{\partial A_o}{\partial x_k} , \qquad (4.3a)$$

$$\Pi_4(x) = i \frac{\partial A_\mu}{\partial x_\mu} . \qquad (4.3b)$$

In particular, we note that the canonical momentum $\Pi_4(x)$ does not vanish identically here. The canonical quantization scheme now gives

$$\left[A_\mu(x), A_\nu(x')\right]\Big|_{x_o=x'_o} = \left[\Pi_\mu(x), \Pi_\nu(x')\right]\Big|_{x_o=x'_o} = 0 , \qquad (4.4a)$$

$$\left[\Pi_\mu(x), A_\nu(x')\right]\Big|_{x_o=x'_o} = - i \delta_{\mu\nu} \delta(\bar{x}-\bar{x}') . \qquad (4.4b)$$

From here one can work out the expansion of the electromagnetic potential in terms of creation and annihilation operators. In this case the result is formally simpler than eqs. (3.10) and reads

$$A_\mu(x) = \frac{1}{\sqrt{V}} \sum_{\bar{p}} \frac{1}{\sqrt{2\omega}} \sum_{\lambda=1}^{4} e_\mu^{(\lambda)}(p)\{e^{ipx}\, a^{(\lambda)}(p) + e^{-ipx}\, a^{*(\lambda)}(p)\} \quad . \qquad (4.5)$$

The polarization index λ here goes from one to four. The first three polarization vectors are the same as in section 3 while the vector $e^{(4)}(p)$ has vanishing space components and a fourth component given by $e_4^{(4)}(p) = 1$ [7] . The annihilation and creation operators in eq. (4.5) fulfill the commutation relations (3.11). At the first moment, this formalism appears deficient first of all because the Lorentz condition is completely absent and, second, because the time component of the vector in the expansion (4.5) appears to have incorrect reality properties. However, in the Gupta-Bleuler method one also has a certain "metric operator" η which has the properties

$$[a^{(\lambda)}(p), \eta] = 0 \qquad \text{for} \quad \lambda \neq 4 \ , \qquad (4.6a)$$

$$\{a^{(4)}(p), \eta\} = a^{(4)}(p)\eta + \eta\, a^{(4)}(p) = 0 \ . \qquad (4.6b)$$

From these two defining equations one finds that the metric operator η is diagonal in the Fock representation of the free field and that its value is given by

$$\langle a|\eta|\, b\rangle = (-1)^{n^{(4)}}\ \delta_{ab} \quad . \qquad (4.7)$$

Here, the symbol $n^{(4)}$ denotes the total number of scalar photons ($\lambda = 4$) in the state under consideration.

In this formalism the expectation value of an arbitrary operator F for the state $|a\rangle$ is defined by

$$\langle F\rangle = \langle a|\eta F|a\rangle \quad . \tag{4.8}$$

One verifies directly from eq. (4.8) and the two equations (4.6) that the expectation values of the operator (4.5) have the correct reality properties. Finally, the Lorentz condition in this formalism reads

$$\frac{\partial A_\mu^{(+)}(x)}{\partial x_\mu}|a\rangle = 0 \quad , \tag{4.9}$$

for each physically allowed state. Here, the plus sign on the operator denotes the positive frequency part [8] .

From here we calculate the expectation value of a product of two fields and find the well-known result

$$\langle 0|\eta A_\mu(x)\ A_\nu(x')|0\rangle = \langle 0|A_\mu(x)\ A_\nu(x')|0\rangle =$$

$$= \frac{1}{(2\pi)^3}\int dp\ e^{ip(x-x')}\ \delta(p^2)\ \Theta(p)\delta_{\mu\nu} \quad . \tag{4.10}$$

A comparison with eq. (3.5) gives

$$G_1(p^2) = 0 \quad , \tag{4.11a}$$

$$G_2(p^2) = \delta(p^2) \quad . \tag{4.11b}$$

Because of eq. (4.11a) the integral on the right-hand side of eq. (2.13) vanishes identically and there is no contradiction between the general result (2.13) and the canonical quantization rule (4.4a). In a way, this is not too surprising because to prove the positive definite character of the function G_1 in eq. (2.15) it was necessary to assume first of all the divergence condition and, second, a positive definite metric in the space of the state vectors. None of these two conditions is literally fulfilled in the Gupta-Bleuler method. Therefore, it is not surprising that no formal contradiction appears in spite of the fact that there is no gradient term in the commutator of the fields.

5. The Electromagnetic Field in Interaction with Spin 1/2 Particles

We now turn to the somewhat more complicated problem where the electromagnetic field is not a free field but interacting with a Dirac field describing spin 1/2 particles. The electromagnetic field itself is treated with the Gupta-Bleuler technique as indicated in the last section. The Lagrangian for the interacting system is given by

$$\mathcal{L} = -\frac{1}{4} F_{\mu\nu} F_{\mu\nu} - \frac{1}{2} \left(\frac{\partial A_\mu}{\partial x_\mu}\right)^2 - \frac{1}{4} \left[\bar{\psi},\left(\gamma \frac{\partial}{\partial x} + m\right)\psi\right] -$$

$$- \frac{1}{4} \left[- \frac{\partial \bar{\psi}}{\partial x} \gamma + m \bar{\psi},\psi\right] + \frac{ie}{2} A_\mu \left[\bar{\psi},\gamma_\mu \psi\right] +$$

$$+ \text{renormalization terms.} \tag{5.1}$$

The explicit form of the renormalization terms does not concern us here, except for the statement that they are all of order e^2 or higher. In this section we are going to be interested in the vacuum expectation value of the commutator of two electromagnetic potentials and, in particular, in the appearance or non-appearance of the gradient term in eq. (2.13). First of all, we remark that the terms appearing in the Lagrangian (5.1) but not in the Lagrangian (4.1) do not contain any derivatives of the electromagnetic potentials (apart, possibly, from the renormalization terms). Consequently, up to and including terms of order e the momenta which are canonically conjugate to the electromagnetic potentials are still given by eqs. (4.3). This implies, in particular, that none of these momenta vanishes. Therefore, we can treat all components of the electromagnetic potential as dynamically independent quantities. Therefore, the canonical formalism tells us that to all orders in e and for all combinations of μ and ν we must have

$$[A_\mu(x), A_\nu(x')]\Big|_{x_o=x'_o} = 0 \quad . \tag{5.2}$$

Consequently, if the canonical quantization is consistent, there must be no gradient term of the type shown in eq. (2.13).

To see whether or not we arrive at any formal contradiction this way, we consider the equation of motion for the electromagnetic field

$$\Box A_\mu(x) = - j_\mu(x) = - \frac{ie}{2} [\bar{\psi}(x), \gamma_\mu \psi(x)] +$$

$$+ \text{renormalization terms.} \tag{5.3}$$

The total current operator $j_\mu(x)$ in electrodynamics is a conserved quantity

$$\frac{\partial j_\mu(x)}{\partial x_\mu} = 0 \quad . \tag{5.4}$$

Consequently, the argument of section 2 tells us that

there exists a weight function $\Pi(p^2)$ defined by the equation

$$<0|j_\mu(x)\; j_\nu(x')|0> =$$

$$= \frac{1}{(2\pi)^3} \int dp\; e^{ip(x-x')}\; (p_\mu p_\nu - \delta_{\mu\nu} p^2)\; \Pi(p^2)\; \Theta(p)\;, \tag{5.5}$$

$$\Pi(p^2) = \frac{V}{-3p^2} \sum_{p^{(n)}=p} <0|j_\mu|n><n|j_\mu|0>\quad . \tag{5.5a}$$

To discuss the states $|n>$ appearing in eq. (5.5a) we have to introduce the concept of "physical" or "incoming" particles in the formalism. For this purpose we rewrite the differential equation of motion (5.3) as an integral equation with a retarded singular function introducing the incoming field $A_\mu^{(in)}(x)$ in the following way

$$A_\mu(x) = A_\mu^{(in)}(x) + \int D_R(x-x')\; j_\mu(x')\; dx'\quad , \tag{5.6}$$

$$D_R(x) = -\;\Theta(x)\;\Delta(x,0)\quad , \tag{5.6a}$$

$$\Box\, D_R(x) = -\;\delta(x)\quad , \tag{5.6b}$$

$$\Box\, A_\mu^{(in)}(x) = 0\quad . \tag{5.6c}$$

Formally, the field $A_\mu^{(in)}(x)$ is the boundary value of the Heisenberg field $A_\mu(x)$ for $x_0 = -\infty$. In a similar way, we introduce an incoming spin 1/2 field defined by the equations

$$\psi(x) = \psi^{(in)}(x) - \int S_R(x-x')\; f(x')dx'\quad , \tag{5.7}$$

$$f(x) = ie\,\gamma\, A(x)\,\psi(x) + \text{renormalization terms}, \tag{5.7a}$$

$$S_R(x) = -\,\Theta(x)\left(\gamma\,\frac{\partial}{\partial x} - m\right)\Delta(x,m^2) =$$

$$= \left(\gamma\,\frac{\partial}{\partial x} - m\right)\Delta_R(x,m^2) \quad , \tag{5.7b}$$

$$\left(\gamma\,\frac{\partial}{\partial x} + m\right) S_R(x) = -\,\delta(x) \quad , \tag{5.7c}$$

$$\left(\gamma\,\frac{\partial}{\partial x} + m\right)\psi^{(in)}(x) = 0 \quad . \tag{5.7d}$$

As the canonical commutation relations are independent of time and as the incoming fields are formally the boundary values of the interacting fields for $x_o = -\infty$, it is reasonable to assume that the incoming fields also fulfill the canonical commutation rules [9]. Consequently, we can introduce annihilation and creation operators essentially as Fourier components of the incoming fields. These operators describe particles which we will call incoming particles. The physical meaning of these objects is that they describe those particles which were present in our system very long ago and before any interaction had taken place. Intuitively, it is also clear that each state is uniquely characterized when the particles which were present very long ago as incoming particles are specified. Consequently, we assume each state $|n>$ in the sum in eq. (5.5a) to be specified by the incoming particles. The matrix operator η discussed in the previous section still has the explicit form (4.7), where $n^{(4)}$ now means the total number of incoming scalar photons in the state under consideration.

We now return to the weight function $\Pi(p^2)$. Instead of eq. (5.5a) we now use an expression for this weight function analogous to eq. (2.15)

$$\Pi(p^2) = \frac{V}{\bar{p}^2 - 3p^2} \sum_{p^{(n)}=p} \langle 0|j_k|n\rangle\langle n|j_k|0\rangle =$$

$$= \frac{V}{\bar{p}^2 - 3p^2} \sum_{p^{(n)}=p} \sum_{k=1}^{3} |\langle 0|j_k|n\rangle|^2 (-1)^{n_n^{(4)}} \quad . \tag{5.8}$$

Because of the indefinite metric, the right-hand side of eq. (5.8) is not a sum of positive definite terms. However, we can show that the total expression (5.8) still is positive definite. To prove this we first remark that all contributions from states which either contain no incoming scalar photon or which contain an even number of incoming scalar photons are positive. Therefore, we are only interested in states $|n\rangle$ where an odd number of scalar incoming photons is present. We discuss explicitly the case when the state $|n\rangle$ contains just one incoming photon together with other particles. We symbolically write this as

$$|n\rangle = |a,\bar{p},\lambda\rangle \quad , \tag{5.9}$$

where the letter a stands for all other incoming particles in the state and the letters $\bar{p}$ and λ for the momentum and polarization of the photon. From the general structure of the formalism it follows that the matrix element of any operator and, in particular, of the current operator j_k must be linear in the polarization vector of the photon, [10],

$$\langle 0|j_k|n\rangle = \langle 0|j_k|a,\bar{p},\lambda\rangle = F_{k\mu}(a,p)\; e_\mu^{(\lambda)}(p) \quad . \tag{5.10}$$

Further, the formalism must be gauge invariant also for scalar photons. This means, according to general calculational rules in quantum electrodynamics, that if we replace the polarization vector of any photon by the energy momentum vector of the same photon, the result must be zero. This implies

$$F_{k\mu}(a,p)p_\mu = 0 \quad . \tag{5.11}$$

Using the fact that the photon mass is exactly zero and, therefore, that the absolute value of the space component of p is the same as the time component, we can also write eq. (5.11) in the form

$$F_{k\mu}(a,p)\, e_\mu^{(3)}(p) = F_{ko}(a,p) = -i\, F_{k\mu}(a,p)e_\mu^{(4)}(p) \quad . \tag{5.12}$$

Returning to eq. (5.10) we find that we can write eq. (5.12) in the alternative form

$$|\langle 0|j_k|a,\bar{p},3\rangle| = |\langle 0|j_k|a,\bar{p},4\rangle| \tag{5.13}$$

Consequently, the negative contributions in the sum (5.8) from terms with one scalar photon are exactly compensated by similar terms with one longitudinal photon. Therefore, the only surviving contributions from the states which have one incoming photon are the contributions from the transverse photons. These terms are all positive. A similar discussion can evidently be done for any state with an odd number of photons in eq. (5.8) and the final conclusion is that in spite of the negative metric of the Gupta-Bleuler formalism, the sum on the right-hand side of eq. (5.8) is actually positive definite because of the gauge invariance condition (5.11).

Next, we remark that the matrix elements of the electromagnetic potentials can be related to the matrix elements of the current operator according to eq. (5.3)

$$\langle 0|A_\mu|n\rangle = \frac{1}{[p^{(n)}]^2} \langle 0|j_\mu|n\rangle \quad , \; [p^{(n)}]^2 \neq 0 \quad . \tag{5.14}$$

Consequently, the spectral functions for the electromagnetic potentials themselves can be written as follows

$$\{[p_\mu p_\nu - \delta_{\mu\nu}p^2]G_1(p^2) + \delta_{\mu\nu}G_2(p^2)\}\ \theta(p) =$$

$$= V \sum_{p^{(n)}=p} \langle 0|A_\mu|n\rangle\langle n|A_\nu|0\rangle = (p_\mu p_\nu - \delta_{\mu\nu}p^2)\ \frac{\Pi(p^2)}{(p^2)^2}\ \theta(p) +$$

$$+ V \sum_{\substack{p^{(\gamma)}=p \\ \text{one photon} \\ \text{states } |\gamma\rangle}} \langle 0|A_\mu|\gamma\rangle\langle\gamma|A_\nu|0\rangle \quad . \tag{5.15}$$

The one photon matrix elements remaining in eq. (5.15) must have a general form analogous to the right-hand side of eq. (5.10). In this case we write

$$\langle 0|A_\mu|\gamma\rangle = F_{\mu\nu}(p)\ \langle 0|A_\nu^{(in)}|\gamma\rangle \quad , \tag{5.16}$$

where p, as before, is the energy momentum vector of the photon. The quantity $F_{\mu\nu}(p)$ must transform as a tensor. Consequently, it is a linear combination of $p_\mu p_\nu$ and $\delta_{\mu\nu}$. The two coefficients multiplying these basic tensors are, in principle, functions only of p^2 but this latter quantity is identically zero. Therefore, the two coefficients are constants. The number multiplying $\delta_{\mu\nu}$ is conventionally made equal to one by charge renormalization and we find

$$\langle 0|A_\mu|\gamma\rangle = (\delta_{\mu\nu} - M\ p_\mu p_\nu)\ \langle 0|A_\nu^{(in)}|\gamma\rangle \quad . \tag{5.17}$$

When this expression is substituted in the last term on the right-hand side of eq. (5.15) we get

$$V \sum_{p^{(\gamma)}=p} \langle 0|A_\mu|\gamma\rangle\langle\gamma|A_\nu|0\rangle = (\delta_{\mu\nu} - 2Mp_\mu p_\nu)\delta(p^2)\,\Theta(p) \,. \tag{5.18}$$

The two weight functions $G_1(p^2)$ and $G_2(p^2)$ in eq. (5.15) are then given by

$$G_1(p^2) = \frac{\Pi(p^2)}{(p^2)^2} - 2M\,\delta(p^2) \quad , \tag{5.19a}$$

$$G_2(p^2) = \delta(p^2) \quad . \tag{5.19b}$$

From eq. (2.13) we then conclude that the vacuum expectation value of the commutator between the time component and a space component of electromagnetic potentials for equal times must have the form

$$\langle 0|\left[A_o(x),\, A_k(x')\right]|0\rangle\Big|_{x_o=x'_o} =$$

$$= -i\,\frac{\partial}{\partial x_k}\,\delta(\bar{x}-\bar{x}')\left[\int_o^\infty \frac{da\,\Pi(-a)}{a^2} - 2M\right] \quad . \tag{5.20}$$

According to the canonical quantization scheme this commutator should, however, vanish. Consequently, the constant M must be given by

$$M = \frac{1}{2}\int_o^\infty \frac{da}{a^2}\,\Pi(-a) > 0 \quad . \tag{5.21}$$

The right-hand side of eq. (5.21) is an integral over a positive definite integrand and, therefore, non-vanishing. Consequently, the constant M in eq. (5.17) must be different from zero if eq. (2.13) should be consistent with the canonical commutation relation (5.2).

The appearance of the constant M in eq. (5.17) may

be somewhat surprising at the first moment. Therefore, we want to discuss it in some more detail [11]. We start by calculating the integral (5.21) in perturbation theory. From the defining equation (5.5a) we find that we can obtain the first nontrivial contribution to $\Pi(p^2)$ which is of order e^2 by considering only terms of order e in the definition of the current operator. From eq. (5.3) we see that we get the current to this order by replacing the spin 1/2 fields by the field without any interaction, i.e., by the incoming fields. Therefore, we have to first non-trivial order

$$\Pi(p^2) \simeq \frac{V\ e^2}{12p^2} \sum_{p^{(n)}=p} \langle 0|[\bar{\psi}^{(in)}, \gamma_\mu \psi^{(in)}]|n\rangle \times$$

$$\times \langle n|[\bar{\psi}^{(in)}, \gamma_\mu \psi^{(in)}]|0\rangle \quad . \quad (5.22)$$

As the incoming fields are free fields, they contain just one creation or annihilation operator of incoming electrons or positrons each. Consequently, the state $|n\rangle$ in eq. (5.22) can only be a state with one incoming electron-positron pair. For such a state, the matrix elements appearing in (5.22) are easily computed and one finds

$$\Pi(p^2) = \frac{e^2}{3p^2}\frac{1}{V} \sum_{q+q'=p} \sum_{pol} \bar{u}^{(-)}(-\bar{q}')\gamma_\mu\ u^{(+)}(\bar{q}) \times$$

$$\times \bar{u}^{(+)}(\bar{q})\gamma_\mu\ u^{(-)}(-\bar{q}') \quad , \quad (5.23)$$

where the notation $u^{(\pm)}(\bar{q})$ has been used for the positive and negative energy plane wave solution of the Dirac equation. The summation over polarizations in eq. (5.23) is performed using standard techniques and gives a trace of a product of factors involving γ-matrices. Further, the summation over the energy momentum vectors can conveniently be replaced by an integration and we get after simple formal rearrangements

$$\Pi(p^2) = \frac{e^2}{3p^2} \frac{1}{(2\pi)^3} \iint dq\, dq'\, \delta(p-q-q')\, \delta(q^2+m^2) \times$$

$$\times \delta(q'^2+m^2)\Theta(q\ )\Theta(q')\ Sp\left[\gamma_\mu(i\gamma q-m)\gamma_\mu(i\gamma q'+m)\right] . \tag{5.24}$$

The calculation of the trace and the integration over the vectors q and q' is elementary and straightforward. We do not enter upon the details here but just write down the result

$$\Pi(p^2) = \frac{e^2}{12\pi^2} (1- \frac{2m^2}{p^2}) \sqrt{1+(4m^2/p^2)}\ \Theta(-p^2-4m^2) . \tag{5.25}$$

The integral in eq. (5.21) can now be calculated in this approximation and we find

$$M \simeq \frac{e^2}{24\pi^2} \int_{4m^2}^{\infty} \frac{da}{a^2} (1+ \frac{2m^2}{a}) \sqrt{1- 4m^2/a} = \frac{e^2}{120\pi^2} \frac{1}{m^2} . \tag{5.26}$$

We note that the integral appearing here is convergent and yields a result which is different from zero.

In view of the appearance of the somewhat unfamiliar constant M it might be worth while to mention that the value given in eq. (5.26) can also be obtained in another way and by solving the equation of motion (5.6) for the electromagnetic potential in perturbation theory. To do this, we evidently have to evaluate the retarded integral which appears in Eq. (5.6) to order e^2. Therefore, the expression used in eq. (5.22) for the current operator is not enough. We also have to include the first correction to the operator $\psi(x)$ as well as some renormalization terms in eq. (5.3). The first order correction to the spin 1/2 field is obtained from eqs. (5.7). More in detail, we have

$$j_\mu(x) = \frac{ie}{2}[\bar{\psi}^{(in)}(x), \gamma_\mu\psi^{(in)}(x)] + \\ + \frac{ie^2}{2}[\bar{\psi}^{(in)}(x), \gamma_\mu\delta\psi(x)] + \\ + \frac{ie^2}{2}[\delta\bar{\psi}(x), \gamma_\mu\psi^{(in)}(x)] + \\ + \text{renormalization terms}, \tag{5.27a}$$

$$\delta\psi(x) = -i\int S_R(x-x')\gamma_\lambda A_\lambda^{(in)}(x')\,\psi^{(in)}(x')dx'. \tag{5.27b}$$

We are now interested in calculating the matrix element between a one photon state with energy momentum vector p and polarization λ and the vacuum of the current operator (5.27a). Evidently, the first term on the right-hand side of eq. (5.27a) does not contribute and we find

$$<0|j_\mu(x)|\gamma> = \frac{e^2}{2}\int dx'\{<0|[\bar{\psi}^{(in)}(x), \gamma_\mu S_R(x-x')\gamma_\lambda \times \\ \times \psi^{(in)}(x')]|0> + <0|[\bar{\psi}^{(in)}(x')\gamma_\lambda S_A(x'-x), \gamma_\mu \times \\ \times \psi^{(in)}(x)]|0>\} \times \\ \times <0|A_\lambda^{(in)}(x')|\gamma> + \text{renormalization terms}. \tag{5.28}$$

The expression inside the curly bracket in eq. (5.28) is the well-known kernel appearing in the elementary first order vacuum polarization [12]. It can be evaluated using standard techniques. Actually, it contains one divergent integral which has to be compensated for by the renormalization term in eq. (5.28). For our purpose, it is advantageous to make explicit use of the fact that the kernel in eq. (5.28) is explicitly retarded in its character, i.e., that it vanishes as soon as $x_o' > x_o$. This tells us that the Fourier transform of the kernel has to fulfill a dispersion relation. The weight function which appears in the dispersion relation can be explicitly calculated and turns out to be exactly the function $\Pi(p^2)$ in eq. (5.25). After a calculation of this kind one finds

$$\frac{e^2}{2}\{<0|[\bar{\psi}^{(in)}(x),\gamma_\mu \; S_R(x-x')\gamma_\lambda \; \psi^{(in)}(x')]|0> +$$

$$+ <0|[\bar{\psi}^{(in)}(x')\gamma_\lambda \; S_A(x'-x),\gamma_\mu \; \psi^{(in)}(x)]|0>\} =$$

$$= \frac{1}{(2\pi)^4} \int dq \; e^{iq(x-x')}(q_\mu q_\lambda - q^2\delta_{\mu\lambda}) \int_0^\infty da \times$$

$$\times \Pi(-a) \left[\frac{1}{(a+q^2)_P} + i\pi \; \delta(q^2+a)\,\varepsilon(q)\right] . \qquad (5.29)$$

The letter P on the denominator in eq. (5.29) indicates that the principle value of the integral has to be taken. The factor $q_\mu q_\nu - q^2\delta_{\mu\nu}$ in eq. (5.29) appears because of current conservation. It guarantees the gauge invariance of eq. (5.28) as well as conservation of the current $j_\mu(x)$. This factor is also consistent with the previous condition (5.11) in the special case considered here [13]. Substituting in eq. (5.28) we now get formally

$$<0|j_\mu(x)|\gamma> = p_\mu \; p_\lambda \; <0|A_\lambda^{(in)}(x)|\gamma> \int_{4m^2}^\infty \frac{da}{a}\,\Pi(-a) +$$

$$+ \text{renormalization terms.} \tag{5.30}$$

In this expression we have used that the photon has zero mass and put $p^2 = 0$ everywhere. On the other hand, we see from eq. (5.17) and the equation of motion (5.3) that the matrix element between the vacuum and the one photon states of the current operator has to vanish identically. This defines the renormalization term in eq. (5.30). However, if we want to use this expression to calculate the difference between the analogous matrix element of the interacting Heisenberg field $A_\mu(x)$ and the incoming field, we see from the integral equation (5.6) that we have to divide by p^2 which is equal to zero. Therefore, we run into an indeterminate expression. Such a quantity can only be defined as a careful limit of factors, each of which is a well defined quantity. A suitable limit of this kind is obtained if we consider the charge e not as a fixed quantity but as a function of the time x_0 according to

$$e \rightarrow e\, e^{-\varepsilon|x_0|} \quad . \tag{5.31}$$

Here, ε is a positive number and we want to take the limit $\varepsilon \rightarrow 0$ in the final result. Clearly, a damping factor of this kind improves the convergence at $x_0 = \pm\infty$ of integrals like (5.6) and (5.7). Therefore, it makes the statement that the incoming field actually is the boundary value of the interacting field a little more precise. With this convention, eq. (5.27b) acquires an extra exponential factor inside the integral and the matrix element of the current in eq. (5.28) is replaced by

$$<0|j_\mu(x)|\gamma> =$$

$$= e^{-\varepsilon|x_0|} \int dx'\, e^{-\varepsilon|x_0'|} <0|A_\lambda^{(in)}(x')|\gamma> \frac{1}{(2\pi)^4} \int dq\, e^{iq(x-x')} \times$$

$$\times\ (q_\mu q_\lambda - \delta_{\mu\lambda} q^2) \int_0^\infty da\Pi(-a)\left[\frac{1}{(a+q^2)_P} + i\pi\delta(q^2+a)\varepsilon(q)\right] +$$

$$+ \text{ renormalization term } =$$

$$= e^{-2\varepsilon|x_o|}(p_\mu p_\lambda - \delta_{\mu\lambda}p^2)\Big|_{p_o\to p_o+i\varepsilon} \times$$

$$\times <0|A_\lambda^{(in)}(x)|\gamma> \int_{4m^2}^\infty da\Pi(-a)\left[\frac{1}{a+\bar{p}^2-(p_o+i\varepsilon)^2} - \frac{1}{a}\right] \quad . \tag{5.32}$$

The last term inside the square bracket in eq. (5.32) is the renormalization terms which is adjusted so as to compensate the first term for $\varepsilon = 0$ and $p^2 = 0$ according to the remark made after eq. (5.30). Keeping only the leading term in ε in eq. (5.32) we find

$$<0|j_\mu(x)|\gamma> = e^{-2\varepsilon|x_o|} p_\mu p_\lambda <0|A_\lambda^{(in)}(x)|\gamma> 2i\varepsilon\, p_o \times$$

$$\times \int_{4m^2}^\infty \frac{da}{a^2}\Pi(-a) \quad . \tag{5.33}$$

By construction this expression vanishes when ε goes to zero as should be the case. The matrix element of the electromagnetic potential now becomes, according to eq. (5.6)

$$<0|A_\mu(x)|\gamma> = <0|A_\mu^{(in)}(x)|\gamma> +$$

$$+ 2i\varepsilon\, p_o \int_{4m^2}^\infty \frac{da}{a^2}\Pi(-a)\, p_\mu p_\lambda \int dx'\, D_R(x-x')\, e^{-2\varepsilon|x_o'|} \times$$

$$\times \langle 0|A_{\lambda}^{(in)}(x')|\gamma\rangle = [\delta_{\mu\gamma} - \frac{1}{2}\int_{4m^2}^{\infty} \frac{da}{a^2}\, \Pi(-a) p_{\mu} p_{\lambda}] e^{-2\varepsilon|x_0|} \times$$

$$\times \langle 0|A^{(in)}(x)\ |\gamma\rangle \quad . \tag{5.34}$$

In the limit when ε tends to zero, this expression is in complete agreement with eqs. (5.17) and (5.21). Consequently, a direct integration of the equations of motion and the canonical formalism gives a result completely consistent with eq. (2.13). The particular reason why no gradient term appears in this case is the same as in section 4, viz. first of all that the divergence condition for the electromagnetic potentials is not fulfilled and, second, that the metric in the space of the state vectors is not positive definite. These two circumstances together generate a compensating mechanism and no gradient term actually appears. It is very essential for this to happen that the term involving the constant M in eq. (5.17) actually exists. Quite evidently, the particular phenomenon which we have discussed here is gauge dependent. Our discussion has been entirely inside the so-called Gupta-Bleuler gauge. This version of the formalism has the advantage that it is explicitly Lorentz covariant and, from every point of view, very suitable for practical calculations. In some other gauge the commutator between two potentials may be entirely different and the last term in eq. (5.17) may be different or even absent.

6. Commutator of the Currents in Spin 1/2 Electrodynamics

In all the examples which we have discussed in the previous sections, the integral over the weight function appearing as coefficient for the gradient term in the equal time commutator was convergent. Consequently, the argument which led to eq. (2.13) was a reasonably strict argument. Therefore, our main concern in the previous examples was to show that the commutator in eq. (2.13) had the same value independently of the way in which it was evaluated. However, we now turn to our last example where the situation is somewhat different. The physical system we consider is the same as in section 5, viz. electrodynamics for spin 1/2 particles. However, instead of the fields describing the electromagnetic potentials we now want to investigate the currents. In this case we have from eq. (5.5) the general formula

$$<0|[j_\mu(x),j_\nu(x')]|0> =$$

$$= i \int_0^\infty da\ \Pi(-a)[\frac{\partial^2}{\partial x_\mu \partial x_\nu} - a\ \delta_{\mu\nu}]\Delta(x'-x,a) \quad , \qquad (6.1)$$

where the function $\Pi(p^2)$ is defined in eq. (5.5a). The explicit form of this function in first non-vanishing order of perturbation theory is given in eq. (5.25). Putting the two times x_o and x_o' equal and using the properties (2.11) of the singular function we find the following gradient term for the current commutator

$$<0|[j_o(x),j_k(x')]|0>\Big|_{x_o=x_o'} =$$

$$= -\ i\ \frac{\partial}{\partial x_k}\ \delta(\bar{x}-\bar{x}')\int_0^\infty da\ \Pi(-a)\ . \qquad (6.2)$$

On the other hand, a straight-forward computation using the canonical formalism, in particular, the canonical anticommutators for the spin 1/2 fields gives

$$[j_o(x),j_k(x')]\Big|_{x_o=x'_o} = i\ e^2\{\bar\psi(x)\gamma_4\psi(x)\bar\psi(x')\gamma_k\psi(x') -$$

$$- \bar\psi(x')\gamma_k\psi(x')\bar\psi(x)\gamma_4\psi(x)\}_{x_o=x'_o} =$$

$$= i\ e^2[\bar\psi(x)\gamma_4\{\psi(x),\bar\psi(x')\}\gamma_k\psi(x') -$$

$$- \bar\psi(x')\gamma_k\{\psi(x')\bar\psi(x)\}\gamma_4\ \psi(x) - \bar\psi(x)\gamma_4\bar\psi(x')\psi(x)\gamma_k\psi(x') +$$

$$+ \bar\psi(x')\ \ \bar\psi(x)\gamma_4\gamma_k\psi(x')\psi(x)]_{x_o=x'_o} =$$

$$= i\ e^2\delta(\bar x-\bar x')[\bar\psi(x)\gamma_k\psi(x') -$$

$$- \bar\psi(x')\gamma_k\ \psi(x)]_{x_o=x'_o} = 0 \quad . \qquad (6.3)$$

Evidently, the two results (6.2) and (6.3) are formally inconsistent as the integrals over the positive definite function $\Pi(-a)$ on the right-hand side of eq. (6.2) can never vanish. This formal contradiction is also the origin of much of the rather confused literature which exists about the gradient terms in eq. (2.13). The origin of the confusion is, evidently, that none of the two calculations leading to eq. (6.2) or eq. (6.3) is really satisfactory. In the first computation where the gradient term appears explicitly on the right-hand side of eq. (6.2) we have interchanged the limit when the two times become equal and the integration over the mass variable a. Such an interchange of limiting processes is allowed provided the expressions one is working with are sufficiently regular. In particular, the interchange has been justified for our previous examples where the integral over the weight function was convergent. However, the explicit form of the weight function $\Pi(p^2)$ given in eq. (5.25) shows that the integral in eq. (6.2) is actually rather strongly divergent. Therefore, the formal computation leading to the result (6.2) is unreliable. However, the formal computation leading to eq. (6.3) is not better. The two terms inside the square bracket multiplying the three-dimensional δ-function in eq. (6.3) both have an infinite vacuum expectation value when the two points x and x' coincide. Therefore, the value of the bracket is of the form $\infty - \infty$ and, therefore, indeterminate. In view of this situation it is clear that the current commutator investigated in this section is a very singular expression and that it cannot be evaluated except, possibly, with the aid of some cut-off procedure. Also it is reasonable to expect that the actual value obtained for such an indeterminate expression depends on the particular cut-off procedure chosen. A priori, it is therefore to be expected that we can find mathematical definitions of this singular commutator where either the result (6.2) or the result (6.3) or, possibly, some entirely different expression is declared to be mathematically the correct value [14]. In our opinion, such mathematical procedures are inherently quite as arbitrary as the more formal ways to evaluate the commutator given here. Before one can decide which mathematical procedure is the correct one, it is essential to get some feeling for the "physical" significance of the integral appearing on the right-hand side of eq. (6.2). Only after such a feeling has been developed is it possible to choose a suitable cut-off procedure and to make this choice in such a way that the result which is expected on the basis of physical intuition really comes out.

The first point which we want to make here is that the integral, which appears in eq. (6.2) actually occurs also in other places in the formalism and then has the significance of a formal self mass of the photon. To substantiate this remark we return to some of the calculations made in the previous section and want to discuss some of the steps there more in detail. A self mass of the photon would appear, e.g., in such a way that the equation of motion for the vacuum to one photon matrix element of the electromagnetic potential reads

$$(\Box - \mu^2) \langle 0|A_\lambda(x)|\gamma\rangle = 0 \quad . \tag{6.4}$$

Comparing this with eq. (5.3) we find

$$\langle 0|j_\lambda(x)|\gamma\rangle = -\mu^2 \langle 0|A_\lambda^{(in)}(x)|\gamma\rangle \quad . \tag{6.5}$$

Our explicit calculation of this matrix element of the current operator as given, e.g., in eqs. (5.32) or (5.33) showed no term of the type indicated in eq. (6.5). Instead we only obtained a term in eq. (5.33) involving $p_\mu p_\lambda$ and this term was responsible for the expression involving M in eq. (5.17). The formal reason for this lies in eq. (5.29) where we have introduced the factor $q_\mu q_\lambda - q^2\delta_{\mu\lambda}$ as an important part of the Fourier transform of the kernel in eq. (5.29). This factor is requested by current conservation and the gauge invariance of the formalism. Therefore, the photon self mass has to vanish in a gauge invariant formulation of electrodynamics, as is rather well-known. We have also remarked that, actually, the integral kernel appearing in eq. (5.29) involves a divergent integral and, therefore, is not entirely well defined. As is also rather well-known and pointed out several times in the literature of 20 years ago, the highly divergent and ambigous integrals in eq. (5.29) can, by formal manipulations, very well yield an extra term which violates gauge invariance and which formally corresponds to a self mass of the photon. We now want to show that this extra term contains exactly the integral on the right-hand side of Eq. (6.2). To demonstrate this, we start by writing the left-hand side of eq. (5.29) as follows

$$K_{\mu\lambda}(x-x') \equiv$$

$$\equiv \frac{e^2}{2}\{<0|[\bar{\psi}^{(in)}(x),\gamma_\mu S_R(x-x')\gamma_\lambda\ \psi^{(in)}(x')]|0> +$$

$$+ <0|[\bar{\psi}^{(in)}(x')\gamma_\lambda S_A(x'-x),\gamma_\mu\psi^{(in)}(x)]|0>\} \equiv$$

$$\equiv\ -\ \Theta(x-x')\ F_{\mu\lambda}(x-x') \quad . \tag{6.6}$$

$$F_{\mu\lambda}(x-x') = \frac{e^2}{2}\{Sp[\gamma_\mu S(x-x')\gamma_\lambda\ S^{(1)}(x'-x)] -$$

$$- Sp[\gamma_\mu\ S^{(1)}(x-x')\gamma_\lambda\ S(x'-x)]\} \quad , \tag{6.6a}$$

$$S^{(1)}(x'-x) = <0|[\bar{\psi}^{(in)}(x),\psi^{(in)}(x')]|0> =$$

$$= \frac{1}{(2\pi)^3}\int dq\ e^{iq(x'-x)}\delta(q^2+m^2)(i\gamma q-m) \quad . \tag{6.6b}$$

The expression in eq. (6.6a) has a Fourier transform

which is given by a convergent integral and which can be computed without difficulty. Straight-forward calculations give

$$F_{\mu\lambda}(x-x') = \frac{-i}{(2\pi)^3} \int dq\, e^{iq(x-x')} \varepsilon(q) [q_\mu q_\lambda - \delta_{\mu\lambda} q^2] \Pi(q^2). \tag{6.7}$$

To proceed from here to a Fourier representation of the functions $K_{\mu\lambda}(x-x')$ in eq. (6.6) we can, e.g., write the step function in eq. (6.6) as a Fourier integral

$$\Theta(x-x') = \frac{1}{2\pi i} \int_{-\infty}^{+\infty} \frac{d\tau\, e^{i\tau(x_o - x'_o)}}{\tau - i\varepsilon} =$$

$$= \frac{1}{2\pi i} \int \frac{dp\, \delta(\bar{p})}{p_o - i\varepsilon} e^{-ip(x-x')} \quad . \tag{6.8}$$

The Fourier transform of the expression in eq. (6.6) is now obtained as a formal convolution integral

$$K_{\mu\lambda}(x-x') = \frac{1}{(2\pi)^4} \int dq\, e^{iq(x-x')} K_{\mu\lambda}(q) \quad , \tag{6.9}$$

$$K_{\mu\lambda}(q) = \int \frac{dp\, \delta(\bar{p})}{p_o - i\varepsilon} \varepsilon(q+p) [(q+p)_\mu (q+p)_\lambda -$$

$$- \delta_{\mu\lambda}(q+p)^2] \, \Pi((q+p)^2) \quad . \tag{6.9a}$$

The integration over the vector p in this last expression can be performed by elementary techniques provided all integrals are treated as convergent. In that way one finds formally

$$K_{\mu\lambda}(q) = [q_\mu q_\lambda - \delta_{\mu\lambda} q^2] [\int_0^\infty \frac{da \Pi(-a)}{(a+q^2)_R} + i\pi\varepsilon(q)\Pi(q^2)] +$$

$$+ (\delta_{\mu\lambda} - \delta_{\mu 4}\, \delta_{\lambda 4}) \int_0^\infty da\; \Pi(-a) \quad . \qquad (6.10)$$

The result exhibited in eq. (6.10) is rather startling. Only the first term here agrees with the expected expression as given in eq. (5.29). The extra term which appears is not Lorentz invariant and also violates the gauge invariance of the formalism. Substituting this expression in eq. (6.5) and adding the renormalization term one finds

$$<0|j_\lambda(x)|\gamma> = (\delta_{\mu\lambda} - \delta_{\mu 4}\delta_{\lambda 4}) <0|A_\lambda^{(in)}(x)|\gamma> \int_0^\infty da\Pi(-a). \qquad (6.11)$$

The expression in (6.11) vanishes only for scalar photons while physical transversal photons as well as longitudinal photons acquire a self mass given by the expression

$$\mu^2 = - \int_0^\infty da\; \Pi(-a) \quad . \qquad (6.12)$$

Such a result is physically impossible [15].

Quite evidently, the explicit calculation which has led to the impossible result (6.10) instead of the physically correct result (5.29) is as unreliable as the calculations leading to eqs. (6.2) and (6.3) because of the somewhat careless handling of divergent integrals. The main point which we would like to make here is that the same integral which appears as a formal photon self mass in this calculation also appears as coefficient of the gradient term in eq. (6.2). As we know that the photon self mass can be avoided if the theory is treated as a sufficiently careful limit of a cut-off theory, we also expect that the gradient term will disappear in such a treatment and, therefore, that the calculation using the canonical formalism leading to the result (6.3) in retrospect is declared to be the correct one. This, of course, does not mean that the argument actually used to derive eq. (6.3) is logically consistent, only that the result obtained by this elementary and formal calculation happens to be physically correct.

There are several cut-off procedures available in the literature which are both gauge invariant and Lorentz covariant. We shall pick one of them here and demonstrate that with this cut-off procedure the limit of the integral in eq. (6.12) is, indeed, zero. We choose the cut-off method of Pauli and Villars, a method which is conventionally referred to as "regularization" [16]. Characteristic of this cut-off procedure is the idea that the cut-off is not introduced by brute force as an upper limit of integration in the formalism but, rather, with the aid of auxiliary heavy particles with masses m_i. These auxiliary particles are coupled to the electromagnetic field with coupling constants $e\sqrt{C_i}$. To achieve convergence of the formalism it is further important that the following conditions are fulfilled

$$\sum C_i = 0 \quad , \tag{6.13a}$$

$$\sum C_i \, m_i^2 = 0 \tag{6.13b}$$

The sums over the index i in these two expressions also include a term corresponding to the real physical electron. By definition we make this term correspond to i=1 and put $C_1 = 1$ and $m_1 = m$. The masses of the other particles are assumed to be very large and effectively act as cut-off parameters in the theory. In the final result we take the limit when all these auxiliary masses go to infinity. Evidently, the relations (6.13) cannot be fulfilled with all constants C_i positive. Therefore, some of the couplings of these heavy particles are purely imaginary. For such high energies that these auxiliary particles can be created, we therefore expect that the unitarity properties of the theory are not allright. However, in the limit when the cut-off masses become infinitely large, the energy at which this difficulty with unitarity occurs is pushed away to infinity. Therefore, unitarity is restored for all observable processes in the limit when the cut-off masses tend to infinity. This cut-off method is explicitly Lorentz and gauge invariant also for finite values of the cut-off parameters. The S-matrix is unitary in the limit when the cut-off parameters go to infinity.

Using this technique and working in first order perturbation theory as we have done so far, it is evident that the weight function $\Pi(p^2)$ is replaced by the following expression

$$\Pi(p^2) \rightarrow \Pi^{reg}(p^2) =$$

$$= \sum_i \frac{C_i e^2}{12\pi^2} \left(1 - \frac{2m_i^2}{p^2}\right) \sqrt{1 + 4m_i^2/p^2}\ \Theta(-p^2 - 4m_i^2) \ . \qquad (6.14)$$

Because of the step functions appearing in this definition, the regularized function $\Pi^{reg}(p^2)$ agrees with $\Pi(p^2)$ for any finite value of p^2 and the cut-off masses sufficiently large. However, the asymptotic behaviour of the two expressions is different. When the number $-p^2$ is much larger than all the cut-off masses, the regularized weight function goes to zero while the unregularized one approaches a constant value. We can now calculate the integral corresponding to the photon self mass in eq. (6.12) or the coefficient of the gradient term in eq. (6.2). The calculation is elementary and gives

$$\int_0^\infty da\ \Pi^{reg}(-a) =$$

$$= \frac{e^2}{12\pi^2} \sum_i C_i \int_{4m_i^2}^\infty da\left[\left(1+\frac{2m_i^2}{a}\right)\sqrt{1-4m_i^2/a}-1\right] =$$

$$= \frac{-e^2}{6\pi^2} \sum_i C_i\, m_i^2 = 0 \quad . \tag{6.15}$$

In the regularized version of the theory, the integral (6.15) as well as the photon self mass and the coefficient of the gradient term in eq. (6.2) is exactly zero even before the limit when the cut-off parameters go to infinity has been taken. Therefore, the canonical evaluation of the current commutator yields the correct result in this version of quantum electrodynamics.

We emphasize that the conclusion just reached has not been obtained on the basis of some arbitrary mathematical definition of which particular limiting process should be used to evaluate the current commutator but, rather, from the observation that the coefficient of the gradient term is related to the photon self mass and, therefore, has to be zero in a consistent formalism.

Finally, we want to remark that the integral appearing as coefficient of the gradient term in section 5 and given, e.g., in eq. (5.21) or (5.26) is not changed by regularization. Indeed, one finds

$$\int_0^\infty da \frac{\Pi^{reg}(-a)}{a^2} = \frac{e^2}{12\pi^2} \sum_i C_i \int_{4m_i^2}^\infty \frac{da}{a^2} \left(1+\frac{2m_i^2}{a}\right) \sqrt{1-4m_i^2/a} =$$

$$= \frac{e^2}{120\pi^2} \sum_i \frac{C_i}{m_i^2} \to \frac{e^2}{120\pi^2} \frac{1}{m^2} , \qquad (6.16)$$

which is the same as the result (5.26) in the limit when the cut-off parameters become very large.

7. Conclusions

With the aid of several quite elementary examples we have discussed the appearance or non-appearance of the gradient term in the vacuum expectation value of the equal time commutator between the time component and the space component of a field operator which transforms like a vector. The field operator in our examples is either a vector field itself or the current belonging to such a field. The standard calculation leading to the gradient term is only well defined when one is dealing with convergent integrals. If the vector field in question is conserved and if the metric in the space of the state vectors is positive definite, the gradient term always exists and is never in contradiction with any other aspect of the formalism as, e.g., the canonical commutation relations. However, when some of these conditions are not fulfilled, the gradient term may be absent. For instance, if the field under consideration is not conserved or if the metric is not positive definite as in the Gupta-Bleuler version of electrodynamics, the gradient term in the commutator between the two fields is absent. Again, as long as this coefficient is finite, no formal contradiction arises between the evaluation of the gradient term using the spectral representation or any other way of calculating the commutator between the two fields. Finally, in the case when the coefficient of the gradient term is represented by a divergent integral, certain formal contradictions can exist as is always the case when one is handling divergent quantities in a formal way. The contradiction is removed if the theory is consistently handled as the limit of a cut-off theory. We have here demonstrated one particular way of doing this, viz. by the use of the Pauli-Villars regulators. In this way of understanding electrodynamics, the gradient term is absent also in the equal time commutator of two current operators.

References

1. The original inventors of this representation are S. Kamefuchi and H. Umezawa, Progr. Theor. Phys. 6, 543 (1951). A few years later, and without any claim of priority, this representation was also used by G. Källén, Helv. Phys. Acta 25, 417 (1952); M. Gell-Mann, F. E. Low, Phys. Rev. 95, 1300 (1954). Later, a pedagogical summary was given by H. Lehmann, Nuovo Cimento 11, 342 (1954).
2. The first authors who explicitly pointed out the result shown in eq. (2.13) were T. Goto and T. Imamura, Progr. Theor. Physics 14, 396 (1955). Later, the same remark was made by J. Schwinger, Phys. Rev. Lett. 3, 296 (1959). The right-hand side of eq. (2.13) is often referred to as a "Schwinger term" in the literature. We shall not follow this conven-

tion here but call this expression a "gradient term".

3. The original discussion of such a field was given by A. Proca, Journal de Physique 7, 347 (1936). A somewhat more readily available reference is the textbook by G. Wentzel, Einführung in die Quantentheorie der Wellenfelder, Franz Deuticke, Wien 1943, chapter III.
4. S. Gupta, Proc. Phys. Soc. London A63, 681 (1950), 64, 850 (1951); K. Bleuler, Helv. Phys. Acta 23, 567 (1950).
5. Cf., e.g., S. Coester, Phys. Rev. 83, 798 (1951), and R. J. Glauber, Progr. Theor. Phys. 9, 295 (1953).
6. Cf., e.g., G. Källén, Quantenelektrodynamik, Handbuch der Physik V_1, Springer-Verlag 1958, esp. p. 194.
7. Note, that we do not put $e_4^{(4)}(p) = i$.
8. For further details about the Gupta-Bleuler method we refer to the original literature quoted in footnote [4] or to various textbooks, e.g., ref. [6], p. 199 .
9. This statement which has here been made on an entirely intuitive basis can be given a precise mathematical meaning but we do not want to enter into these details here. Cf., however, the discussion following eq. (5.31) below.
10. This is essentially trivial to each order in perturbation theory where the annihilation operator of the incoming photon is always accompanied by a polarization vector. More generally, the same result can be proved using reduction formula techniques. We do not want to enter into details here but refer to the existent literature, e.g., ref. [6], esp. p. 351.
11. The existence of this constant has previously been mentioned in the literature; G. Källén, Helv. Phys. Acta 25, 417 (1952).
12. Cf., e.g., ref. [6], p. 280, eq. (29.4).

13. The detailed calculations leading to the result (5.29) can be found, e.g., in ref. [6], pp. 280-284.
14. A mathematical procedure where the result (6.2) is declared to be the correct one has been given, e.g., by K. Johnson, Nucl. Physics 25, 431 (1961). Cf. also W. S. Hellman, P. Roman, Nuovo Cim. 52, ser. X, 1341 (1967) and the literature quoted there.
15. As has been mentioned earlier, the fact that a self mass of the photon may appear in this way has been pointed out repeatedly in the literature. Cf., e.g., G. Wentzel, Phys. Rev. 74, 1070 (1948). The fact that some relation exists between the photon self mass and the gradient term in eq. (2.13) was mentioned very briefly by L. S. Brown, Phys. Rev. 150B, 1338 (1966), esp. p. 1345.
16. W. Pauli, F. Villars, Rev. Mod. Phys. 21, 434 (1949).

MODELS OF LOCAL CURRENT ALGEBRA†

By

H. LEUTWYLER

Institut für theoretische Physik, Universität
Bern, Switzerland

Abstract

We point out that the infinite momentum limit has a simple interpretation in terms of quantities associated with a lightlike surface and derive the "local current algebra relations at infinite momentum" without making use of the infinite momentum limit. Two models satisfying these relations are then described: A trivial solution with degenerate mass spectrum and a solution that exhibits mass splitting. In particular we discuss the significance of ghost states with spacelike momentum arising in non-trivial solutions of the local current algebra relations.

---

† Lecture given at the VII. Internationale Universitätswochen f. Kernphysik, Schladming, February 26-March 9, 1968.

1. Charge Algebra at Infinite Momentum

The basic objects involved in current algebra are a set of currents $j_i^\mu(x)$ and their associated charges

$$T_i = \int d^3x \; j_i^0(x) \tag{1.1}$$

The set may include e.g. the conserved vector currents whose charges are the generators of the isospin symmetry group, strangeness changing currents such as the vector and axial vector currents associated with weak semileptonic decays etc. Current algebra is an attempt to formulate simple properties for these currents and their charges. A particularly attractive assumption is the hypothesis that the charges T_i generate a Lie algebra

$$[T_i, T_k] = i f_{ik\ell} T_\ell \tag{1.2}$$

Of course, if the currents $j_i^\mu(x)$ are conserved, then the generators T_i are independent of time, i.e. they commute with the Hamiltonian of the system and we have a genuine symmetry algebra. However, it has been pointed out by Gell-Mann [1], that even if the currents are not conserved and the generators are not independent of time, the algebra may still be valid at any fixed instant of time. Moreover, one may hope that the commutation rules (1.2) which could in principle be satisfied exactly, are saturated approximately if one restricts himself to one particle states at rest and takes care of many particle contributions in the intermediate states on the left by including a small set of low-lying resonances. However, Coleman [2] has shown that these assumptions are not compatible, unless the currents are conserved and the algebra (1.2) represents a symmetry algebra.

Fubini and Furlan [3], [4] pointed a way out of this difficulty: they suggested to use states at infinite momentum rather than states at rest. This technique avoids the inconsistencies pointed out by Coleman and even apart from the one particle saturation hypothesis the infinite momentum limit is a very convenient device, used e.g. in the derivation of the Adler-Weisberger relation [5].

In the first part of this lecture we want to show that the infinite momentum limit is not only a convenient device, but has a very simple meaning. The prescription that the matrix elements should be evaluated at infinite momentum simply means that we are not really dealing with the charges (1.1) contained in a spacelike surface x^o = const., but instead with the charges belonging to a lightlike surface [6]

$$T_i = \int_\Sigma d\sigma_\mu j_i^\mu(x) \quad ; \quad \Sigma : x^\mu n_\mu = \text{const.} \qquad (1.3)$$

where the normal n^μ is a lightlike vector, $n^2 = 0$.

We claim that the assumptions:

- the generators T_i are identified with integrals over a lightlike surface according to (1.3)
- the commutation rules (1.2) are satisfied by the generators associated with one and the same surface Σ

are equivalent to the infinite momentum limit of the equal time charge algebra.

Moreover, if we add the assumption:

- the algebra is saturated by a set of one particle states (at finite momentum)

we reproduce the integrated current algebra relations

derived from saturation of the equal time algebra by one particle states at infinite momentum.

The intuitive reason behind this connection between the infinite momentum limit and current algebra on a lightlike surface is clear: The infinite momentum limit corresponds to a frame of reference which moves with the velocity of light compared to a frame in which the momentum is finite. If the corresponding Lorentz transformation is applied to the generators T_i defined by an integral over the surface x^o = const. we obtain an analogous integral over a hypersurface whose normal approaches the light cone as the velocity tends to the velocity of light.

There is thus a simple lesson to be learned from the virtues of the infinite momentum limit: Instead of dealing with quantities associated with a fixed instant of time, one should work with quantities belonging to a fixed lightlike hypersurface. In particular, the Adler-Weisberger relation does not test the equal time commutation rules of the generators, but rather the analogous commutation rules for the charges contained in a lightlike surface. The validity of the Adler-Weisberger relation supports the hypothesis that the corresponding generators associated with a lightlike surface satisfy an exact algebra of the type (1.2).

A detailed proof that the charge algebra on a lightlike surface is equivalent to the infinite momentum limit of the equal time charge algebra is given in the Appendix. We point out in the next section that our statements do not only apply to the algebra of the generators, but may be generalized immediately to local current algebra.

2. Local Current Algebra on a Lightlike Surface

The charge algebra relations discussed in the last section of course contain only a very limited amount of information about the matrix elements of the currents, since the matrix elements of the generators belonging to a lightlike surface involve only processes with vanishing invariant momentum transfer t. To obtain information about the form factors which describe the t-dependence of the matrix elements, Gell-Mann [7] has proposed to make use of locality which tells us that the currents $j_i^\mu(x)$, $j_k^\nu(y)$ commute whenever the points x and y are separated by a spacelike distance. More specifically, the quark model suggests that the charge densities satisfy the equal time commutation rules

$$[j_i^o(x),\ j_k^o(y)]_{x^o=y^o} = i\delta^3(\vec{x}-\vec{y})f_{ik\ell}j_\ell^o(x) \qquad (2.1)$$

which are both compatible with locality and with the equal time commutation rules of the generators.

Actually it is not these commutation rules that are used in local current algebra models, which make use of the infinite momentum limit [8] and a test of these models does not provide a test of the above commutation rules in much the same way as the Adler-Weisberger relation does not test the commutation rules of the generators associated with a spacelike but rather with a lightlike surface. Analogously, the infinite momentum limit of the local algebra (2.1) involves the current densities belonging to a lightlike surface. In order to formulate an appropriate set of commutation rules for these densities we observe that locality does <u>not</u> imply that the currents $j_i^\mu(x)$, $j_k^\nu(y)$ associated with the same lightlike surface $x, y \in \Sigma$, commute whenever $x \neq y$. This is because the surface Σ touches the light cone emerging from x along the line L

$$L : \xi^{\mu} = x^{\mu} + \lambda n^{\mu} \tag{2.2}$$

where n^{μ} denotes the normal of the surface Σ. If y is situated on this line the two currents need not commute. The generators belonging to Σ, defined in (1.3) are therefore not given by an integral over mutually commuting densities. However, we may split up the integration into a line integral along the light-like world lines mentioned above and a two dimensional integral perpendicular to these lines.

Let us for simplicity suppose that the normal n^{μ} is given by

$$n^{\mu} = (1,0,0,-1) \tag{2.3}$$

We then write the general point $x \in \Sigma$ in the form

$$x^{\mu} = (\sigma + \lambda,\ x^1,\ x^2,\ \sigma - \lambda) \qquad \mu = 0,1,2,3 \tag{2.4}$$

where σ characterizes the position of the surface Σ and is to be hold fixed, whereas λ and $\underline{x} = (x^1, x^2)$ vary, as the point x runs over Σ . The surface element then takes the form [9]

$$d\sigma_{\mu} = n_{\mu} d\lambda d^2\underline{x} \tag{2.4}$$

The line integrals

$$\tilde{j}_i(\underline{x}) = \int_L j_i^{\mu}(x)\ dx_{\mu} = \int_L j_i^{\mu}(x)\ n_{\mu} d\lambda \tag{2.5}$$

are local with respect to each other

$$[\tilde{j}_i(\underline{x}),\ \tilde{j}_k(\underline{y})] = 0 \qquad \underline{x} \neq \underline{y} \tag{2.7}$$

and the generators are given by

$$T_i = \int d^2\underline{x}\ \tilde{j}_i(\underline{x}) \tag{2.8}$$

We have suppressed the dependence of the new currents $\tilde{j}_i(\underline{x})$ on the parameter σ, since we shall deal always with one and the same surface Σ.

A simple analog of the local commutation rules (2.1) presents itself:

$$[\tilde{j}_i(\underline{x}),\ \tilde{j}_k(\underline{y})] = i\delta^2(\underline{x}-\underline{y})\ f_{ik\ell}\tilde{j}_\ell(\underline{x}) \tag{2.9}$$

In fact this algebra is equivalent to the infinite momentum limit of the equal time commutation rules (2.1). In particular the assumption that the equal time commutation rules (2.1) are saturated by one particle states at <u>infinite</u> momentum is equivalent to the assumption that the commutation rules (2.9) for the new currents $\tilde{j}_i(\underline{x})$ are saturated by one particle states at <u>finite</u> momentum.

We avoid to make use of the infinite momentum limit at all in this lecture [8]. The proof that the commutation rules (2.9) are indeed equivalent to the infinite momentum limit of the equal time algebra of the charge densities is given in the Appendix, together with a detailed analysis of the kinematics of the infinite momentum limit.

3. Momentum Space

We proceed by translating the commutation rules (2.9) into momentum space. Let us denote the state vectors of the system by $|\vec{p},N\rangle$ where $\vec{p}$ stands for the total momentum and N denotes the remaining quantum numbers. There is no need to restrict ourselves to one particle states at this point. The moving states are obtained from states at rest by

$$|\vec{p},N\rangle = U\left[L(\vec{p}/M)\right]|0,N\rangle (M/p^{o})^{1/2} \tag{3.1}$$

where $L(\vec{p}/M)$ is the pure Lorentz transformation which takes the vector $M^{\mu} = M(1,0,0,0)$ into p^{μ} . (The angles belonging to this transformation involve the ratio $\vec{p}/M$.) From the definition (2.6) of the currents $\tilde{j}_i(\underline{x})$ we obtain

$$\langle\vec{p}',N'|\tilde{j}_i(\underline{x})|\vec{p},N\rangle = 2\pi\, e^{i(p'-p)x}\, \delta(p'n-pn) \times$$

$$\times \langle\vec{p}',N'|j^{\mu}_{i}(o)n_{\mu}|\vec{p},N\rangle \tag{3.2}$$

The currents connect only states with the same value of the projection pn, say

$$p^{\mu}n_{\mu} = m_{o} \tag{3.3}$$

where $m_o > 0$ is some fixed constant; i.e. the currents involve only those momenta p^{μ} that belong to the intersection I of the mass shell $p^2 = M^2$, $p^{o} > 0$ with the plane $pn = m_o$. This is a two parameter set of momenta and it suffices to know the value of the transverse momentum $\underline{p} = (p^1,p^2)$ to determine the vector p^{μ} completely. Moreover, Lorentz invariance implies that the matrix elements $<\vec{p}',N'|j_i^\mu(o)n_\mu|\vec{p},N>$ depend, for a fixed value of m_o, essentially only on the difference $\underline{p}' - \underline{p}$ of the transverse momenta. This may be seen as follows. Let us denote by π^μ the particular momentum that lies in the intersection I and has no transverse components, $\pi^1 = \pi^2 = 0$. This momentum is determined uniquely by M and m_o [10]:

$$\pi^\mu = \frac{1}{2} m_o^{-1}(m_o^2 + M^2, 0,0, m_o^2 - M^2) \qquad (3.4)$$

Suppose now that p is an arbitrary momentum in the intersection I. Since p and π have the same projection onto n they are connected by a Lorentz transformation which belongs to the little group of the lightlike vector n

$$p^\mu = \Lambda^\mu{}_\nu \pi^\nu \qquad ; \qquad \Lambda^\mu{}_\nu n^\nu = n^\mu \qquad (3.5)$$

The generators of this little group are J_3, F_1, F_2 where J_3 generates rotations around the third axis and the commuting operators F_1 and F_2 are given by

$$F_1 = N_1 + J_2$$

$$F_2 = N_2 - J_1 \qquad (3.6)$$

in terms of the generators $\vec{J}$, $\vec{N}$ of the homogeneous Lorentz group. Since rotations around the third axis leave π invariant, we may restrict ourselves to a transformation of the type $\Lambda = F(\underline{\alpha})$

$$F(\underline{\alpha}) = \exp(-i\underline{\alpha} \cdot \underline{F}) \qquad (3.7)$$

In fact it is shown in the Appendix that the angles α^1, α^2 are related in a simple fashion to the transverse momentum $\underline{p}$:

$$\underline{\alpha} = \underline{p}/m_o \qquad (3.8)$$

and we have the result:

$$p^\mu = F(\underline{p}/m_o)^\mu{}_\nu \; \pi^\nu \qquad (3.9)$$

The pure Lorentz transformation which takes the state at rest to a moving state with momentum p may therefore be written in the form

$$L(\vec{p}/M) = F(\underline{p}/m_o) \; L(\vec{\pi}/M)R \qquad (3.10)$$

where R is a rotation. This implies the representation

$$|\vec{p},N\rangle = U[F(\underline{p}/m_o)] \, |\vec{\pi},N'\rangle D(R)^{N'}{}_N (\pi^o/p^o)^{1/2} \qquad (3.11)$$

where $D(R)^{N'}{}_N$ stands for the representation of the rotation group induced on the states at rest. Finally we observe that the operators $U[F]$ and $j_i^\mu(o)n_\mu$ commute, because the transformation F belongs to the little group of n and furthermore

$$F(\underline{p}/m_o) \; F(\underline{p}'/m_o) = F(\underline{p} + \underline{p}'/m_o) \qquad (3.12)$$

because the generators F_1 and F_2 commute. This leads to the following representation of the current matrix elements:

$$\langle\vec{p}',N'|\tilde{j}_i(\underline{x})|\vec{p},N\rangle = (2\pi)^{-2} e^{i(p'-p)x} \; \delta(p'n-pn) \times$$

$$\times \; (m_o^2/p^{o\prime}p^o)^{1/2} \; D^*(R')^{N_1'}{}_{N'} \; D(R)^{N_1}{}_N (N_1'|J_i(\underline{p}'-\underline{p})|N_1) \qquad (3.13)$$

$$(N'|J_i(\underline{p})|N) = (2\pi)^3(\pi^{0\prime}\pi^0/m_o^2)^{1/2}\langle\vec{\pi}',N'|U^+[F(\underline{p}/m_o)] \times$$

$$\times \quad j_i^\mu(o)n_\mu|\vec{\pi},N\rangle \qquad (3.14)$$

The states $|N)$ are introduced for convenience to represent the result in the form of a matrix [11]. These states have nothing to do with the state vectors $|\vec{p},N\rangle$. Accordingly the quantity $J_i(\underline{p})$ is not an operator acting in the space of physical state vectors like, say $j_i^\mu(x)$. It acts instead in the "space of quantum numbers" spanned by the vectors $|N)$.

The result (3.13) confirms our statement that the matrix elements of the currents $\tilde{j}_i(\underline{x})$ involve, apart from kinematical factors, only the difference of the transverse momenta.

It is now a simple matter to translate the commutation rules (2.9) into momentum space, by inserting a complete set of intermediate states on the left. The result reads

$$[J_i(\underline{p}'),\ J_k(\underline{p})] = i\ f_{ik\ell}\ J_\ell(\underline{p}'+\underline{p}) \quad . \qquad (3.15)$$

In this equation the products $J_i(\underline{p})J_k(\underline{p}')$ on the left hand side are to be understood in the sense of matrix multiplication in the space of quantum numbers

$$(N'|J_i(\underline{p})J_k(\underline{p}')|N) = \sum_{N''} (N'|J_i(\underline{p})|N'')(N''|J_k(\underline{p}')|N)$$

Eq. (3.15) is the content of the local algebra of the operators $\tilde{j}_i(\underline{x})$. Note that we have not assumed so far that the algebra is saturated by one particle states. This special case results if one restricts the set of states $|\vec{p},N\rangle$ to one particle states and analogously for the set $|N)$. The result (3.15) remains the same.

In the literature the relation (3.15) is derived

under the assumption that the equal time algebra of the charge densities (2.1) is saturated by one particle states in the infinite momentum limit [10], [11]. This establishes the equivalence of the local algebra of the currents $j_i(\underline{x})$ with the equal time algebra of the charge densities at infinite momentum. We refer the reader to the Appendix for a detailed account of this equivalence.

4. Relativistic Invariance

We now turn to the problem of solving the algebra (3.15). The following question then immediately arises: Suppose that we have a solution of this algebra, i.e. suppose that the matrix elements $(N'|J_i(\underline{p})|N)$ are given explicitly as functions of $\underline{p}$. Is it always possible to interpret this solution in terms of matrix elements of a relativistic current operator $j_i^\mu(x)$ according to (3.14)? The answer is no. In fact, suppose that we are given the matrix elements of one particular component, say $(N'|J_1(\underline{p})|N)$ between two states of definite charge and spin j,j' respectively. Let us for simplicity assume j' = 0. We are thus given a total of 2j+1 quantities. On the other hand relativity implies that the matrix elements of the current $j_i^\mu(x)$ are not independent, but involve only a few invariant form factors. Their number in the present case is 4 (we assume $j \geq 1$), two form factors with positive parity and two with negative parity. We will have trouble fitting the given numbers as soon as $j \geq 2$.

The conditions that must be imposed on the matrix elements $(N'|J_i(\underline{p})|N)$ in order that they belong to a covariant current $j_i^\mu(x)$ are referred to in the literature as angular conditions [8],[10],[12]. They can be derived in a straightforward fashion from the explicit expression (3.14). We quote the infinitesimal form of these conditions [13]:

$$[L_3, J_i(\underline{p})] = i\underline{p} \times \underline{\nabla} J_i(\underline{p}) \tag{4.1}$$

$$\{I,\{I,\{I,J_i(\underline{p})\}\}\} =$$

$$= \frac{1}{4}[M^2,[M^2,\{I,J_i(\underline{p})\}]] + \frac{1}{2}\,\underline{p}^2[M^2,\{I,J_i(\underline{p})\}]_+ +$$

$$+ \frac{1}{4}(\underline{p}^2)^2\{I,J_i(\underline{p})\} \tag{4.2}$$

Here the symbol $\{I,X\}$ stands for

$$\{I,X\} = \frac{1}{2}[M^2,[L_3,X]] - \frac{1}{2}\,\underline{p}^2[L_3,X]_+ - [\underline{p}\cdot M\underline{L},X] \tag{4.3}$$

Note that these relations involve operators acting in the space of quantum numbers spanned by the vectors $|N)$. In particular, M is the mass operator whose eigenvalue is the mass of the particular state and the angular momentum operators L_1, L_2, L_3 are the generators of the representation u(R) of the rotation group defined by

$$u(R)|N) = |N')D(R)^{N'}{}_{N}\ . \tag{4.4}$$

In order that a solution $J_i(\underline{p})$ of the algebra (3.15) be acceptable we must therefore at the same time provide operators M and L_1, L_2, L_3 such that the angular conditions are satisfied. In particular we are interested in a solution that corresponds to saturation by one particle states, such that the operator M has only a discrete spectrum. Clearly the spectrum of M is intimately related with the properties of the current $J_i(\underline{p})$ through the angular conditions and we may thus hope to get information on the mass spectrum from current algebra.

5. A Model with Degenerate Mass Spectrum

Recently explicit, nontrivial solutions of the problem posed in the last section have been given by Gell-Mann, Horn and Weyers [14] and by Leutwyler [13]. Before we turn to these models it is instructive to look at a solution with degenerate mass spectrum which has been known for some time [15].

We restrict ourselves to a model of the isovector current algebra

$$[J_i(\underline{p}),J_k(\underline{p}')] = i\,\varepsilon_{ik\ell}\,J_\ell(\underline{p} + \underline{p}') \tag{5.1}$$

and furthermore assume that the algebra is saturated by one particle states of one given isospin T, e.g. T = 1/2, and label the states by the eigenvalue of T_3: $|N) \to |\hat{N},T_3)$. We assume that isospin is conserved and therefore the dependence of the matrix elements on the isospin quantum numbers factors out

$$(\hat{N}',T_3'|J_i(\underline{p})|\hat{N},T_3) = {t_i}^{T_3'}{}_{T_3}\,(\hat{N}'|J(\underline{p})|\hat{N}) \; . \tag{5.2}$$

The algebra (5.1) is satisfied provided the reduced matrix element satisfies

$$J(\underline{p})\;J(\underline{p}') = J(\underline{p}+\underline{p}') \tag{5.3}$$

We thus look for a solution of this relation together with a set of operators M, L_1,L_2,L_3 that satisfy the angular conditions. It is very easy to solve this problem if one is willing to accept that all particles have the same mass. In this case the angular conditions simplify considerably. This is because we may in this case choose the parameter m_o to coincide with the common value of the mass of the particles, such that (3.14) reduces to

$$(N'|J_i(\underline{p})|N) = (2\pi)^3 \langle 0,N'|U^+[F(\underline{p}/M)]\, j_i^\mu(o)n_\mu|0,N\rangle \tag{5.4}$$

We recall that the matrices $F(\underline{p}/M)$ satisfy the relation (3.12) and therefore the operators $U^+[F(\underline{p}/M)]$ constitute a representation of the current algebra relations (5.3). The origin of the angular conditions is the fact that the representation $U^+[F]$ is not only a representation of the subgroup of matrices F but in fact a representation of the full homogeneous Lorentz group. This suggests that we try the ansatz

$$J(\underline{p}) = u^+[F(\underline{p}/M)] \tag{5.5}$$

where $u(\Lambda)$ is a representation of the homogeneous Lorentz group. The representation $u(\Lambda)$ then not only induces a representation of the subgroup of matrices $F(\underline{p}/M)$, but at the same time it induces a representa tion of the subgroup of rotations in such a fashion that the corresponding generators L_1, L_2, L_3 satisfy the angular conditions. Note that the representation $u(\Lambda)$ must be unitary to guarantee that the currents $j_i^\mu(x)$ are real. The unitary representations of the homogeneous Lorentz group are well known [16]. There is the trivial representation which corresponds to $J(p)=1$, i.e. constant form factors associated with point charges. The nontrivial representations are all infinite dimensional and correspond to infinitely many particles with spin $j_o, j_o+1, j_o+2, \ldots$.

If we choose e.g. $j_o = 1/2$ we can work out the current matrix elements explicitly and relate the result to the conventional electric and magnetic isovector form factors. As an example one finds for the electric form factor [10]

$$F_1^V(t) = (1- t/4m^2)^{-3/2} \tag{5.6}$$

where M is the common mass of the infinite sequence of particles.

6. A Model with Mass Splitting

In this section we consider a model with non-degenerate mass spectrum [13]. We start with the observation that the current algebra relation (5.3) is solved with the ansatz

$$J(\underline{p}) = \exp(i\, \underline{p}.\underline{x}) \tag{6.1}$$

where x^1 and x^2 are two commuting operators. In order that the currents $j_i^\mu(x)$ be real, x^1 and x^2 must be hermitian. What remains to be found are operators M, L_1, L_2, L_3, with the property that the angular conditions (4.1) and (4.2) are satisfied for all $\underline{p}$. The angular conditions may be developed in a power series with respect to the momentum $\underline{p}$ and each coefficient provides a relation to be satisfied by the set of six operators M, $\vec{L}$, $\underline{x}$. At first sight this wealth of conditions seems to be too much for our six operators. It turns out however, that these conditions are not independent. In fact they admit of a two-parameter family of solutions of the following type. We assume that there exist operators P^1 and P^2 conjugate to x^1 and x^2

$$[P^a, x^b] = -i^{\;ab} \qquad a,b = 1,2 \tag{6.2}$$

Furthermore, the representation of the algebra of the operators $\underline{x}, \underline{P}$ is assumed to be irreducible in the sense that an operator commuting with both $\underline{x}$ and $\underline{P}$ is a multiple of the identity. This means that we try to make the space of vectors $|N)$ as small as possible; the models described by Gell-Mann, Horn and Weyers [14] are not irreducible in this sense. In terms of the operators $\underline{P}$ introduced above the solution reads

$$L_3 = \underline{x} \times \underline{P}$$

$$M^2 = \tfrac{1}{2}[\alpha, \underline{P}^2]_+ + \beta \tag{6.3}$$

where α and β are functions of the operator $r = |\underline{x}|$

$$\alpha = \mu r(1-\mu r)^{-1}$$

$$\beta = \frac{1}{4}\,\alpha r^{-2}(1+\alpha)^2 + \kappa(1+\alpha) \tag{6.4}$$

Note that the solution involves the two parameters μ and κ which represent the only degrees of freedom in our model. We shall not make use of the operators L_1 and L_2 for which analogous expressions may be given.

It is not very instructive to describe the heuristic way that led us to this solution and it is quite tedious to check that the above expressions really constitute a solution of the angular conditions.

There is an elegant independent way to arrive at the same model by using a relativistically invariant infinite component field equation [17]. The existence of such an underlying field equation explains why the infinite set of angular conditions admits of a non-trivial solution. On the other hand, the direct approach shows that our model represents in a sense the simplest solution, characterized by the fact that the representation of the operators $\underline{P}$, $\underline{x}$ is irreducible. For more details the reader is referred to the literature [13],[17].

We shall in the following take it for granted that the above expressions constitute a relativistically invariant solution of the local current algebra relations and restrict ourselves to a discussion of the properties of this model.

7. Physical States

Since the algebra of operators $\underline{x},\underline{P}$ is irreducible there exists a complete set of basis vectors $|\underline{x})$ which diagonalize the operator $\underline{x}$:

$$x^a|\underline{x}') = x^{a'}|\underline{x}')$$

$$P^a|\underline{x}') = +\, i\, \frac{\partial}{\partial x^{a'}}|\underline{x}') \qquad a=1,2 \qquad (7.1)$$

and it is convenient to make use of this basis in order to analyze the properties of the model. We denote the x-representation of the vectors $|\hat{N})$ by

$$(\underline{x}|\hat{N}) = \psi_N(\underline{x}) \qquad (7.2)$$

We require that the states $|\hat{N})$ be eigenstates of the square of the mass operator. The explicit expression for this operator then leads to the following eigenvalue equation:

$$\{-\frac{1}{2}[\alpha,\Delta]_+ + \beta\}\psi_N(\underline{x}) = M^2\ \psi_N(\underline{x}) \qquad (7.3)$$

where Δ stands for $(\partial/\partial x^1)^2 + (\partial/\partial x^2)^2$ and M^2 denotes the eigenvalue. We may assume that simultaneously the states $|\hat{N})$ are also eigenstates of L_3

$$L_3|\hat{N}) = m|\hat{N}) \qquad (7.4)$$

Expressed in terms of polar coordinates, $\underline{x}=r(\cos\phi,\sin\phi)$ the eigenfunctions of L_3 are of the form

$$\psi_N(\underline{x}) = (r/\alpha)^{1/2}\ R(r)\ \exp(im\phi) \qquad (7.5)$$

where we have extracted the normalization factor $[r/\alpha(r)]^{1/2}$ to simplify the radial equation that determines the mass eigenvalue:

$$\left[\frac{d^2}{dr^2} + \frac{2}{r}\frac{d}{dr} - \frac{1}{r^2}(m^2 - \frac{1}{4}) + \frac{1}{\mu r}(M^2-\kappa) - M^2\right] R(r) = 0$$

$$(7.6)$$

This is precisely the radial equation for the wave functions of the hydrogen atom. Putting the reduced mass of the electron equal to 1/2 the usual radial equation reads

$$\left[\frac{d^2}{dr^2} + \frac{2}{r}\frac{d}{dr} - \frac{1}{r^2}\ell(\ell+1) + \frac{e^2}{r} + E\right]R(r) = 0 \qquad (7.7)$$

with the bound state eigenvalues

$$E = - e^4/4n^2 \quad , \; n=1,2,... \tag{7.8}$$

Note that in our model the charge e^2 depends on the energy

$$e^2 = \mu^{-1}(M^2-\kappa) \; ; \quad E=-M^2 \tag{7.9}$$

Inserting these expressions in the eigenvalue condition (7.8) we obtain

$$(M^2 - \kappa)^2 = 4n^2 \mu^2 M^2 \tag{7.10}$$

We assume $\mu > 0$ and in this case we must have $M^2 > \kappa$ in order that the potential be attractive. If we furthermore suppose $\kappa > 0$ then the only acceptable solution of (7.10) is

$$M = \mu n + (\kappa + \mu^2 n^2)^{1/2} \tag{7.11}$$

The values of the quantum number ℓ that belong to a given value of n are $\ell = 0,1,2,...,n-1$ [18]. Comparison of the two wave equations (7.6) and (7.7) shows that m is related to ℓ by $|m| = \ell + 1/2$.

To a given value of the mass there thus belongs a set of 2n wave functions labelled by the eigenvalue m of L_3 which varies in the range $-(n-1/2) \leq m \leq (n-1/2)$. This is the spectrum of L_3 expected for a particle of spin $j=n-1/2$. We conclude that the model describes an infinite set of particles with spin $j = 1/2, 3/2, 5/2, ...$. Each of these values of the spin occurs once; the mass of the corresponding particle is given by

$$M = \mu(j + \tfrac{1}{2}) + \left[\kappa + \mu^2(j + \tfrac{1}{2})^2\right]^{1/2} \tag{7.12}$$

For large values of j we see that M grows linearly

with j in accordance with a general theorem on current algebra due to Grodsky, Martinis and Święcki [19]. The parameter μ measures the splitting between consecutive levels. If μ tends to zero, the mass spectrum degenerates into the point $M = \kappa^{1/2}$ and we recover one of the degenerate solutions described in section 5.

The radial wave function corresponding to the state j, m is essentially a Laguerre polynomial:

$$R_{j,m}(r) = N(Mr)^{|m|-1/2}\, e^{-Mr}\, L^{2|m|}_{j+|m|}(2Mr) \qquad (7.13)$$

where M is the mass of the state and N denotes normalization constant. Once these wave functions are known the matrix elements of the current may be evaluated in a straightforward fashion

$$\begin{aligned}(j',m'|J(\underline{p})|j,m) &= \int d^2x (j',m'|\underline{x})\, e^{i\underline{p}\underline{x}}\, (\underline{x}|j,m) = \\ &= \int d^2x r\alpha(r)^{-1}\, R_{j',m'}(r)\, R_{j,m}(r)\, \times \\ &\quad \times e^{i(m-m')\phi}\, e^{i\underline{p}\underline{x}} \qquad (7.14)\end{aligned}$$

As an example we quote the expectation value in the ground state: $j = j' = 1/2$, $m = m'$, which represents the form factor $F_1^V(t)$ at $t = -\underline{p}^2$:

$$\begin{aligned}(\tfrac{1}{2},m|J(\underline{p})|\tfrac{1}{2},m) &= (1+\underline{p}^2/4M^2)^{-3/2}(M-\mu)^{-1} \times \qquad (7.15)\\ &\quad \times \left\{M + \tfrac{1}{2}\mu - \tfrac{3}{2}\mu(1+\underline{p}^2/4M^2)^{-1}\right\}\end{aligned}$$

where we have adjusted the normalization constant such that $J(0) = 1$. As the mass splitting parameter μ tends to zero, this expression reduces to the result quoted in section 5 for the degenerate solution.

It is amusing to consider the Fourier transform of the form factor

$$\rho(r) = (2\pi)^{-3} \int d^3p\, F(-\vec{p}^{\,2})\exp(-i\vec{p}\vec{x}) \tag{7.16}$$

which essentially represents the corresponding charge distribution. Let us insert in this expression the particular form factor given by the expectation value

$$F(-\underline{p}^2) = (j,m|J(\underline{p})|j,m) \tag{7.17}$$

According to (7.14) this is the (two-dimensional) Fourier transform of the square of the wave function $\psi_{j,m}(\underline{x})$. But $\rho(r)$ is the (three-dimensional) Fourier-transform of $F(-\vec{p}^{\,2})$. Therefore the charge distribution is essentially given by the square of the wave function, more precisely:

$$\rho(r) = (\pi r)^{-1} \frac{d}{dr} \int_r^\infty dr' r'^2 \alpha(r')^{-1}(r'^2-r^2)^{-1/2}|R_{j,m}(r)|^2 \tag{7.18}$$

For distances large compared to the Compton wave length of the particle this expression simplifies to [20]

$$\rho(r) \simeq -(Mr/\pi)^{1/2}\, \alpha(r)^{-1}|R_{j,m}(r)|^2 \tag{7.19}$$

This means that the charge distribution of these particles is essentially the same as the charge distribution of the corresponding radial electron wave function in the hydrogen atom, roughly $\rho(r) \sim \exp(-2Mr)$ where M is the mass of the particle.

8. Ghosts

In the last section we considered the bound state solutions of our wave equation and ignored the fact that this equation also admits of scattering solutions, belonging to positive values of the energy, i.e. negative values of M^2. These solutions cannot be disregarded, because the current causes transitions between bound states and scattering states. The fact that these states are associated with negative values of M^2 means that they represent states with spacelike momentum p, $p^2 < 0$. It is difficult to interpret these states physically [21] and they must be regarded as a shortcoming of our model. The occurrence of ghosts with spacelike momenta is however by no means characteristic for our model, but is shared by the other known solutions of local current algebra [22]. In view of the connection between local current algebra and infinite component field equations, this is not astonishing, since spacelike solutions have been known to be a disease of infinite component field equations since they were invented by Majorana [23] in 1932 .

In order to clarify the significance of these ghosts it is instructive to study the behaviour of the solution as the mass splitting parameter μ tends to zero. From the explicit expression (7.15) it is clear that the particular current matrix element $(\frac{1}{2},m|J(\underline{p})|\frac{1}{2},m)$ is an analytic function of μ near $\mu = 0$ and the same is true of the general matrix element $(j',m'|J(\underline{p})|j,m)$. The limit $\mu \to 0$ is well-defined and in fact the bound states approach smoothly the corresponding wave functions of the degenerate solution as μ tends to zero.

On the other hand the degenerate solution does not possess ghosts and in fact the scattering states of our model do not have a well-defined limit as $\mu \to 0$.

This may be seen from the fact that the potential given by $-(M^2-\kappa)/\mu r$ is repulsive if $M^2 < 0$, and as μ approaches zero the corresponding wave functions are pushed away completely. It is not astonishing that the transition matrix elements between bound and scattering states are not analytic in μ near $\mu = 0$.

Let us now suppose that we want to verify the current algebra relations

$$\sum_{j'',m''} (j',m'|J(\underline{p})|j'',m'')(j'',m''|J(\underline{p}')|j,m) =$$

$$= (j',m'|J(\underline{p}+\underline{p}')|j,m) \quad (8.1)$$

where we have restricted the summation over the intermediate states to bound states only. If $\mu = 0$ the relation is satisfied since there are no ghosts. For finite values of μ on the other hand, the relation is not satisfied, since the transitions to scattering states may not be ignored. This might seem surprising, since the matrix elements involved are analytic in μ at $\mu = 0$. To add transition matrix elements to ghost states means to add functions that are not analytic in μ and the result is to be analytic, since the right hand side is. The point is that the domain of analyticity of the matrix elements in the sum shrinks as j tends to infinity. These matrix elements possess branch points in μ arising from the square root in the expression (7.12) for M. These branch points are located at

$$\mu = \pm \frac{i\kappa^{1/2}}{j+1/2} \quad (8.2)$$

and come closer and closer to the origin as j tends to infinity. The ghosts serve precisely to compensate the singularity resulting from this accumulation of branch points in the sum.

This shows that the ghosts have to do with the fact that infinitely many states are involved, in particular their existence is closely related to the behaviour of the matrix elements as the spin j tends to infinity, i. e. to the high energy part of the spectrum. Our model is not expected to be reliable in that region anyway, since it treats resonances like stable particles with a sharp value of the mass, a very poor approximation in the high energy region. The question is therefore: What properties of the model may be expected to be insensitive to the failure at the high energy end of the spectrum? An observation which bears on this question is the following. If one expands the matrix elements $(j',m'|J(\underline{p})|j,m)$ into a Taylor series in μ and keeps only terms to some given order then it is possible to show [17], that this expansion satisfies the current algebra relations - with bound intermediate states only - to the same order in μ. In other words, the ghosts do not contribute in such an expansion to any finite order in μ. We may therefore expect that the matrix elements $(j',m'|J(\underline{p})|j,m)$ are reliable in as much as they agree with their Taylor expansion. The validity of the Taylor expansion is restricted by the branch points mentioned above. For a given value of the mass splitting μ, we may expect that the solution is not too sensitive to the inadequacies of the model for those matrix elements which satisfy $j,j' < \kappa^{1/2}/\mu$. If the mass splitting is small enough compared to the lowest mass, we may hope that the solution represents a good approximation for the lowest lying states, whereas it certainly fails for states of high spin.

Appendix:

Kinematics of the Infinite Momentum Limit

The infinite momentum limit of the current matrix elements is defined by

$$L_i^\mu = \lim_{p^3, p^{3\prime}\to\infty} \langle \vec{p}\,', N' | j_i^\mu(0) | \vec{p}, N\rangle \qquad (A.1)$$

where in the limiting process the difference $p^{3\prime} - p^3$ is to be kept fixed. A convenient way of writing this limit is [11]

$$L_i^\mu = \lim_{\kappa\to\infty} \langle \vec{p}\,' + \kappa\vec{a}, N' | j_i^\mu(0) | \vec{p} + \kappa\vec{a}, N\rangle \qquad (A.2)$$

where $\vec{a}$ is a unit vector pointing in the direction of the third axis. In view of the definition (3.1) of the states $|\vec{p}, N\rangle$ the above limit involves a pure Lorentz transformation of the type $L(\vec{p} + \kappa\vec{a}/M)$. As κ tends to infinity this transformation becomes essentially a boost in the direction of the third axis, with a velocity approaching the velocity of light. It is important to know how this limit is approached and for this purpose we write

$$L(\vec{p} + \kappa\vec{a}/M) = L(\kappa\vec{a}/M)\Lambda \, .$$

To determine Λ it is convenient to make use of the 2x2 representation of the homogeneous Lorentz group. We write

$$L(\vec{p}/M) = \exp(\tfrac{1}{2}\,\vec{\chi}\cdot\vec{\tau}) \qquad (A.4)$$

where $\vec{\chi}$ is a vector in the direction of $\vec{p}$, whose length is given by

$$\sinh|\vec{\chi}| = |\vec{p}|/M \tag{A.5}$$

and τ_1, τ_2, τ_3 are the Pauli matrices. The 2x2 matrix Λ is easily worked out for large values of κ. One finds that it approaches a constant matrix F

$$\Lambda = F(\underline{p}/M) + O(\kappa^{-1}) \tag{A.6}$$

given by

$$F(\underline{p}/M) = \begin{pmatrix} 1 & 0 \\ (p^1+ip^2)/M & 1 \end{pmatrix} \tag{A.7}$$

This matrix involves only the transverse part of the momentum, $\underline{p} = (p^1, p^2)$, and is the 2x2 representation of the object introduced in (3.7). Note that the 2×2 representation of the generators of the Lorentz group is $\vec{J} = \frac{1}{2}\vec{\tau}$, $\vec{N} = \frac{1}{2} i\vec{\tau}$. The matrix $L(\kappa\vec{a}/M)$ is a pure Lorentz transformation in the direction of the third axis. In the 2x2 representation we have

$$L(\kappa\vec{a}/M) = B(\kappa/"M) + O(\kappa^{-1/2}) \tag{A.8}$$

where the matrix $B(\alpha)$ stands for the boost

$$B(\alpha) = \begin{pmatrix} \alpha^{1/2} & 0 \\ 0 & \alpha^{-1/2} \end{pmatrix} \tag{A.9}$$

We can go further and separate explicitly the dependence of these matrices on the mass M of the state:

$$L(\vec{p} + \kappa\vec{a}/M) = B(\kappa/2m_o)F(\underline{p}/m_o)B(m_o/M) + O(\kappa^{-1/2}) \tag{A.10}$$

Here $m_o > 0$ is a parameter whose value may be fixed once and for all, independent of the particular state under consideration. For large values of κ we thus have the result that the pure Lorentz transformation $L(\vec{p} + \kappa\vec{a}/M)$ is a product of three factors:

- a finite boost $B(m_o/M)$ in the direction of the third axis, which accounts for the fact that the mass spectrum of the system is not assumed to be degenerate,
- a transformation $F(\underline{p}/m_o)$ which involves only the transverse momentum $\underline{p}$ and is independent of the mass,
- a boost $B(\kappa/2m_o)$ which represents the main effect of the infinite momentum limit.

We now return to the current matrix elements and consider the limit L_i^μ defined by (A.2). Making use of the definition (3.1) we obtain

$$L_i^\mu = \lim_{\kappa\to\infty} (MM'/p^o p^{o\prime})^{1/2} \langle\langle \underline{p}', m_o, N'|U^+[B(\kappa/2m_o)] \times$$

$$\times \; j_i^\mu(o)U[B(\kappa/2m_o)]|\underline{p}, m_o, N\rangle\rangle \tag{A.11}$$

where we have introduced the abbreviation

$$|\underline{p}, m_o, N\rangle\rangle = U[F(\underline{p}/m_o)B(m_o/M)]|O,N\rangle \quad . \tag{A.12}$$

Lorentz invariance requires

$$U^+[B(\kappa/2m_o)]j_i^\mu(o)U[B(\kappa/2m_o)] = B^\mu_\nu \; j_i^\nu(o) \tag{A.13}$$

Here B^μ_ν stands for the 4x4 representation of the boost

$B(\kappa/2m_o)$. In the limit of very large κ this representation degenerates into

$$B^\mu_\nu = \kappa/m_o \; a^\mu \bar{a}_\nu + O(1) \tag{A.14}$$

where the vectors a^μ, $\bar{a}^\mu$ are defined by [11]

$$a^\mu = (1,0,0,1) \; ; \quad \bar{a}^\mu = (1,0,0,-1) \tag{A.15}$$

The normalization factor $(MM'/p^o p^{o'})^{1/2}$ behaves like $(MM')^{1/2}\, \kappa^{-1}$ as κ tends to infinity and therefore

$$L^\mu_i = a^\mu (MM'/m_o^2)^{1/2} \langle\langle \underline{p}', m_o, N' | \bar{a}_\nu j^\nu_i(o) | \underline{p}, m_o, N \rangle\rangle \tag{A.16}$$

In this result we recognize the following well-known facts about the infinite momentum limit:

- In the infinite momentum limit the current matrix elements become proportional to the vector a^μ, i. e. the first and second components of the current vanish whereas the components L^o_i and L^3_i become equal.
- The limit involves only the transverse momenta $\underline{p}$ and $\underline{p}'$; in fact the matrix elements depend only on the difference $\underline{q} = \underline{p}' - \underline{p}$.

This last statement may be verified by relating the infinite momentum limit to the matrix elements of the local current associated with a lightlike surface. Note that the boost $B(m_o/M)$ takes the rest momentum $(M,0,0,0)$ precisely into the momentum π^μ introduced in section 3, i.e .

$$B(m_o/M) = L(\vec{p}/M) \; . \tag{A.17}$$

The states $|\underline{p}, m_o, N\rangle\rangle$ may therefore be written

$$|\underline{p},m_o,N\rangle\rangle = U[F(\underline{p}/m_o)]\,|\vec{\pi},N\rangle(\pi^o/M)^{1/2} \qquad (A.18)$$

and the result of the infinite momentum limit, (A.16), may be expressed in terms of the matrix elements $(N'|J_i(\underline{p})|N)$ introduced in section 3 as follows

$$L_i^\mu = a^\mu(2\pi)^{-3}\,(N'|J_i(\underline{p})|N) \qquad (A.19)$$

This shows that the matrix elements of the local currents $\tilde{j}_i(\underline{x})$ associated with a lightlike surface are essentially the infinite momentum limits of the charge densities.

We finally consider matrix elements of the local equal time commutation rules of the charge densities in the infinite momentum limit, in particular

$$L_{ik} = \lim_{p^{3\prime},p^3\to\infty} \langle\vec{p}\,',N'|j_i^o(x)j_k^o(y)|\vec{p},N\rangle_{x^o=y^o} \qquad (A.20)$$

Inserting a complete set of intermediate states and making use of the expressions for the infinite momentum limit derived above, we obtain, interchanging limit and sum over intermediate states

$$L_{ik} = (2\pi)^{-6}\sum_{N''}\int d^3p''(N'|J_i(\underline{p}'-\underline{p}'')|N'')(N''|J_k(\underline{p}''-\underline{p})|N)\times$$

$$\times\exp\{i(p'-p'')x+i(p''-p)y\} \qquad (A.21)$$

This result minus an analogous expression obtained by interchanging i,k and x, y must equal the infinite momentum limit of the right hand side in (2.1). This implies the relation (3.15) and establishes the equivalence of the two methods.

Footnotes

1. M. Gell-Mann, Phys. Rev. 125, 1067 (1962).
2. S. Coleman, Phys. Letters 19, 144 (1965). For a very instructive discussion of Coleman's Theorems see the lectures by J. S. Bell in Recent Developments in Particle Symmetries, International School of Physics "Ettore Majorana", 1965.
3. S. Fubini and G. Furlan, Physics 4, 229 (1965).
4. The virtues of the infinite momentum limit arc discussed by R. F. Dashen and M. Gell-Mann, Proc. 3 rd Coral Gables Conf. on Symmetry Principles at High Energy (W. H. Freeman Co., San Francisco,1966).
5. S. L. Adler, Phys. Rev. Letters 14, 1051 (1965); W. I. Weisberger, Phys. Rev. Letters 14, 1047 (1965).
6. The connection of the infinite momentum limit with lightlike surfaces is briefly mentioned in ref. 4 .
7. M. Gell-Mann, Phys. Rev. 125, 1067 (1962); Physics 1, 63 (1964).
8. For a detailed account of local current algebra at infinite momentum see the lectures by M. Gell-Mann in Strong and Weak Interactions, International School of Physics "Ettore Majorana", 1966.
9. We use the metric $g_{oo} = - g_{11} = - g_{22} = - g_{33} = 1$.
10. H. Bebié and H. Leutwyler, Phys. Rev. Letters 19, 618 (1967).
11. F. Coester and G. Roepstorff, Phys. Rev. 155, B1583 (1967). This paper contains a very readable analysis of the properties of the infinite momentum limit.
12. J. Weyers, SLAC-PUB-281, 1967.
13. H. Leutwyler, Current Algebra: A Simple Model with Nontrivial Mass Spectrum, Phys. Rev. Letters, March 1968.
14. M. Gell-Mann, D. Horn and J. Weyers, amplified ver-

sion of the report delivered at the Heidelberg Conference on High Energy Physics and Elementary Particles, 1967 (Preprint, Institute for Advanced Studies, Princeton, October 1967).

15. R. F. Dashen and M. Gell-Mann, Phys. Letters 17, 340 (1966); S. Fubini in Proc. 4th Coral Gables Conf. on Symmetry Principles at High Energy (W. H. Freeman Co., San Francisco, 1967). See also refs. 8 and 10.
16. M. A. Naimark, Amer. Math. Soc. Transl., Ser. 2, Vol. 6, 379 (1957).
17. H. Bebié, F. Ghielmetti, V. Gorgé and H. Leutwyler, to be published.
18. There exists a second similar solution with half integer values of ℓ and n, corresponding to particles with integer spin. F. Ghielmetti, to be publ.
19. I. T. Grodsky, M. Martinis and M. Święcki, Phys. Rev. Letters 19, 332 (1967).
20. Note that if $\mu > 0$, $\alpha(r)$ is negative for large values of r.
21. For an attempt to interpret the physical significance of states with spacelike momenta see G. Feinberg, Phys. Rev. 159, 1089 (1967); M. E. Arons and E. C. G. Sudarshan, Lorentz Invariance, Local Field Theory and Faster than Light Particles, Preprint Syracuse University, 1967. This paper contains references to the literature on this subject.
22. In particular, close examination of the models considered by Gell-Mann, Horn and Weyers reveals that they too contain states with spacelike momenta. The same seems to be true for a model proposed recently by H. Kleinert, Saturation of Current Algebra by Non-Degenerate O(4,2) Multiplet, preprint University of Colorado, November 1967.
23. E. Majorana, Nuovo Cim. 9, 335 (1932). Spacelike

solutions of infinite component field equations are described by V. Bargman, Math. Rev. 10,584 (1949); E. Abers, I. T. Grodsky and R. E. Norton, Phys. Rev. 159, 1222 (1967); W. Rühl, Preprint Rockefeller University, 1967.

ROLE OF STRONG INTERACTIONS AND THE INTERMEDIATE BOSON IN THE RADIATIVE CORRECTIONS TO WEAK PROCESSES†

By

A. SIRLIN

Department of Physics, New York University
New York
and
International Centre for Theoretical Physics, Trieste, Italy

I. Introduction

The problem of radiative corrections to weak interactions is a rather old one, as some of the main results have been known for a long time [1] - [6]. In the last year or so there has been considerable interest on the subject and, perhaps, the main reason for the recent attention is that new approaches have been developed which attempt to take into account the effects of the strong interactions.

In these two lecture hours we will restrict ourselves to a discussion of two methods. One of these ap-

---

† Lecture given at the VII. Internationale Universitätswochen f.Kernphysik,Schladming,February 26-March 9,1968.

proaches assumes the validity of current algebra at small distances and, within this context, we will single out two models: one assumes a special form for the space-space part of the current algebra, the other assumes the existence of the intermediate boson. These two models are particularly interesting because they give, with some qualifications to be explained later, finite second order radiative corrections to the ratio $G_V/G_\mu = \cos\Theta$. The other approach does not assume current algebra and attempts to show some general properties of the second order corrections which are apparently independent of the details of the strong interactions and the possible existence of the intermediate boson. This approach consists essentially in separating out in a finite and gauge invariant manner the contributions of order 1/k as $k_\mu \to 0$ (k_μ is the photon four momentum) in the hadronic covariants and for this reason we will call it, for want of a better name, the 1/k method. As we will see this latter approach is not sufficiently powerful to permit a complete calculation of the corrections to G_V, but it isolates a large and interesting part of the corrections and, in particular, it determines within good approximations the second order corrections to some observables such as the allowed spectrum in β-decay. Although the 1/k method is less powerful than the current algebra approach, it has the obvious advantage that its assumptions are much more general.

Our presentation is not intended to be exhaustive. In particular, we will not discuss the interesting work of G. Källén in which the effect of strong interactions is described by the introduction of phenomenological form factors without assuming the validity of current algebra. This approach was already described in great detail by Prof. Källén in the Schladming schools of the past two years and in a published paper [7].

The plan of these two lectures will be roughly as follows: We will first recall some of the main old results concerning the radiative corrections to μ and β decay. This will perhaps show in a pragmatic way why we need these corrections. It will also permit us to describe in a simple manner the results of last year as they are very similar to the old ones. After this, we will explain very briefly the idea behind the 1/k method and we will then turn our attention to the current algebra approach. In this last method, for simplicity of presentation, we will discuss mainly the divergent contributions, but we will also give the results of the finite contributions. Finally, we will examine the present situation regarding the universality of the weak interactions, using the latest analysis of the Cabibbo theory.

We would like to emphasize that aside from the particular models discussed here, there may be other speculative and perhaps more profound methods to obtain finite second order corrections. The study of such methods may shed light in our understanding of physical phenomena at very short distances. This is a challenge and people here, particularly the younger ones, are urged to explore these unforeseen paths.

II. Muon Decay

Turning our attention to the old results, let us first consider the μ decay. This is one of the best understood weak processes as it is not affected by the strong interactions. Assuming the two component theory of the neutrino, the energy-angle distribution of the electrons from a completely polarized μ meson can be written up to terms of order α in the following way [3]

$$\Delta P\, d^3p = \frac{m_\mu^5 A^2 x^2 dx\, d\Omega}{96(2\pi)^4} \{3 - 2x + 6\zeta \frac{m_e}{m_\mu} (\frac{1-x}{x}) + \frac{\alpha}{2\pi} f(x) +$$

$$+ \xi \cos\Theta [1-2x + \frac{\alpha}{2\pi} g(x)]\} \qquad (1)$$

where $x = 2|\vec{p}_e|/m_\mu$, $\vec{p}_e$ is the electron momentum and Θ is the angle between $\vec{p}_e$ and the muon momentum. Here f(x) and g(x) are two complicated and finite functions of x given in ref. 3.

They arise from the contributions of the second order radiative corrections corresponding to the Feynman diagrams of figs. 1 and 2.

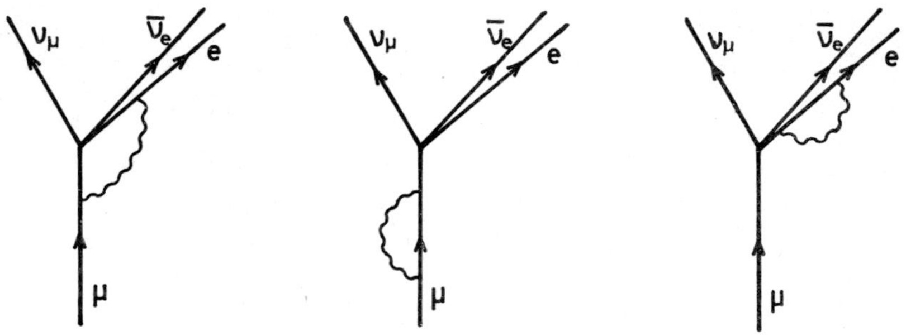

Fig. 1

Virtual Radiative Corrections to μ Decay

In eq. (1) we have neglected terms of order $\alpha\frac{m_e}{m_\mu}$ and $\alpha(m_e/E_e)$ but they have been calculated also. The three parameters A, ζ, ξ are functions of the coupling constants.

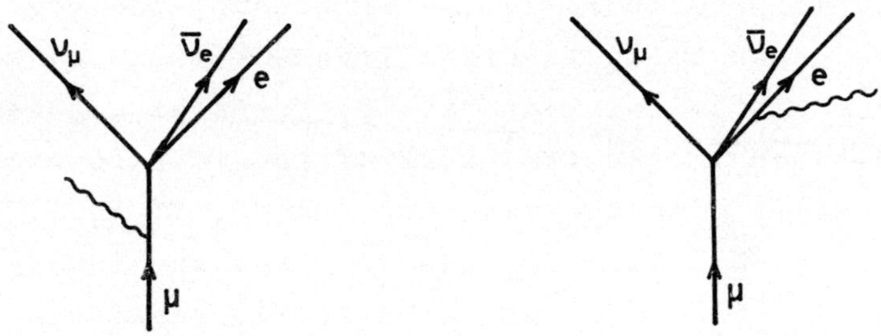

Fig. 2

Inner Bremsstrahlung Diagrams in μ Decay

In the V-A theory, $A^2 = 4G_\mu^2$, $\xi=1$ and $\zeta=0$ so that we get a particularly simple and beautiful expression depending on only one parameter, G_μ , which can be determined from the μ meson lifetime.

In this connection we note that the function f(x) contributes a correction to the muon lifetime given by [3]

$$\left(\frac{\Delta\tau}{\tau}\right)_\mu = \frac{\alpha}{2\pi}\left(\pi^2 - \frac{25}{4}\right) \quad . \qquad (2)$$

We will not go here into a detailed discussion of the comparison between theory and experiment, because that alone may take up a good part of my hour. It is sufficient to say that the agreement between the experimental and theoretical energy-angle distributions is excellent if one includes the functions $\frac{\alpha}{2\pi}$ f(x) and $\frac{\alpha}{2\pi}$ g(x). These corrections are in fact necessary to describe μ decay at the present level of experimental accuracy.

Before leaving the discussion of μ decay it is perhaps useful to make a parenthetical remark, mainly for the people interested in the details of the weak interactions. Muon decay is still frequently discussed, perhaps for historical reasons, in terms of a general formula derived in the framework of the four component neutrino theory with the ten couplings S, V, A, T, P, S', V', A', T', P'. This formula involves the famous parameters ρ, δ, ξ and ζ and is perfectly consistent to zero order in α. However, to determine these parameters at the present level of accuracy one needs the second order corrections and these are not finite and well defined for all the couplings (they are only finite for the V, V', A, A' couplings in the charge retention order). For the purpose of comparing theory and experiment, a more logical approach, in the opinion of this lecturer, would be therefore to forget the general formula (and the parameters ρ and δ), and to specialize to those theories for which the corrections are well defined, such as the two component theory or the V-A theory. In this approach one would simply compare eq. (1) with experiment and see how good the fits are, rather than determining the parameters ρ and δ.

III. β Decay (Summary of Old Results)

The next point in the development of the subject was the study of the second order corrections to the β decay of a "bare" nucleon, that is to say a hypothetical nucleon with the physical mass of the nucleon but somehow devoid of strong interactions. The quantity of main interest are the corrections to the Fermi part and is given by the expression [5]

$$\Delta P\, d^3p = \frac{\alpha}{2\pi} P^{0} d^3p \{3(1+\rho)\ln \frac{\Lambda}{m_p} + \frac{9}{4}\rho + g(E,E_m,m_e)\} \tag{3}$$

$$g(E,E_m,m_e) = 3\ln\frac{m_p}{m_e} - \frac{3}{4} + 4\left[\frac{\tanh^{-1}\beta}{\beta} - 1\right] \times$$

$$\left[\frac{E_m - E}{3E} - \frac{3}{2} + \ln\left(2\frac{E_m-E}{m_e}\right)\right] +$$

$$+ \frac{4}{\beta} L\left(\frac{2\beta}{1+\beta}\right) + \frac{1}{\beta}\tanh^{-1}\beta\left[2(1+\beta^2) + \frac{(E_m-E)^2}{6E^2} - 4\tanh^{-1}\beta\right] \tag{4}$$

where $\rho = -G_A/G_V$, $\beta = p/E$; P, E, E_m are the momentum, energy and end point energy of the electron and L(x) is the Spence function:

$$L(x) = \int_0^x \frac{dt}{t}\ln(1-t)$$

Very small terms of order $\alpha E/m_p$ and $\alpha E/m_p \ln(m_p/E)$ have been neglected in eqs. (3) and (4) but they can be included if desired, in this particular model.

Eqs. (3) and (4) present several interesting features: in the first place we see that the corrections are divergent (Λ is an ultraviolet cutoff) if we use either the experimental value of $\rho = -G_A/G_V = 1.23\pm0.02$ or the bare V-A value $\rho = +1$. The reason why μ decay is finite and neutron β decay is infinite in the V-A theory without strong interactions has been known for a long time: it essentially amounts to the fact that the axial current gives rise to an infinite contribution to the vector coupling which enters asymmetrically, i.e. with opposite relative signs, in μ decay and in β decay. In the first decay it cancels the divergence coming from the vector current, while in β decay it doubles this divergent contribution. In a hypothetical world without strong interactions and without axial currents the constants G_V and G_μ would be renormalized by the same universal divergent quantity. As we will see later these simple observations are closely connected to the developments of last year. Eqs. (3) and (4) contain another interesting feature: if we set $\rho = 0$ so that we isolate the contributions of the vector current, all reference to strongly interacting particles disappears from the formula except perhaps for the cutoff Λ , which may or may not have to do with the strong interactions.

A final observation is that eqs. (3) and (4) give not only a correction to the decay lifetime but also to the electron spectrum. To order α the corrections to the shape of the electron spectrum are finite: this can be seen by noting that the divergent contribution is a constant multiplying the uncorrected spectrum $P^o d^3p$ so that it simply renormalizes the coupling constant. Alternatively, writing $\int (P^o + \Delta P) d^3p = 1/\tau$ where τ is the lifetime, one can eliminate the coupling constant in P^o in terms of the observed lifetime and one verifies that the resulting correction to the electron spectrum is finite to order α. Numerically this correction is significant, particularly when applied to nuclear β decays with $E_m \gg m_e$. For example, it plays a significant role in the analysis of the spectra of the B^{12} and N^{12} β-decays. As you probably remember, these decays provided one of the basic tests of the conserved vector theory.

IV. β Decay

(Summary of Some of the New Developments)

We can now anticipate the results of the 1/k method and the current algebra approach. The 1/k method [8] shows essentially that in the presence of strong interactions the corrections are given again by the function $\frac{\alpha}{2\pi} g(E,E_m,m_e)$ plus undetermined constants of order α, provided one neglects terms of order

$$\frac{\alpha}{2\pi}\frac{E}{M}\,,\ \frac{\alpha}{2\pi}\frac{E}{M}\ln\frac{M}{E}\,,\ \frac{\alpha}{2\pi}\frac{q}{M}$$

(M is some hadronic mass and q is the momentum transfer to the leptons) and provided that some small matrix elements of order $\alpha/2\pi$ are evaluated in the allowed approximation. These results hold for both the Fermi and the Gamow-Teller parts of the transition probability and are independent of the possible existence or absence of the intermediate boson. On the other hand, this method does not permit an evaluation of all the constant terms of order α and is unable to tell whether they are finite or infinite. The general results of this method have been checked "a posteriori" with a number of calculations of radiative corrections to n β-decay and π β - decay, some of which assume the existence of the intermediate boson, some of which introduce, as in ref. 7, the effect of the strong interactions by means of form factors. According to the above mentioned results all of these calculations should give a result such as eq. (3) with the same function $\frac{\alpha}{2\pi} g(E,E_m,m)$ as in eq. (4) plus constant terms of order α . This is, in fact, the case. The method can also be used to study corrections to some of the other observables, such as the longitudinal polarization of electrons in allowed β decay [8], but this is mainly of academic interest at the present time.

The current algebra approach [9] - [11] has been used to study the corrections of order α to the Fermi part of the decay probability. Here one assumes essentially that the relevant matrix elements are given by the Fourier transforms of the appropriate time ordered products of currents and the validity of the equal time commutation relation

$$\delta(x_o)[j_\mu^{el}(x),\ V_o^+(o)] = \delta^4(x)\ V_\mu^+(o) \qquad (5)$$

where j_μ^{el} and V_μ^+ are the electromagnetic and isospin currents. As we mentioned before, there are two classes of contributions in the second order radiative corrections to the Fermi part of the decay probability: one part originates in the matrix elements involving the vector current, the other part arises from the axial vector current contributions. Using the above assumptions and again neglecting terms of order $\alpha\,\frac{E}{M}$, $\alpha\frac{E}{M}\ln(\frac{M}{E})$, $\alpha\frac{q}{M}$ etc., it has been shown by Bjorken [9], Abers, Dicus, Norton and Quinn [10],[11] that the entire contribution of the vector current is model independent and, therefore, identical to that given in eqs. (3) and (4) with $\rho = 0$. There is a slight qualification to this statement: there is a small finite term which depends on the commutation relation $\delta(x_o)[j_\mu^{el}(x),V_\mu^+(o)]$ (sum over μ). The exact equality with eqs. (3) and (4) for $\rho = 0$ holds if this particular commutator is the one given, for example, by the quark algebra. In any case, this "model dependent" term is very small: for the quark algebra it gives a correction for the decay probability of -0.06%. The contributions from the axial vector current are instead model dependent and give rise, in general, to divergent contributions which depend on the space-space part of the current algebra.

The fact that under the above assumptions the entire contribution from the vector current is the same with or without strong interactions is rather remarkable. The fact that one obtains the function $g(E,E_m,m_e)$ is not in itself surprising. It should be so according to the 1/k method. What is in fact rather surprising is that the constant terms, including the logarithmic divergence, are the same as in the "bare particle" calculations.

In order to solve the "divergence problem" two approaches have been studied in considerable detail.

In one approach the space-space part of the current algebra is chosen in such a manner that the divergent contribution from the axial current cancels that arising from the vector current. This approach has been proposed by Cabibbo, Maiani and Preparata [12],[13] and by Low, Johnson and Suura [14]. The first three authors discuss in great detail a model of fundamental fields, the "SUB" model, which may be regarded as the underlying fields of the proposed algebra [13]. The second approach was discussed by Abers, Dicus, Norton and Quinn in ref. 11 and by this lecturer in ref. 15. It involves assuming that the weak interactions proceed via an intermediate boson with electromagnetic properties described by the Proca-Wentzel Lagrangian. In this approach the vector current gives the same model independent answer as before, but the contribution of the axial current is cut off, so to speak, by the boson mass, Thus, the answer for the Fermi part is now logarithmically divergent. However, in this model the same phenomenon occurs in muon decay; the contribution of the axial current to the vector coupling and the contribution of the vector current to the axial coupling are now finite, and to order α

one obtains the same divergent factor in μ decay and in the Fermi part of β decay. This universal divergent contribution can then be interpreted as an unobservable universal renormalization of the weak interaction constant. The above statements are exact if $q=o$ (q is the momentum transferred to the leptons). If $q \neq o$ there are some qualifications to be explained later.

Thus these two methods attempt to solve the problem by taking out of the picture the original culprit of the "bare particle calculations": that is, the divergent contributions of the axial current. One method does this by requiring that in the presence of the strong interactions the divergent contributions from the axial current should have the opposite sign to that which obtains in the "bare particle" calculations. The second method, instead, renders finite the asymmetrical contributions of the axial current, by introducing the W meson.

V. The 1/k Method [8]
(A brief description)

Consider for simplicity n β-decay and let us concentrate our attention in the diagrams depicted in fig. (3b):

The matrix element for this diagram can be written in the form

$$M_b = \frac{G_V}{\sqrt{2}} \frac{\alpha}{4\pi^3 i} \int \frac{d^4k \, D_{\mu\nu}(k)}{(k^2 + 2\ell . k + i\varepsilon)} \times$$

$$\times \left| \bar{u}_e (2\ell_\mu + \gamma_\mu \not{k}) \gamma_\rho (1+\gamma_5) v_\nu \right| \times$$

$$\times \left\{ \frac{2p_{2\nu}-k_\nu}{k^2-2p_2.k+i\varepsilon} <p_2|j^W_\rho(o)|p_1>+ P_{\nu\rho}(p_2,p_1,k)\right\} \tag{6}$$

Here $D_{\mu\nu}(k)$ is the photon propagator, $j^W_\rho(o)$ is the total weak interaction current which consists of a vector and an axial vector part and $P_{\nu\rho}$ is a combination of tensors and pseudotensors which satisfies $k_\nu P_{\nu\rho}(p_2,p_1,k) = 0$.

A simple but important property of eq. (6) is that all the contributions of order 1/k (as $k \to 0$) to the hadronic part of the matrix elements are contained in the first term between curly brackets in eq. (6). The quantity $P_{\nu\rho}(p_2,p_1,k)$ is regular as $k \to 0$. Physically the terms of order 1/k (as $k \to o$) come from the graphs in which the photon line is attached to the external proton line. Mathematically this can be shown, for example, by considering

$$T_{\nu\rho}(p_1,p_2,k) = i \int e^{-ikx}<p_2|T[j^{el}_\nu(x)\ j^W_\rho(o)]|p_1>d^4x \tag{7}$$

which, with some qualifications to be explained in section VI, describes the hadronic part of the matrix element. By inserting a complete set of intermediate states we can verify that only the one proton state gives rise to terms of order 1/k (as $k \to 0$) and that the residue of this singularity coincides with the first term in the curly brackets of eq. (6). There are other ways of obtaining this result: some of them are discussed in ref. 8. The term involving k_μ, of course, is not of order 1/k: it has been separated out for reasons of convenience.

We now notice that in the expression proportional to $<p|j^W_\rho(o)|n>$, the k integration can be explicitly

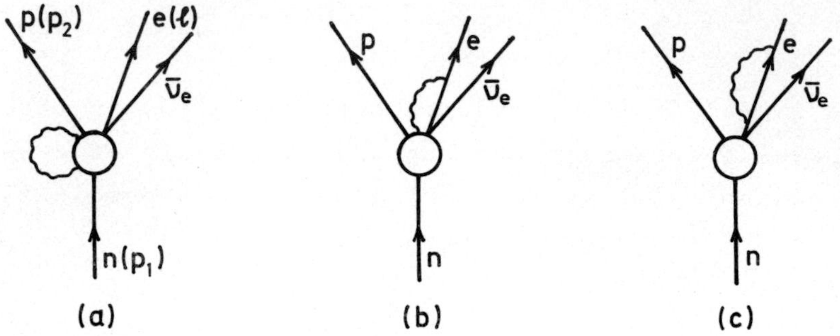

Fig. 3

Radiative corrections to n β-decay in the local Fermi-theor

performed. Let us call $M_b^{(1)}$ the resulting quantity. If we choose for the photon propagator the Landau gauge, $M_b^{(1)}$ is finite. In any other gauge we would simply regularize the integral. The contribution $M_b^{(1)}$ is clearly not invariant under gauge transformations of the photon propagator and is moreover infrared divergent. Next we consider the graphs depicted in fig. (3a) and isolate the infrared divergent part. This part originates in graphs in which the photon is emitted and absorbed by the outgoing proton and is contained in the following expression [16]:

$$M_a^{inf} = -\frac{\alpha}{8\pi^3 i} M_o \int d^4k \frac{D_{\mu\nu}(k)(2p_2-k)_\mu(2p_2-k)_\nu}{(k^2-2p_2.k+i\varepsilon)^2} \tag{8}$$

where

$$M_o = \frac{G_V}{\sqrt{2}} \langle p_2|j_\rho^W(o)|p_1\rangle \bar{u}_e\gamma_\rho(1+\gamma_5)v_\nu \tag{9}$$

is the zeroth order matrix element. Again, for convenience, we have included some terms which are not infrared divergent and we note that M_a^{inf} is obviously finite in the Landau gauge.

Finally we compute diagram (c) with the usual perturbation techniques (we will not write the expression here but it can be found in ref. 8) and we consider the sum $M_a^{inf} + M_b^{(1)} + M_c$. We first verify that the sum is invariant under gauge transformations of the photon propagator: namely we replace

$$D_{\mu\nu}(k) \rightarrow C(k^2)k_\mu k_\nu$$

and verify that the sum of the three new quantities vanishes for arbitrary $C(k^2)$. Next we note that the sum is finite: the easiest way is to specialize to the Landau gauge. It is well known that M_c is finite in this gauge and we have already seen that the same is true for M_a^{inf} and $M_b^{(1)}$. Finally we compute the contribution of $M_a^{inf} + M_b^{(1)} + M_c$ and add that arising from the inner bremsstrahlung. After neglecting terms of order $\alpha E/m_p$, $\alpha E/m_p \ln(m_p/E)$, $\alpha q/M$ and evaluating some small matrix elements not proportional to M_o in the allowed approximation, we find that these contributions are identical to the function $(\alpha/2\pi)\, g(E,E_m,m)$ in eqs. (3) and (4).

What can we say about the terms involving $P_{\nu\rho}$ in eq. (6)? To give an idea of the method consider the integral:

$$\alpha \int \frac{d^4k\, D_{\mu\nu}(k)}{k^2+2\ell.k+i\varepsilon}\, \bar{u}_e \gamma_\mu \not{k} \gamma_\rho (1+\gamma_5) v_\nu P_{\nu\rho}(p_2,p_1,k) \quad (10a)$$

and write

$$[k^2+2\ell.k+i\varepsilon]^{-1} = [k^2+i\varepsilon]^{-1} - 2\ell.k[k^2+i\varepsilon]^{-1}[k^2+2\ell.k+i\varepsilon]^{-1} \quad (10b)$$

Using the fact that $P_{\nu\rho}$ is regular as $k \to 0$, one shows that the term proportional to $\ell.k$ in eq. (10b) gives rise to terms of order $\alpha\frac{E}{M}$ and $\alpha\frac{E}{M}\ln\frac{M}{E}$ where M is some hadronic mass [8] . We argue that these terms are probably very small as they involve the product of two small quantities and we neglect them. Next we use the identity

$$\gamma_\mu\gamma_\lambda\gamma_\rho = g_{\mu\lambda}\gamma_\rho - g_{\mu\rho}\gamma_\lambda + g_{\lambda\rho}\gamma_\mu - i\varepsilon_{\mu\rho\lambda\alpha}\gamma_\alpha\gamma_5 \quad (11)$$

and transform eq. (10a) into a sum of four integrals, of which

$$-\alpha\int \frac{d^4k\ D_{\mu\nu}(k)}{k^2+i\varepsilon} k_\lambda\ P_{\nu\mu}\ \bar{u}_e\gamma_\lambda\ (1+\gamma_5)v_\nu \quad (12)$$

is a typical one. The k integral is a sum of a vector and a pseudovector. If we neglect terms of order $\alpha q/M$, the only possible such quantities are $\bar{u}_p\gamma_\mu u_n$ and $\bar{u}_p\gamma_\mu\gamma_5 u_n$. As the proportionality constants are unknown in this approach, we can say that these contributions are of the form

$$\alpha\left[c\bar{u}_p\gamma_\lambda u_n + d\rho\bar{u}_p\gamma_\lambda\gamma_5 u_n\right]\bar{u}_e\gamma_\lambda(1+\gamma_5)v_\nu \quad (13)$$

where c and d are undetermined constants independent of E and m_e. The same is true for the other integrals involving $P_{\nu\mu}$ and for the contributions of diagram (3b) not contained in M_b^{inf}. To order α , contributions such as eq. (12) simply renormalize the coupling constants G_V and G_A. Therefore, to this order they will not affect those observables which to zero order in α (and in a V-A theory) are independent of the actual values of G_V and G_A. Examples are the shape of the allowed spectrum and the longitudinal polarization of the electron. To order α one finds, for example, for the allowed spectrum [17]:

$$Pdp = \frac{\tilde{\xi}[E_m - E]^2}{2\pi^3} p^2 dp\left[1 + \frac{\alpha}{2\pi} g(E, E_m, m)\right] \tag{14a}$$

$$\tilde{\xi} = |G_V|^2 (1 + \frac{\alpha}{\pi} \text{Re } c)|M_F|^2 + |G_A|^2 (1 + \frac{\alpha}{\pi} \text{Re } d)|M_{GT}|^2 \tag{14b}$$

Thus the correction to the shape of the allowed spectrum is given by a single universal function g which is the same for the Fermi and Gamow-Teller parts of the decay probability. This is very nice and rather unexpected. Moreover, to our approximation, g contains the entire dependence of the radiative corrections on E and m_e. In particular, g has the following remarkable property: although it contains terms which diverge logarithmically as $m_e \to 0$ for fixed $\vec{p}_e$ (mass singularity) such logarithmic terms disappear if we first integrate over all values of E (in the β decay case $\ln m_e$ is replaced by $\ln E_m$). Such behaviour was first suggested as a general property in ref. 3 and later on discussed in detail by Kinoshita [18] and by Lee and Nauenberg [19]. The function g, aside from giving the corrections to the spectrum, also gives a significant correction to the lifetime. For example, if eq. (14) is applied to the decay of O^{14}, we find that the contribution of g decreases the lifetime by 1.3%. This is rather close to the result of the bare V-A calculation when a cutoff $\Lambda = m_p$ is used, and is very close to the result of the current algebra models involving special forms of the space-space commutators, which give $\sim$ 1.23 %. The reason is easy enough to understand. The function g involves very large logarithmic terms such as $3(\alpha/2\pi)\ \ln(m_p/m_e)$. If the mechanism that makes the corrections finite involves a cutoff of the order of m_p, one expects this contribution to be the dominant one. However, it should be stressed that this method does not give a complete calculation of the corrections to the lifetime as the constants c and d in eq. (14b) have not been determined.

Finally, it should be mentioned that although we have described the 1/k method for the case of n β-decay, it can be extended to nuclear β decay in a simple manner. The extension to the case in which the weak interactions occur via an intermediate boson is based on the simple observation that for the purpose of separating the terms of order 1/k, the intermediate boson may be regarded as part of the weak vertex. As was mentioned in section III, the predictions embodied in eqs. (14) have been "verified" by comparison with a number of calculations of radiative corrections to n and π β-decays which assume either the intermediate boson or structure effects of the strong interactions.

VI. The Current Algebra Approach

The general discussion of this approach was developed mainly by Bjorken [9] and by Abers, Dicus, Norton and Quinn [10],[11]. The study of particular models for which the corrections to G_V/G_μ are finite is the work of many authors and is contained in refs. 11 to 15.

Here we will give a simplified version of the general approach and restrict ourselves to study the divergent contributions in the ultraviolet region, although the discussion can be extended to the finite terms. We consider again figs. (3a), (3b) and (3c) and regard the initial and final hadrons of momenta p_1 and p_2 as belonging to the same isomultiplet.

In the Feynman gauge, (3b) can be written as

$$M_b = \tag{15a}$$

$$= - \frac{i\alpha G_V}{4\pi^3\sqrt{2}} \int \frac{d^4k}{k^2+i\varepsilon} [\bar{u}_e \gamma_\mu \frac{1}{\not{\ell}+\not{k}-m_e} \gamma_\rho (1+\gamma_5)v_\nu] T_{\mu\rho}(p_1,p_2,k)$$

In eq. (15a) $T_{\mu\rho}(p_1,p_2,k)$ is the hadronic part of the matrix element and is a combination of second rank tensors and pseudotensors:

$$T_{\mu\rho}(p_1,p_2,k) = V_{\mu\rho} + A_{\mu\rho} \tag{15b}$$

where $V_{\mu\rho}$ and $A_{\mu\rho}$ involve the vector and axial vector currents V_ρ^+ and A_ρ^+, respectively. In order to discuss the divergent parts we can set $\ell = m_e = 0$ in eq. (15a) and obtain

$$M_b^{div} = - \frac{i\alpha G_V}{4\pi^3\sqrt{2}} \int \frac{d^4k}{(k^2+i\varepsilon)^2} k_\lambda [\bar{u}_e \gamma_\mu \gamma_\lambda \gamma_\rho (1+\gamma_5) v_\nu] T_{\mu\rho} \tag{15c}$$

Using eq. (11) this expression can be transformed to:

$$M_b^{div} = - \frac{i\alpha G_V}{4\pi^3\sqrt{2}} \int \frac{d^4k}{(k^2+i\varepsilon)^2} \bar{u}_e [k_\mu T_{\mu\rho}\gamma_\rho + \gamma_\mu T_{\mu\rho} k_\rho - \not{k} T_{\mu\mu} - i\varepsilon_{\mu\lambda\rho\alpha}\gamma_\alpha T_{\mu\rho} k_\lambda](1+\gamma_5)v_\nu \tag{15d}$$

At this stage we specialize our study to the corrections to the Fermi part of the decay probability, as this is the most interesting quantity in discussing the universality of the weak interactions. After we have performed the k integration we must obtain a vector in the hadronic part of the matrix element. Therefore, the first three terms contribute to the Fermi part via $V_{\mu\rho}$, the last one through $A_{\mu\rho}$. Furthermore, we use the equations

$$k_\mu \; V_{\mu\rho} = \langle p_2 | V_\rho^+ | p_1 \rangle \tag{16a}$$

$$k_\rho \; V_{\mu\rho} = \langle p_2 | V_\mu^+ | p_1 \rangle + O(\alpha) \tag{16b}$$

and obtain the following simplified expression:

$$M_b^{div} = \frac{i\alpha G_V}{4\pi^3\sqrt{2}} \, L_\rho \int \frac{d^4k}{(k^2+i\varepsilon)^2} \;\times$$

$$\times \left\{- 2\langle p_2 | V_\rho^+ | p_1 \rangle + k_\rho V_{\mu\mu} + i\varepsilon_{\mu\lambda\alpha\rho} A_{\mu\alpha} k_\lambda \right\} \tag{17a}$$

where

$$L_\rho = \bar{u}_e \, \gamma_\rho (1+\gamma_5) v_\nu \tag{17b}$$

is the leptonic part of the matrix element. In eq. (17a) we have neglected terms of order α^2 .

Eq. (16a) is very general: it can be derived by inserting a photon of four momentum k_μ and polarization ε_μ in all possible lines of the zeroth order matrix element (including the lepton line) and demanding that the total matrix element should vanish when $\varepsilon_\mu \to k_\mu$ (gauge invariance).

Let us assume a theory in which the interaction of the hadrons with the electromagnetic field is linear in $\mathcal{A}_\mu$ and given by

$$\mathcal{L} = e\, j_\mu^{el} \mathcal{A}_\mu \tag{18}$$

This implies that j_μ^{el} does not involve derivatives of fields, otherwise we would get terms of higher order in $\mathcal{A}_\mu$. Let us further assume that $V_{\mu\rho}$ can be represented by

$$V_{\mu\rho}(k, p_1, p_2) = i\int d^4x\, e^{-ikx} \langle p_2 | T[j_\mu^{el}(x) V_\rho^+(o)] | p_1 \rangle \tag{19}$$

Using the equal time commutation relation of eq. (5) and observing that the integral

$$\int d^4x\, e^{-ikx} \langle p_2 | T[j_\mu^{el}(x) \partial_\rho V_\rho^+(o)] | p_1 \rangle$$

is of order α, we readily obtain eq. (16b).

Next we turn our attention to the matrix element of fig. (3a) which in the Feynman gauge reads:

$$M_a = -\frac{i\alpha G_V}{8\pi^3\sqrt{2}} L_\lambda \int \frac{d^4k}{k^2+i\varepsilon} V_{\lambda\mu\mu}(k, q = p_1 - p_2, p_1, p_2) \tag{20a}$$

Let us again assume that we can represent $V_{\lambda\mu\nu}$ as the Fourier transform of the appropriate time ordered products of currents:

$$V_{\lambda\mu\nu}(k, q, p_1, p_2) =$$

$$= \int d^4x \int d^4y\, e^{iq.x} e^{ik.y} \langle p_2 | T[V_\lambda^+(x) j_\mu^{el}(y) j_\nu^{el}(o)] | p_1 \rangle - \\ - \delta M_{\lambda\mu\nu} \tag{20b}$$

In eq. (20b) $\delta M_{\lambda\mu\nu}$ is a tensor such that when inserted in eq. (20a) gives the electromagnetic contribution to the mass of the initial and final hadrons. It plays the role of the usual mass counterterms in quantum electrodynamics and it cancels the 1/q poles (at $q = p_1 - p_2$) of the time ordered product. Therefore $V_{\lambda\mu\nu}$ is free from poles at $q = p_1 - p_2$. Next we make use of a very interesting relation which states that [9],[10],[11]

$$V_{\lambda\mu\nu}(k,p_1-p_2,p_1,p_2) = \frac{\partial}{\partial k_\lambda} V_{\nu\mu}(k,p_1,p_2) + O(\alpha) \quad (21)$$

and which will be discussed later. Inserting eq. (21) into (20a), performing a partial integration and neglecting terms of order α^2:

$$M_a = - \frac{i\alpha G_V}{4\pi^3\sqrt{2}} L_\rho \int \frac{d^4k}{(k^2+i\varepsilon)^2} k_\rho V_{\mu\mu} + \text{finite term} \quad (22)$$

The finite term in eq. (22) is a "surface term" that can be computed by properly regularizing the integral [11],[20].

Comparing eqs. (17a) and (22) we find that the divergent terms involving $V_{\mu\mu}$, which clearly depend on the details of the strong interactions, exactly cancel. This is a crucial point in the argument. Let us now call $M_b^{(V)}$ the contribution to M_b arising solely from the vector current. Adding the divergent part of $M_b^{(V)}$ which is given by the first two terms in eq. (17a), the divergent part of M_a (eq. (22)) and diagram (3c) which is computed with the usual perturbation methods of quantum electrodynamics, we find

$$(M_a + M_b^{(V)} + M_c)^{div} = \frac{3\alpha M_0}{8\pi^3 i} \int \frac{d^4k}{(k^2+i\varepsilon)^2} \sim \frac{3\alpha}{4\pi} M_0 \ln \Lambda \quad (23)$$

where Λ is a cut-off [20].

Thus we obtain the very interesting result that the

contribution from V_ρ^+ contains a divergent contribution which is independent of the details of the strong interactions. As we mentioned in section IV, this result can be extended to the finite terms: the entire contribution of V_ρ^+ is model independent and, therefore, given by eqs. (3) and (4) with $\rho=0$, provided that we neglect terms of order $\alpha E/M$, $\alpha E/M \ln(M/E)$, $\alpha q/M$ and evaluate small matrix elements of order α in the allowed approximation. There is a slight qualification: the finite "surface term" of eq. (22) is slightly model dependent. The exact equality with eqs. (3) and (4) holds for those models in which the commutator

$$\delta(x_o)[j_\mu^{el}(x),\ V_\mu^+(o)]$$

is equal to the value one obtains in the quark model. In any case, this particular term is very small in those models.

We now turn our attention to eq. (21) and briefly sketch its derivation by means of a divergence condition method [11]. We contract eq. (20b) with q_λ and perform a partial integration on the variable x. Using the relation

$$\begin{aligned}\frac{\partial}{\partial x_\lambda} T[V_\lambda^+(x)\ j_\mu^{el}(y) j_\nu^{el}(o)] &= T[\partial_\lambda V_\lambda^+(x)\ j_\mu^{el}(y) j_\nu^{el}(o)] + \\ &+ \delta(x_o-y_o)\{\Theta(x_o)[V_o^+(x), j_\mu^{el}(y)] j_\nu^{el}(o) + \\ &+ \Theta(-x_o) j_\nu^{el}(o)[V_o^+(x), j_\mu^{el}(y)]\} + \\ &+ \delta(x_o)\{\Theta(x_o-y_o)[V_o^+(x), j_\nu^{el}(o)] j_\mu^{el}(y) + \\ &+ \Theta(y_o-x_o) j_\mu^{el}(y)[V_o^+(x), j_\nu^{el}(o)]\}\end{aligned} \qquad (24)$$

which is obtained by straightforward differentiation, and the equal time commutation relation of eq. (5), one obtains

$$q_\lambda V_{\lambda\mu\nu} = V_{\mu\nu}(-k,p_1,p_2) +$$

$$+ i\int d^4y\, e^{i(k+q).y} \langle p_2|T[V^+_\mu(y)\, j^{el}_\nu(o)]|p_1\rangle +$$

$$+ M_{\mu\nu} - q_\lambda \delta M_{\lambda\mu\nu} \qquad (25a)$$

where

$$M_{\mu\nu} =$$

$$= - i\int d^4x\, d^4y\, e^{iq.x}\, e^{ik.y} \langle p_2|T[\partial_\lambda V^+_\lambda(x) j^{el}_\mu(y)\, \times$$

$$\times\, j^{el}_\nu(o)]|p_1\rangle \qquad (25b)$$

Differentiating with respect to q_α :

$$V_{\alpha\mu\nu} = - q_\lambda \frac{\partial V_{\lambda\mu\nu}}{\partial q_\alpha} +$$

$$+ i \frac{\partial}{\partial k_\alpha} \int d^4y\, e^{i(k+q).y} \langle p_2|T[V^+_\mu(y) j^{el}_\nu(o)]|p_1\rangle +$$

$$+ \frac{\partial}{\partial q_\alpha} (M_{\mu\nu} - q_\lambda \delta M_{\lambda\mu\nu}) \qquad (25c)$$

As we mentioned before, $V_{\lambda\mu\nu}$ does not contain poles at $q = p_1 - p_2$ and the same holds for $M_{\mu\nu} - q_\lambda \delta M_{\lambda\mu\nu}$. Setting $q = p_1 - p_2$ we see that the first term on the right hand side of (25c) is $O(q)$ and that the last term is $O(\alpha)$ as it involves $\partial_\rho V^+_\rho$. Neglecting these terms, making use of translational invariance and remembering eq. (19), we readily obtain eq. (21).

In the derivation of eqs. (16b) and (21) we made use of eq. (5) and we assume that the hadronic parts of the matrix elements can be represented by the Fourier transforms of the appropriate time-ordered products of currents, as in eqs. (19) and (20b). If the vector currents explicitly contain the electromagnetic field $\mathcal{A}_\mu$ (as would be in general the case if the currents involve derivatives of fields), one must add to the time ordered products certain terms involving the functional derivatives of the currents with respect to the electromagnetic field. But then the current commutators are also altered by operator Schwinger terms. The derivations of eqs. (16b) and (21) have been studied in a model in which the vector currents obtain contributions from a triplet of spin zero fields (π-mesons for example) and the relations are found to hold still [11].

We want to emphasize that what is crucial to the argument is the validity of eqs. (16b) and (21) (or 25a). These are actually divergence conditions, not unlike the Takahashi-Ward identities. Whether or not there are operator Schwinger terms, the essential requirement for the argument is that in the evaluation of the divergence relations these terms should be cancelled by the contributions to the matrix elements not contained in the T products. As has been emphasized by Bell, such situation naturally occurs if gauge properties rather than commutation relations are "abstracted" from the models constructed with the underlying fields [21].

VII. Models Involving Special Assumptions about the Space-Space Part of the Current Algebra

We turn our attention to the term involving $A_{\mu\alpha}$ in eqs. (17a) and assume that $A_{\mu\alpha}$ admits a representation analogous to eq. (19) with $V^+_\rho(o) \to A^+_\rho(o)$. According to an argument due to Bjorken [9], the behaviour of $A_{\mu\alpha}$ as $k_o \to \infty$ is given by

$$A_{\mu\alpha} = \frac{1}{k_o} \int d^3x \langle p_2 | [j_\mu(\vec{x},o), A^+_\alpha(o)] | p_1 \rangle e^{i\vec{k}.\vec{x}} + O(\frac{1}{k^2}) \tag{26}$$

This eq. can be obtained, for example, by inserting a complete set of intermediate states, performing the x_o integration and assuming that one can restrict oneself to intermediate states of finite energies. (In this case one can pull out k_o from the energy denominators.)

To visualize what eq. (26) implies, it is convenient to introduce models of underlying spin 1/2 fields ψ in which the axial vector and electromagnetic currents are given by

$$A^+_\alpha = \psi^+ \gamma_o \gamma_\alpha \gamma_5 T^+ \psi \tag{27a}$$

$$j^{el}_\mu = \psi^+ \gamma_o \gamma_\mu Q \psi \tag{27b}$$

where T^+ and Q are the isospin and charge operators. As $A_{\mu\alpha}$ is contracted with $\varepsilon_{\mu\lambda\alpha\rho}$ in eq. (17a), only the antisymmetric part of $A_{\mu\alpha}$ is relevant. Therefore we compute

$$[j^{el}_\mu(\vec{x},o), A^+_\alpha(o)] - \mu \leftrightarrow \alpha = - 2i\delta^3(\vec{x}) \times \times \varepsilon_{o\alpha\mu\nu} \psi^+ \gamma_o \gamma_\nu [T^+, Q]_+ \psi \tag{27c}$$

Let us further assume that ψ contains one isodoublet but no other isomultiplets except isoscalars. Then using the Gell-Mann-Nishijima formula we see that $[T^+,Q]_+ = 2\bar{Q}\, T^+$ where $\bar{Q}$ is the average charge of the underlying fields which belong to the fundamental isodoublet. Inserting then eq. (27c) into eq. (26) and writing the result in a covariant manner one obtains:

$$\frac{1}{2}[A_{\mu\alpha}-A_{\alpha\mu}] \to -\, 2i\, \frac{\bar{Q}\varepsilon_{\sigma\alpha\mu\nu}k_\sigma}{k^2} <p_2|V_\nu|p_1> + O(\frac{1}{k^2}) \quad (27d)$$

When this expression is introduced into eq. (17a) and its contribution added to eq. (23) we obtain a total divergent contribution

$$(M_a+M_o^{(V)}+M_b^{(A)}+M_c)^{div} = \frac{3\alpha}{8\pi^3 i}\, \overset{o}{M}\, [1+2\bar{Q}] \int \frac{d^4k}{(k^2+i\varepsilon)^2} \quad (28)$$

For the quark algebra $\bar{Q} = \frac{1}{6}$ and for the algebra of fields [22] $\bar{Q} = 0$ (this is so because the space-space commutators cancel in this model) and we get a divergent answer for these two models, at least in the "local" Fermi theory. In refs. 12), 13) and 14) models of underlying particles with $\bar{Q} = -1/2$, in particular the SUB-model involving three triplets of integer charge particles, have been extensively discussed. These models give rise to finite second order radiative corrections.

The pseudotensor $A_{\mu\alpha}$ also contributes finite terms which have been evaluated approximatively in ref. 11). Fortunately, these model dependent contributions from $A_{\mu\alpha}$ turn out to be very small. Adding these contributions to those arising from the vector current one obtains

$$\Delta P\, d^3p = P^o d^3p\, \{\frac{\alpha}{2\pi}\, [3\, \ln \frac{\Lambda}{m_p} + g(E,E_m,m_e)] +$$

$$+ \frac{\alpha}{2\pi} 6 \bar{Q} \ln \frac{\Lambda}{m_{A_1}} - 1.2 \times 10^{-3} \} \qquad (29)$$

where $g(E,E_m,m_e)$ is the universal function defined in eq . (4) and m_{A_1} is the mass of the A_1 meson.

In equation (29) the expression between square brackets is the model independent contribution of the vector current while the last two terms are the contribution of the axial current. We note that eq. (29) has the structure required by the 1/k method, namely the corrections contain the function $g(E,E_m,m_e)$ plus constant terms. For $\bar{Q} = -1/2$, eq. (29) gives a finite answer which is completely dominated by the contribution of g. For example, for O^{14} and $\bar{Q} = -1/2$, eq. (29) gives a correction to the lifetime of -1.23% while the function g gives a correction of -1.30%. These values are also close to the ones obtained with the old "bare particle calculation" of ref. 3 when a cutoff $\Lambda = m_p$ is used in the latter. In fact, such calculation gives (when small terms of order $\alpha\ m_p/\Lambda$ are not included) a correction of - 1.56%. The reason, of course, is the large contribution of the universal function g.

VIII. Models Involving the Intermediate Boson

This model has been recently discussed in connection with the radiative corrections in refs. 11 and 15.

We assume that the weak interactions are mediated by a vector boson, and concentrate our attention in those diagrams involving the second order radiative corrections and the strong interactions. These are depicted in figs. (4a), (4b) and (4c).

The matrix element for diagram (4b) is

$$\tilde{M}_b = \frac{-i\alpha G_V M_W^2}{4\pi^3 \sqrt{2}} \int \frac{d^4k}{k^2+i\varepsilon} T_{\mu\rho} \frac{[-g_{\rho\lambda}+(k_\rho k_\lambda/M_W^2)]}{k^2-M_W^2 + i\varepsilon} \times$$

$$\times [\bar{u}_e \gamma_\mu \frac{1}{\not{l}+\not{k}-m_e+i\varepsilon} \gamma_\lambda (1+\gamma_5)\, v_\nu] \qquad (30a)$$

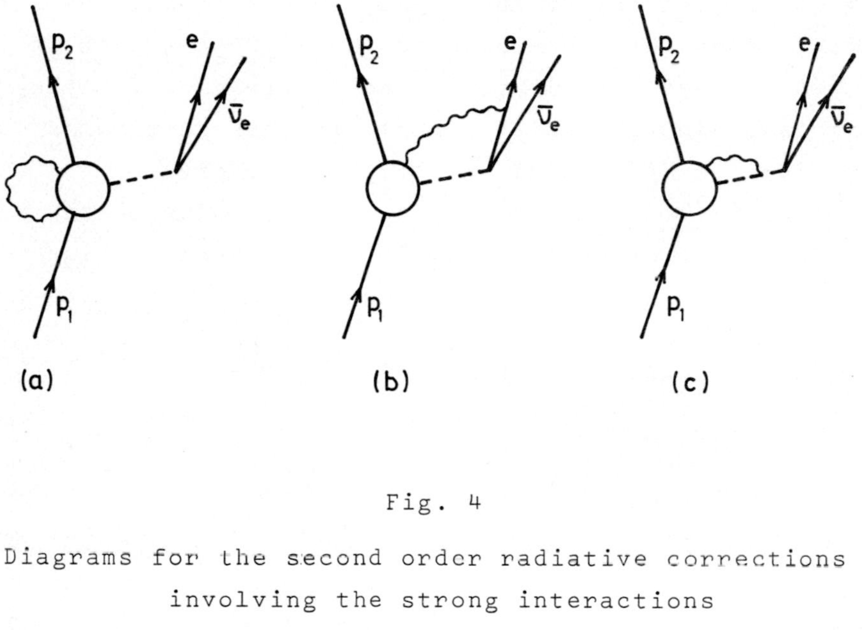

Fig. 4

Diagrams for the second order radiative corrections involving the strong interactions

In eq. (30a) $T_{\mu\rho}$ is again the hadronic part of the matrix element as discussed in section VI. We have also set q=0 where q is the total momentum transfer to the leptons, as terms of order αq are neglected in the current algebra approach. Assuming that $T_{\mu\rho}$ tends at most to a constant as $k_o \to \infty$, the divergent part arises from the term involving $k_\rho k_\lambda$ in the W propagator. The divergent part of eq. (30a) is

$$\tilde{M}_b^{div} = \frac{-i\alpha G_V}{4\pi^3\sqrt{2}} L_\mu \int \frac{d^4k}{k^2+i\varepsilon} \frac{T_{\mu\rho}\, k_\rho}{k^2-M_W^2+i\varepsilon} \qquad (30b)$$

Only $V_{\mu\rho}$ contributes here to the Fermi part of the amplitude and using eq. (16b) we obtain up to terms of order α:

$$\tilde{M}_b^{div} = -\frac{i\alpha}{4\pi^3} M_o \int \frac{d^4k}{k^2+i\varepsilon} \frac{1}{k^2-M_W^2+i\varepsilon} \qquad (30c)$$

Thus, from diagram (4b) we obtain a divergent integral independent of the strong interactions. The axial current does not contribute to the divergent part of $\tilde{M}_b$ but it does contribute to the finite part via the term involving $g_{\rho\lambda}$ in eq. (30a). In this particular contribution M_W acts as an effective cutoff.

Diagram (4c) involves the electrodynamics of the vector boson. Here we use an interaction Lagrangian obtained from the free Lagrangian

$$\mathcal{L}^W_{free} = -m^2 W^+_\mu W_\mu - \frac{1}{2} G_{\mu\nu} G^+_{\mu\nu} \qquad (31a)$$

where W_μ is the boson field and

$$G_{\mu\nu} = \partial_\mu W_\nu - \partial_\nu W_\mu \qquad (31b)$$

by means of the replacement [23]

$$i\,\partial_\mu W_\nu \rightarrow (i\,\partial_\mu - e\mathcal{A}_\mu) W_\nu \qquad (31c)$$

where e is the charge of the particle annihilated by W_ν. We will call the interaction Lagrangian obtained in this manner the Proca-Wentzel Lagrangian [24].

With this interaction and again setting $q = 0$, diagram (4c) reads

$$\tilde{M}_c = \frac{i\alpha G_V}{4\pi^3\sqrt{2}} \int \frac{d^4k}{k^2+i\varepsilon} \frac{T_{\mu\rho}}{k^2-M_W^2+i\varepsilon} \left[- g_{\rho\lambda} + \frac{k_\rho k_\lambda}{M_W^2} \right] \times$$

$$\times \left[k_\mu\ g_{\lambda\alpha} - g_{\lambda\mu}\ k_\alpha\right] L_\alpha \qquad (32a)$$

Now the term $k_\rho k_\lambda$ gives zero contribution and we obtain

$$\tilde{M}_c = - \frac{i\alpha G_V}{4\pi^3\sqrt{2}} \int \frac{d^4k}{k^2+i\varepsilon} \frac{1}{k^2-M_W^2+i\varepsilon} \{\langle p_2 | V_\rho | p_1\rangle -$$

$$- V_{\mu\mu}\ k_\rho\} L_\rho \qquad (32b)$$

Diagram (4a) in the limit $q = 0$ is still given by eq. (22).

The divergent part of $\tilde{M}_a$ involves the tensor $V_{\mu\mu}$: by inspection of eqs. (22) and (32b) we see that this part combines with the second term in eq. (32b) to give a finite contribution. Therefore, adding eqs. (22), (30c) and (32b) we find

$$\tilde{M}_a^{div} + \tilde{M}_b^{div} + \tilde{M}_c^{div} = \frac{\alpha}{2\pi^3 i} M_o \int \frac{d^4k}{k^2+i\varepsilon} \frac{1}{k^2-M_W^2+i\varepsilon} \qquad (33)$$

Again we obtain the rather remarkable result that the diagrams involving the strong interactions give rise to a model independent divergent integral. There is an important difference with the "local" Fermi case: The axial vector current does not give divergent contributions although it contributes to the finite parts.
To check the universality of the divergent parts we now examine the three diagrams in μ decay which correspond to fig. 4. These are

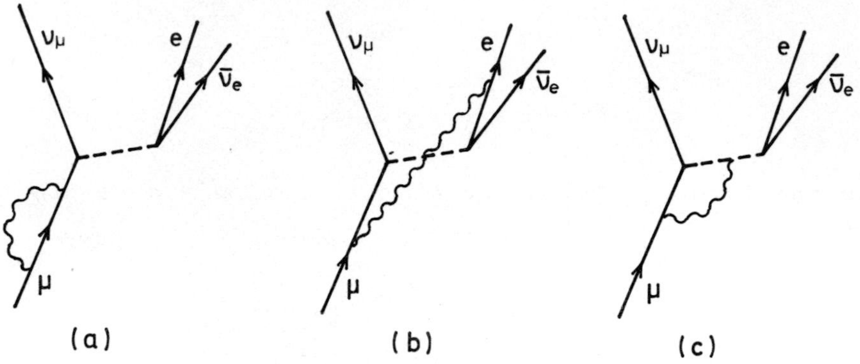

Fig. 5

Diagrams in μ decay corresponding to those of fig. 4 .

We compute these three diagrams in perturbation theory and we find that the sum of the divergent parts is in fact the zeroth order matrix element for μ decay multiplied by the same divergent integral as in eq. (33). Note that in treating μ decay we must consider the corrections to both the vector and axial vector couplings, as these are unseparable in this process.

Aside from the diagrams depicted in figs. 4 and 5 there are others in which the photon is emitted and absorbed by W, or emitted by W and absorbed by the electron, or emitted and absorbed by the electron. These diagrams have nothing to do with the strong interactions and, moreover, they are clearly identical in β and μ decay if we neglect terms of order αq in both processes. When one considers the terms of order αq in μ decay, there appear according to the calculations of Bailin [25] some uncompensated logarithmic divergences of the form $(\alpha/2\pi)(q^2/M_W^2)\ \ln(\Lambda/M_W)$ where Λ is a cutoff. It is widely believed that at momenta of the order $\Lambda \simeq$ 300 BeV the present description of weak interactions will be considerably modified, possibly by the emergence of higher order effects. In μ decay $q^2 \leqq m_\mu^2$ and we also know that $M_W > 2$BeV. Assuming Λ = 300 BeV we see that these terms are smaller than $(\alpha/2\pi)10^{-2}$, or about three orders of magnitude smaller than the finite corrections to $(G_V/G_\mu)^2$ which are of order $15(\alpha/2\pi)$. Putting this in other words, we would need a value of $\Lambda \sim 10^{300}\ M_W$ to make these terms one tenth as large as the finite terms. Therefore if seems very plausible that these contributions do not play an important quantitative role and we neglect them. It is to be hoped that future theory will clarify the nature of these terms and the mechanism that renders them finite.

The discussion can be extended to discuss all the finite terms. Provided that terms of order $\alpha m_\mu^2/M_W^2$ are neglected in μ decay and terms of order $\alpha E/M, \alpha E/M\ \ln M/E$ and $\alpha q/M$ are neglected in β decay, the second order corrections from the vector current are identical to those obtained in the "local" theory. The corrections from the axial current are essentially the same with the important proviso that $\ln\Lambda$ is replaced by $\ln M_W$ which corresponds to the fact that these corrections are now finite. It is then easy to see that the corrections of order α to the ratio $(G_V/G_\mu)^2 = \cos^2\theta$ can be simply obtained by using the old "local" calculation for the μ decay lifetime (eq.(2)) and using for β decay equation (29) with $\Lambda \to M_W$, that is

$$\Delta P\ d^3p = P^0 d^3p\{\frac{\alpha}{2\pi}[3\ \ln \frac{M_W}{m_p} + g(E,E_m,m_e)] + \\ + \frac{\alpha}{2\pi}\ 6\ \bar{Q}\ (\frac{M_W}{m_{A_1}}) - 1.2 \times 10^{-3}\} \qquad (34)$$

We emphasize that eq. (34) is in agreement with the general properties determined with the 1/k method.

IX. Present Situation Regarding the Universality of the Weak Interactions

We briefly review the experimental situation regarding the universality of the weak interactions in the framework of the models of radiative corrections explained in sections VII and VIII All the numbers are taken from ref. 26.

Using the latest data on the $\tilde{f}t$ value of O^{14} and the radiative corrections calculated in the $\bar{Q} = -1/2$ model of section VII we find

$$\cos\Theta = 0.9778 \pm 0.0018 \quad (O^{14},\ \bar{Q} = -1/2) \tag{35a}$$

or

$$\sin\Theta = 0.2095 \pm 0.0086 \quad (O^{14},\ \bar{Q} = -1/2) \tag{35b}$$

To check the universality of the weak interactions one attempts to determine Θ from K_{e3} decays. Unfortunately, the best one can do experimentally is to determine $f_+(o)\sin\Theta$:

$$f_+(o)\sin\Theta = 0.21 \pm 0.01 \quad (K_{e3}) \tag{35c}$$

Here $f_+(o)$ is the form factor that governs K_{e3} decays evaluated at $q^2=0$. The quantity $f_+(o) - 1$ is of second order in SU(3) breaking. Assuming that this is negligible, we would have $f_+(o) = 1$ and $\sin\Theta = 0.21 \pm 0.01$. The agreement between this value and eq. (35b) is as good and spectacular as one can possibly ask. However, this agreement may turn out to be misleading. For one reason it is not clear that $f_+(o) - 1$ is negligible: as Prof. Glashow will no doubt tell you, there are recent calculations of symmetry breaking [27] which give $f_+(o) = 0.85$. Assuming this value one would get from eq. (35c) $\sin\Theta = 0.247 \pm 0.012$ and the very good agreement with (35b) would be lost.

In the models involving W (section VIII) we do not get a definite number for $\cos\Theta$ from β decay because we do not know M_W and we must choose $\bar{Q}$. The best one can do at present is to choose a reasonable algebra and a reasonable range of values for M_W/m_p. As an example, for the quark algebra ($\bar{Q} = 1/6$) we can obtain, using the $\tilde{f}t$ value of O^{14}, the following table for the W models

M_W/m_p	$\sin\Theta$ (O^{14}, $\bar{Q} = 1/6$, W model)
2.72	0.219
7.39	0.230
20.1	0.240
54.6	0.249

For simplicity we have not included the experimental error here: it is about 4 %.

We see that the effect of W in the model of section VIII is to increase $\sin\Theta$. We emphasize that if $f_+(o)$ is as low as 0.85, we can now obtain consistency with universality if we use the W model with reasonable values of M_W.

It is clear that, insofar as we can use the data from O^{14}, the two models of sections VII and VIII and universality are perfectly compatible with the experimental data. In order to make the comparisons more meaningful it is necessary to solve now some problems which have nothing to do with the electromagnetic corrections. One is to clarify the situation regarding second order SU(3) breaking corrections. Another one, as emphasized in ref. 26, is to explain the low $\tilde{f}t$ value of Al^{26}.

I am indebted to S. L. Glashow, M. Nauenberg and F. Capra for very stimulating conversations.

References

1. R. E. Behrends, R. J. Finkelstein and A. Sirlin, Phys. Rev. 101, 866 (1956).
2. S. M. Berman, Phys. Rev. 112, 267 (1958).
3. T. Kinoshita and A. Sirlin, Phys. Rev. 113, 1652 (1959).
4. R. P. Feynman, Proceedings of the 1960 Annual International Conference on High-Energy Physics at Rochester (Interscience Publ. N. Y., 1960), p.501, and Symmetries in Elementary Particle Physics (Academic Press, N. Y., 1965), p. 111.
5. S. M. Berman and A. Sirlin, Ann. Phys. 20, 20 (1962). Eqs. (3) and (4) look simpler than the original expressions of ref. 5 (and also of ref. 3 which specialized to an exact V-A "bare calculation"). The reason is that one can simplify the formula by using a relationship between Spence functions as explained, for example, in ref. 8.
6. Ref. 5 contains more references to the early papers. The list of references given in these lectures is by no means complete.
7. G. Källén, Nuclear Physics B1, 225 (1967).

8. A. Sirlin, Phys. Rev. 164, 1767 (1967).
9. J. D. Bjorken, Phys. Rev. 148 , 1467 (1966) and Lectures at International School of Physics "Enrico Fermi", XLI Course, Varenna, Italy (1967).
10. A. S. Abers, R. E. Norton and D. A. Dicus, Phys. Rev. Lett. 18, 676 (1967).
11. E. S. Abers, D. A. Dicus, R. E. Norton and H. R. Quinn, "Radiative Corrections to the Fermi Part of Strangeness Conserving ß-decay", preprint (1967).
12. N. Cabibbo, L. Maiani and G. Preparata, Phys. Lett. 25B, 31 (1967).
13. N. Cabibbo, L. Maiani and G. Preparata, Phys. Lett. 25B, 132 (1967).
14. K. Johnson, F. E. Low and H. Suura, Phys. Rev. Lett. 18, 1224 (1967).
15. A. Sirlin, Phys. Rev. Lett. 19, 877 (1967).
16. For a rather general discussion of infrared divergence, we refer to the papers of D. R. Yennie, S. C. Frautschi and H. Suura, Ann. of Physics 13, 379 (1961), and N. Meister and D. R. Yennie, Phys. Rev. 130, 1210 (1962). Eq. (8) is a convection-convection contribution in the classification introduced in the latter paper.
17. Most of the contributions in $M_a^{inf} + M_b^{(1)} + M_c$ are proportional to M_o. However, there are some small induced terms of order α which behave as scalars and tensors. In the allowed approximation their interference with M_0 can be rigorously computed and have been included in the function g. It is rather remarkable that the coefficients of these terms are such that they do not prevent the emergence of a single universal function g in the corrections to the spectrum.

18. T. Kinoshita, Journal of Math. Phys. 3, 650 (1962).
19. T. D. Lee and M. Nauenberg, Phys. Rev. 133B, 1549 (1964).
20. All the divergent quantities in these notes are to be regarded as integrals whose precise mathematical definition is obtained by regularization and, if necessary, by the introduction of a photon mass.
21. J. S. Bell, Nuovo Cim. 50, 129 (1967).
22. T. D. Lee, S. Weinberg and B. Zumino, Phys. Rev. Lett. 18, 1029 (1967).
23. The indices μ run from 0 to 3. The scalar product of two four-vectors is given by $a.b = a_o b_o - a_i b_i$. The symbol $\partial_\mu \equiv \frac{\partial}{\partial x^\mu}$ (where $x^o = ct$ is real). The tensor $g_{\mu\lambda}$ has the following non-vanishing components: $g_{oo} = -g_{ii} = 1$ (no summation over i).
24. A. Proca, J. Phys. Radium 7, 347 (1936); G. Wentzel, Quantum Theory of Fields (Interscience Publishers, Inc., New York 1949).
25. D. Bailin, Phys. Rev. 135, B166 (1964); Nuovo Cim. 40A, 822 (1965).
26. N. Brene, M. Roos and A. Sirlin, "Status of the Cabibbo Theory", Nuclear Physics (to be published).
27. S. L. Glashow and S. Weinberg, Phys. Rev. Lett. 20, 224 (1968).

FIELD THEORY OF GRAVITATION†

By

W. THIRRING

Institute for Theoretical Physics, University of Vienna, Austria

Although we do not have much detailed experimental information about gravitation [1] it is clear that it is an amazing and unique interaction. The essential feature of gravitation appears in Newton's legendary vision under the apple tree. He realized a certain universality of gravitation: We, the apple, the moon, all are subjected to it. As we shall see this universality goes far beyond the universality of other interactions and leads to the most fascinating consequences. Gravitation is the weakest interaction we know. Between two protons the gravitational attraction is 8.10^{-36} less than the electric force:

$$\kappa m_p^2 = 8.10^{-36}\ e^2 \ . \tag{1}$$

---

† Lecture given at the VII. Internationale Universitätswochen f.Kernphysik,Schladming,February 26-March 9, 1968.

Thus in units where $\hbar = c = m_p = 1$, the electric interactions are characterized by the fine structure constant 1/137, the corresponding coupling constant for the weak interactions is 10^{-5}, and for gravitation 5.10^{-38}. This enormous difference in orders of magnitude can be illustrated in various ways. Thus the Bohr-radius due to gravitation of two neutrons would be somewhat larger than the radius of the universe. Similarly the time a neutron in a nucleus needs to emit a graviton would be of the order of the lifetime of the universe. This makes it clear why quantum features of gravitation (if they exist) have sofar escaped detection and we shall not discuss them further.

Although powerless on the elementary particle level, gravitation becomes dominant on the cosmic scale. This is due to its universal nature by which the gravitational effects of all particles (about 10^{58}) in a star add up. These accumulative effects are impressively demonstrated by the "gravitational collapse": If a star cools down the gravitational attraction is counterbalanced mainly by the zero-point pressure of the electrons which usually become relativistic at this stage. In this case their kinetic energy is of the order (in our units) [2]

$$E_K = \frac{[N(\text{number of electrons})]^{4/3}}{R(\text{radius of the star})} \tag{2}$$

whereas the gravitational energy is

$$E_G = -\frac{N^2 \ \kappa \ m_p^2}{R} \ . \tag{3}$$

Thus for $N^{2/3} > 10^{38}$, or stars somewhat larger than the sun, gravitation is winning out and the electrons are not going to stop the contraction due to gravitation. The next stage where one might hope that the collapse comes to rest is the one where nuclear density is reached. Although at 10^{-13} cm the nuclear forces may have both signs, at about half this distance nucleons seem to repel each other violently (hard core). If we assume that the hard core is 1 GeV $\approx m_p$ high and the nucleons are so compressed that each is inside the hard core of its neighbour we get a repulsive energy

$$E_C = m_p N \quad . \tag{4}$$

Comparing with (3) we see again that for about as many particles as are in a normal star the hard core energy is of the order of the gravitational energy. For larger objects the latter dominates and the thing will contract further [3]. Then the nucleons will become extrem-relativistic and, which then becomes stronger, depends on the relativistic nature of the hard core. Since this is not known for sure this question is open to speculation.

We shall not further pursue these quantitative features but discuss one consequence of the universality of gravitation which seemed so fascinating to some people and outrageous to others: it determines the geometry of space-time. To see how this comes about and which other interactions would share this property we have to gradually sharpen the somewhat vague notion of universality. One formulation would refer to Galileo's experiment. If one drops two bodies at the same time from the same point they will always stay together. In other words the trajectories of particles in a gravitational field depend only on the initial conditions and are the same for particles of the same internal structure. (What this last qualification means will turn out later). We shall denote this property by strong universality since it requires a little more than the kind of universality as we find in weak or electric interactions. For instance, electrodynamics would possess strong universality only if all charges were proportional to the masses of the respective particles. More generally any theory where the coupling constant is proportional to the mass will have strong universality since the mass will cancel out in the equations of motion.

The empirical basis for strong universality of gravitation has now become exceedingly accurate because the $K^o \to 2\pi$ decay provides us with an excellent means for measuring any deviation from it. The argument which is due to M. L. Good [4] runs as follows: If the gravitational potentials were different between K^o and its antiparticle $\bar{K}^o$ they would introduce transitions between $K_1 = \frac{1}{\sqrt{2}}(K^o-\bar{K}^o)$ and $K_2 = \frac{1}{\sqrt{2}}(K^o+\bar{K}^o)$. This would result in a rapid decay of K_2 into 2π and from the observed $K_2 \to 2\pi$ decay and its velocity dependence one gets a limit for the assumed difference in gravitational interaction of K^o and $\bar{K}^o$. Because of the small mass-difference of K_1 and K_2 they can be mixed very easily. Therefore this test is far more sensitive than Dicke's improvement of Eötvös' experiment, which we shall discuss subsequently. To work out Good's argument quantitatively we have to agree on what kind of gravitational theories we shall discuss. I shall only consider as reasonable candidates theories which are build according to the pattern of electrodynamics. That is to say I am looking for Lorentz-covariant theories where the motion of bodies described by $Z^i(S)$ and the field equations are derived from the same Lagrangian [5]. The latter consists of a field part L_f, a matter part

$$L_m = \int dS \, \frac{m}{2} \, \dot{Z}^i(S) \, Z_i(S) \tag{5}$$

and an interaction part L_i. For a scalar field ϕ we would choose, e.g.,

$$L_i = g \int \phi(Z(S)) \, dS \tag{6}$$

whereas in electrodynamics we have

$$L_i = e \int dS \, \dot{Z}^i(S) \, A_i(Z(S)). \tag{7}$$

We shall see that there is considerable evidence that the gravitational field is a tensor ψ_{ik} coupled to the energy momentum tensor

$$L_i = \kappa m \int dS \, \dot{Z}^i(S) \, \dot{Z}^k(S) \, \psi_{ik}(Z(S)) \tag{8}$$

and we will use (8) as basis of our discussion. Assume that for K^o the coupling constant is $\kappa m(1+\delta/2)$ and $\kappa m(1-\delta/2)$ for $\bar{K}^o$. Then in a gravitational field of the form we think we have here on earth ($V = \frac{M\kappa}{R}$),

$$\psi_{ik} = V \begin{pmatrix} 1 & & & \\ & 1 & & \\ & & 1 & \\ & & & 1 \end{pmatrix} , \quad L_m + L_i = \frac{m}{2} \int ds \, \dot{Z}^i \, \dot{Z}^k \, g_{ik} \, ,$$

$$g_{ik} = \bar{g}_{ik} + 2\psi_{ik} = \begin{pmatrix} 1-2V & & & \\ & -1-2V & & \\ & & -1-2V & \\ & & & -1-2V \end{pmatrix} \tag{9}$$

$$\bar{g}_{ik} = \begin{pmatrix} 1 & & & \\ & -1 & & \\ & & -1 & \\ & & & -1 \end{pmatrix} ,$$

we obtain in a K^o-$\bar{K}^o$ basis

$$L_m + L_i = \left(m \begin{pmatrix} 1 & 0 \\ 0 & 1 \end{pmatrix} + \frac{\Delta m}{2} \begin{pmatrix} 0 & 1 \\ 1 & 0 \end{pmatrix} \right) \times$$

$$\times \left\{\dot{Z}^{o2}\left[(1+2V)\begin{pmatrix}1&0\\0&1\end{pmatrix} + V\delta\begin{pmatrix}1&0\\0&-1\end{pmatrix}\right] - \right.$$

$$\left. - \dot{\vec{Z}}^2\left[(1-2V)\begin{pmatrix}1&0\\0&1\end{pmatrix} - V\delta\begin{pmatrix}1&0\\0&-1\end{pmatrix}\right]\right\}. \quad (10)$$

Δ is the K_1-K_2 mass difference. Actually the situation is a little more complicated because they are unstable particles with different lifetime so that complex masses should be used. The difference in the imaginary parts is of the order of the difference of the real parts; hence orders of magnitudes are not changed by forgetting about the imaginary parts. Diagonalizing (10) for $V\delta \ll \frac{\Delta m}{m} \ll 1$ we see that not K_2 but $K_2 + \varepsilon K_1$, $\varepsilon = \delta\frac{mV}{\Delta m}(\dot{Z}^{c2} + \dot{\vec{Z}}^2)$ [6], is the correct linear combination. Now the essential point is the factor $\dot{Z}^{o2} + \dot{\vec{Z}}^2 = \frac{1+v^2}{1-v^2}$ which gives a strong increase of ε with energy. Experimentally one knows that $\varepsilon \simeq 2.10^{-3}$ between 1 and 10 GeV. Since it is energy independent it must be due to another interaction and thus we can get a limit for δ.

For the gravitational field of the earth mV = 1eV, for the sun 6eV and for our galaxy 500 eV. Using the measured $K \to 2\pi$ -rates [7] we obtain

$$\delta \ll 10^{-16} .$$

The classical Dicke-Eötvös experiment is less accurate but tests a somewhat stronger form of universality. There one sees whether the resultant of centrifugal and gravitational force has the same direction for different substances [8]. From this one deduces that the ratio of gravitational to inertial mass is the same for many substances within 1 part in 10^{11}. To assess the significance of this result one has to realize first that for reasonable theories both inertial mass and gravitational mass are additive. Hence their ratio for N protons should be the same as for 1 proton. If one considers proton and neutron as essentially the same particle, then this experiment tests only that for the electrons or the meson glue in the nucleus this ratio is the same as for nucleons. But these former contribute only $\sim$1% in mass so that the test is actually only 10^9 in accuracy. However since it tests that the acceleration given to nucleons moving inside the nucleus is the same as for a proton at rest it demonstrates a stronger property than strong universality. For this distinction we need two more refined concepts, weak equivalence and strong equivalence; the weak equivalence principle means that the effect of a constant gravitational field is exactly the same as going into a suitable accelerated coordinate system provided we ignore the mutual gravitational interaction of the particles in the system under consideration. Strong equivalence is the same property without the proviso of the mutual gravitational interaction [9]. Clearly equivalence implies weak equivalence and weak equivalence strong universality, but the converse is not true. For instance electrodynamics, where all charges are proportional to the masses, satisfies strong universality but not weak equivalence. Take for instance a particle Larmor-precessing in a constant magnetic field B and the same kind of particle at rest. Now switch on a constant electric field in direction of B which will accelerate both particles. One can easily figure out that the particle at rest will follow the field faster. Thus you cannot find a suitable accelerated frame where the effect of the electric field disappears since now the particles will separate if they were initially together, whereas without electric field they would meet over and over again. You will have noticed that the precessing particle represents the nucleon moving in the nucleus and thus the Eötvös-experiment would show an effect if the gravitational field were of the same nature as the electric field. This is also the reason why we had to add this remark about the internal structure at the formulation of strong universality. If the internal constituents of one object have different relative velocity than in another the objects as a whole may experience different forces even if strong universality holds. For K and $\bar{K}$, however different they are, we expect because of TCP the same internal structure in this sense. This is why Good tests only strong universality whereas Eötvös weak equivalence. One can see easily that the scalar theory (6) satisfies strong universality if $g \propto m$ and weak equivalence.

We shall next discuss a large class of theories which satisfy weak equivalence and see that they all have the property that they change the real geometry of space-time from its original pseudo-euclidean form to a Riemannian geometry. They are of the form (8) if ψ_{ik} couples to the energy-momentum tensor of all particles and fields. We will find a one-parameter family of theories which have Newton's theory as non-relativistic limit and pass the test of the Eötvös-experiment since they contain weak equivalence. For two values of the parameters one has even strong equivalence, one of the values of which corresponds to Einstein's general relativity [10].

Theories of the type (8) have some peculiarities which do not appear in electrodynamics. First of all the limiting velocity to which a particle can be accelerated is not everywhere the same but depends on the potential, whereas in electrodynamics it is everywhere the same and in our units one. This is an immediate consequence of the equations of motion which for (8) or (7) predict, for a suitable normalization of S,

$$\dot{z}^i \dot{z}_i + 2\dot{z}^i \dot{z}^k \psi_{ik} = 1 \qquad (10)$$

or

$$\dot{z}^i \dot{z}_i = 1 \quad . \qquad (11)$$

(11) can be rewritten as

$$v^2 = \left(\frac{dx}{dt}\right)^2 = 1 - \left(\frac{dS}{dt}\right)^2 < 1 \qquad (12)$$

whereas (10) gives for ψ of the form (9) e.g.

$$v^2 = \left(\frac{dx}{dt}\right)^2 = \frac{1}{1+2V}\left[1 - 2V - \left(\frac{dS}{dt}\right)^2\right] < \frac{1-2V}{1+2V} \quad . \qquad (13)$$

It may seem surprising that the velocity of a particle can become larger than "3.10^{10} cm/sec". However it is to be anticipated that the photon as the limit of a particle with $m \to 0$ has always this velocity. This expectation is actually borne out if one couples ψ_{ik} to the energy-momentum tensor of the electromagnetic field and looks at the resulting wave equation. Thus there are no acausalities in the sense that particles go faster than light but there is a common limiting velocity which depends on ψ_{ik} and therefore is different in different space-time points. The interesting fact is that it depends on ψ_{ik}, the potential, and not on its derivative, the field strength. Therefore, whereas in electrodynamics or in theories like (6) a constant potential does not enter into the equation of motion, here it seems to be dynamically effective. However for constant ψ the solution of the equations of motion is quite trivial because, by a suitable linear transformation (with constant coefficients) of the coordinates, L_m+L_i can be reduced to L_m. The same transformation on the A_i also transform away the interaction of the electromagnetic field with gravitation. For instance , if ψ is of the form (9), if $\bar{Z}^o = x^o\sqrt{1-2V}$, $\bar{Z}^{1,2,3} = x^{1,2,3}\sqrt{1-2V}$, we have $\dot{Z}^i\dot{Z}^k g_{ik} = \dot{\bar{Z}}^i\dot{\bar{Z}}^k \bar{g}_{ik}$. That is to say that by a change of scale we get the free equation of motion. This applies not only to a single particle but also to several particles interacting via electric- or other fields. The effect of a constant gravitational potential generally amounts only to a change of scale. This means automatically that e.g. the size of the hydrogen atom is $\hbar^2/me^2$ or its frequency $\frac{1}{2}mc^2.(e^2/\hbar c)^2$ when measured in the $\bar{Z}$-coordinates. From the point of view of the original Z-coordinate system all motions will be slowed down by the factor $\sqrt{1-2V}$, the size of any object will shrink by the same factor. The limiting velocity is 1-2V and hence has its normal value when measured with real clocks and measuring rods. Clearly the same will happen if ψ is not constant but does not vary much over the space-time region of the object. But

then the change of units will be different in different space-time regions and that is why the geometry is changed. The distance between two neighbouring space-time points as measured by real measuring devices will be $d\bar{Z}^i d\bar{Z}^k \bar{g}_{ik} = dZ^i dZ^k g_{ik}$ and not the original pseudo-euclidian $dZ^i dZ^k \bar{g}_{ik}$.

Coming back to the Dicke-Eötvös experiment we shall now show that the effect of a constant gravitational field strength, i.e. ψ varying proportional to the coordinates, can be transformed away locally by a quadratic coordinate transformation. Nonrelativistically this corresponds e.g. to $x \to \bar{x} = x + \frac{g}{2} t^2$, that is to say going into Einstein's freely falling elevator. Mathematically the statement is if

$$\bar{Z}^i = Z^i + \frac{1}{2} a^i_{kl} Z^k Z^l ,$$

$$\dot{\bar{Z}}^i \dot{\bar{Z}}^j \bar{g}_{ij} = \dot{Z}^i \dot{Z}^j g_{ij} \tag{14}$$

$$g_{ij} = \bar{g}_{ij} + a^k_{il} \bar{g}_{kj} + \bar{g}_{ir} a^r_{js} Z^s + a^k_{il} Z^l \bar{g}_{kr} a^r_{js} Z^s .$$

In general this transformation is a little more complicated than the one indicated above and transforms the gravitational field away only locally, i.e. not to order Z^2. But within this approximation again all equations are reduced to the ones without gravitation. Note that the key is the universal nature of the coupling of ψ to the energy-momentum tensor of all particles and fields with the same coupling constant. Only in this case will all objects in the elevator behave as in gravitationless space, irrespective of their internal interactions (safe gravitational ones) and mo tions. This also guarantees the negative outcome of the Dicke-Eötvös experiment. If different particles, originally at rest, stay together in the elevator, the forces acting on them in the laboratory system must have the same direction.

Sofar we have not demonstrated anything about the strong equivalence principle because we have not specified L_f or how the gravitational field is generated by masses. Indeed weak equivalence is independent of the field equations of the gravitational field. Therefore all theories where L_i is of the form (8) create a Riemannian space provided the measuring instruments are not changed by their internal gravitational interactions. The ψ_{ik} could be derivatives of a scalar: $\psi_{ik} = \phi_{,ik}$, or a vector: $\psi_{ik} = A_{i,k} + A_{k,i}$, or a non-linear expression $\psi_{ik} = A_i A_k$. However, such non-linear or derivative couplings do not give Newton's law in the non-relativistic limit and are therefore out. Even with a linear coupling to a symmetric tensor the various choices of L_f lead to different theories. Field-theoretically speaking ψ_{ik} contains one scalar $\psi^i{}_i$, one vector $\psi^k_{i,k}$ and the rest corresponds to a spin 2 particle. Different L_f correspond to different coupling strength of the various irreducible components to the energy-momentum tensor T^{ik}. However, the source term $T^{ik}{}_{,k}$ of the vector part vanishes identically. Thus there remains only one essential parameter α, the ratio between the couplings of the spin 0 and the spin two parts. The sum of the couplings is of course determined by the gravitational constant κ. Strong equivalence now comes down to the question of whether the whole system: field equation + equations of motion is invariant by adding a constant gravitational field and making the transformations (14). It turns out that this happens only for $\alpha = 0$ or ∞, that is to say for a pure spin 2 or pure spin zero theory. In the latter case

$$\psi_{ik} = \bar{g}_{ik}\,\phi(x) \quad . \tag{15}$$

There is no direct empirical evidence for the strong equivalence principle since the Eötvös experiment tests only weak equivalence. Strong equivalence would require, for instance, repeating the Cavendish experiment in a freely falling elevator and comparing it with the laboratory result. To get further one has to compare detailed predictions of these theories with more subtle effects like the 3 classical tests of general relativity. The form (9) of the ψ_{ik} corresponds to Einstein's theory or pure spin 2, $\alpha = 0$. In the other extreme, $\alpha = 1$, according to (15), the metric is

$$g_{ik} = \begin{pmatrix} 1+2V & & & \\ & -1+2V & & \\ & & -1+2V & \\ & & & -1+2V \end{pmatrix} \quad (16)$$

Thus this gravitational potential shows clocks down but swells measuring rods by the same factor so that the velocity of light stays constant. This leads to a characteristic difference: The trajectory of a particle with mass in the gravitational field is a time-like geodesic in the space with metric g_{ik}. In other words everybody tries to stay as young as possible by going from one space-time point to the other. E.g. one falls down in a gravitational field because further down everything goes slower as shown by the gravitational redshift and one is ageing more slowly. In this respect theories with different α agree. However,for massless particles, for which there is no proper time, Fermats principle holds. They want to go from one space-time point to the other as fast as possible. Thus light tries to stay away from massive bodies for a spin 2 gravitational field because this slows it down. On the other hand there is no light deflection by a spin 0 gravitational field since this does not change the velocity of light. Thus the observed light deflection shows a dominant spin 2 component but the experiments are not accurate enough to exclude a small scalar admixture. The perihelion shift of Mercury is a very subtle effect which comes out just right for pure spin 2 (general relativity) but there are so many effects to be considered that a small change in α

cannot be excluded. Before contemplating these facts we shall collect them in the following table:

Strong Universality	Weak Equiv.	Strong Equiv.	Effect
L=m(indep. of m)	$L_i = \kappa T_{ik}\psi^{ik}$	$\alpha = 0,1$	
10^{-16}			Good
10^{-11}	10^{-9}		Dicke
no	yes	yes	determ. Metric
no	yes	yes	determ. Redshift
no	no	no, only if $\alpha = 1$	determ. light defl.
no	no	no, only $\alpha=1$ and right non-linearities.	determ. Perihelion motion

If professional relativists are exposed to these kinds of consideration their reaction usually is: "Why do you want to change something in general relativity, it is so beautiful? ". The school of Dicke replies that strong equivalence contradicts Mach's principle and is therefore not satisfactory. This may sound strange because Einstein claimed that general relativity contains Mach's principle, at least in a rudimentary form. The situation is as follows: We have seen that near a massive body clocks slow down and rods shrink. If one looks at the equations of motion of, say, an electron moving around a proton one can see how this comes about. The effect of the gravitational field is the same as increasing the mass of the electron and decreasing its charge. "So you see", says Einstein, "that Mach's principle results: The inertial mass of a particle is increased by a nearly massive object". "But", replies Dicke, "this effect is completely renormalized away if you measure with real rods and clocks. In these physical units the acceleration which one particle receives due to the gravitation of another particle is exactly the same near or far away from a massive body. If Mach's principle is to have any real meaning the Cavendish experiment must show a different result under the influence of the gravitation of a third body". This clearly contradicts strong equivalence and can also be expressed by a variable gravitational constant $\kappa(x)$. This yields then a theory with an additional scalar field $\kappa(x)$, or $\alpha \neq 0$. I don't want to decide whether one should give strong equivalence or this version of Mach's principle an a priori philosophical preference. Let me conclude with a remark on an argument of a similar nature, namely the one about simplicity. Certainly pure spin 2 seems simpler than some mixture. If one looks at how well simplicity arguments work for other interactions one has to admit the following: For weak interactions a pure V-A theory also would be simpler than a small admixture of something CP-violating and yet we know the latter is present. On the other hand for electric interactions there seem to be plenty of possibilities for non-minimal couplings but they are not there. Thus the game for simplicity arguments is 1:1 at the moment and therefore I think we have to leave it to the experimentalist to settle this question for gravitation.

References

1. For a survey on the experimental status see e.g. B. Bertotti, D. Brill, R. Krotkov in L. Witten (ed.) Gravitation, Wiley 1962; R. H. Dicke, The Theoretical Significance of Experimental Relativity, Gordon and Breach 1964.
2. L. D. Landau, E. M. Lifshitz, Statistical Mechanics, Pergamon Press 1958, p.160ff and 330ff.
3. For details see e.g. J. A. Wheeler in H. Y. Chiu (ed.) Gravitation and Relativity, Benjamin 1964; E. E. Salpeter in I. Robinson (ed.) Quasistellar Sources and Gravitational Collapse 1965. An instructive survey is given by K. S. Thorne, Sci. Am. 217, Nr. 5, 88 (1967).
4. M. L. Good, Phys. Rev. 121, 311 (1961).
5. W. Thirring, Forts. Physik, 7, 79 (1959); W. Thirring, Ann. Phys. (N.Y.) 16, 96 (1961).
6. cf. J. Bernstein, N. Cabibbo, T. D. Lee, P. L. 12, 146 (1964);J. S. Bell, Phys. Rev. Letters 13, 348 (1964), who made similar calculations.
7. See for instance J. M. Christenson, J. W. Cronin, V. R. Fitch and R. Turlay, Phys. Rev. Lett. 13, 138 (1964).
8. R. H. Dicke, Sci. Am. 205, 84 (1961).
9. R. H. Dicke, in Verifiche delle teorie gravitazionali, Academic Press 1961.
10. A detailed discussion on renormalization is given in ref. 5. For the following considerations see also R. U. Sexl, Forts. Physik, 15, 269 (1967).

ON HIGH ENERGY COLLISIONS OF HADRONS AT VERY LARGE MOMENTUM TRANSFERS[†]

By

H. A. KASTRUP

Universität München, Sektion Physik
München, Germany

Content

---

† Lectures given at the VII. Internationale Universitätswochen f. Kernphysik, Schladming, February 26-March 9, 1968.

I. Introduction

Under collisions with large momentum transfers we understand in the following elastic and inelastic scattering reactions where the absolute values of the quantities $t = (p_1 - p_1')^2 < 0$ and $u = (p_1 - p_2')^2 < 0$ are larger than the square of the nucleon mass M. Here p_1, p_2, p_1' and p_2' are the 4-momenta [1] of the two primary particles before and after the collision (we shall consider only those processes in which the two initial particles occur in the final state, too).

In other words, we shall not discuss forward and backward scattering here, the structure of which seems to be dominated by diffraction-like mechanisms and resonances in the direct and crossed channels [2].

Whereas a large amount of very interesting theoretical and experimental work has been done in order to analyze the features of forward and backward scattering, the large momentum transfer collisions - in particular the inelastic ones - are rather poorly investigated and understood in comparison [3]. There are several reasons for this:

On the theoretical side we are no longer near the region of known low energy resonances in the crossed t- and u-channels of the (hopefully!) analytic scattering amplitudes. Thus we can no longer assume that these resonances dominate the physical region we are interested in. Furthermore, large momentum transfer reactions are usually accompanied by the emission of many secondaries, because very large momentum transfers tend to break up the primary particles. Our theoretical knowledge of many-particle inelastic scattering amplitudes - which in turn strongly influence the elastic amplitudes because of the unitarity relations - is, however, extremely poor.

On the experimental side the cross sections for the individual channels, for instance the elastic one, become extremely small for large momentum transfers (down to 10^{-34} cm^2/ster in elastic p-p scattering [4]), implying very small counting rates.

As for the inelastic channels it is difficult to obtain a detailed kinematical analysis of all the different particles in the final state.

Although our present theoretical and experimental knowledge of large momentum transfer collisions is very unsatisfactory, these reactions are of principal importance for at least three main reasons:

1. They are supposed to yield information about the short-range properties of the various interactions. The usual "uncertainty" argument for this goes as follows: If R is the interaction range to be tested, then we have approximately $R|t|^{1/2} \simeq 1$, where t is the momentum transfer squared. In other words, the more "central" the collision, the larger is the probability that the particles are being scattered with large momentum transfers.

A slightly different argument characterizing large momentum transfer reactions, or equivalently, high energy collisions with scattering angles $\Theta \neq 0,\pi$, goes as follows: If ℓ is the relative orbital angular momentum between the two incoming particles in the c.m. system and b their impact parameter, then short range collisions will have a small b. Because of $\ell = bp$, where p is the c.m. momentum of the incoming particles, this means that short-range interactions are mainly characterized by the lower partial waves of the scattering amplitude, if p is kept fixed. However, this argument fails in the extreme relativistic limit $p \to \infty$. In this case not only low partial waves might become important even for $b \to 0$.

2. Because the rest masses of particles become negligible in comparison to the variables $s=(p_1+p_2)^2$, t and u etc., we hope to learn something about the realm of validity of approximate symmetries like SU(3), SU(2) $\otimes$ SU(2) etc. which are assumed to be the better the more mass differences or even rest masses themselves become negligible. Thus one suspects large momentum transfer reactions to provide a suitable testing ground for unitary and chiral symmetries.

3. In the framework of axiomatic field theories it is possible to derive lower and upper bounds for the high energy large momentum transfer behaviour of scattering amplitudes from some very general assumptions [5]. An experimental violation of these bounds would be very interesting! No such violations have been observed so far. We shall not deal with this aspect of large momentum transfer collisions.

I have described the first two motivations for analyzing large momentum transfer collisions rather uncritically. In what follows I shall try to give a critical analysis in the framework of a rather elementary and simplified model.

There are other interesting models for large momentum transfer reactions [6], which we shall not discuss here.

The model to be considered is more or less a phenomenological one. Qualitatively it employs the rather old bremsstrahlung picture for meson production, keeping an eye or two on our basic interests listed above and adding several new features [7] which provide interesting experimental predictions.

Acknowledgments

I benefited a great deal from conversations with Dr. G. Mack. A part of the final version of these lectures was written during a visit at DESY, Hamburg. I am very much indebted to Professor H. Joos and several other physicists for their very kind hospitality and for stimulating discussions.

II. A Statistical Soft Meson Model for Elastic and Inelastic Nucleon-Nucleon Scattering

1. Description of the model

We start this section II by giving a critical discussion of the four basic assumptions concerning the model to be employed. The assumptions are as follows:

a) We consider only laboratory energies E_{lab} larger than 10 GeV. At such energies of the incoming nucleons many collision channels are open and <u>we assume the features of a given channel to be dominated by what is going on in all the other channels.</u> This assumption is, of course, applicable to all high energy collisions, not merely to those with large momentum transfers. A typical relation expressing this overall property is the Optical Theorem which says that the imaginary part of the elastic forward scattering amplitude is proportional to the total cross section at the same energy.

More general: what we are concerned with in our first assumption are the consequences of the unitarity of the S-matrix, which relates the different channel amplitudes to each other. As it is impossible, at least at the moment , to take into account and to exploit in detail all unitarity restrictions in the case of many-particle final states, one has to introduce some statistical hypotheses concerning the contribution of all open channels to an individual one. We shall specify these assumptions under c).

The general moral to be drawn from our first hypothesis is that we should try to get information pertaining to our basic interests in large momentum transfer collisions from the combined inspection of <u>all</u> open channels, not only of a single, for instance the elastic, one. We shall sharpen this assertion later on.

b) Our second assumption is typical for large momentum transfer scattering: At very high energies the hadronic fields - we neglect the electromagnetic and weak interactions in this section - of the two incoming particles are strongly Lorentz-contracted in the c.m. system. The same holds for the outgoing particles. For large momentum transfer reactions the contraction necessitates a strong rearrangement of the long-range parts of the hadronic fields. This strong rearrangement is very probably accompanied by the emission of secondaries.* <u>Our second assumption therefore is that the main bulk of the secondaries is a consequence of the strong rearrangement of the hadronic field tails in large momentum transfer collisions.</u>

A first qualitative conclusion from this hypothesis is that the secondaries should be mostly pions because they constitute the tail of the long-range parts of the hadronic fields. This is supported by cosmic ray and accelerator experiments which show about 80% of the secondaries to be pions [8] . The rest con-

---

* By "secondaries" we always mean the particles produced in addition to the outgoing "primary" particles which initiated the reaction.

sists mainly of kaons. However, it is at present not well-known how many of the pions are decay products of, for instance, vector mesons which were produced in the first place. Furthermore, one does not know the variation of the ratio of the different kinds of secondaries as a function of the momentum transfer of the primary nucleons.

What is very much needed is an experiment which determines the identities and the kinematics of all or most particles - primaries and secondaries - in high momentum transfer collisions with multiparticle final states.

A main conclusion to be drawn from the second assumption is that we shall not get much information about the short-range properties of the hadronic fields if we are not able to subtract the long-range effects. This was first pointed out by Wu and Yang in a slightly different context [9].

c) The last pessimistic conclusion from our second assumption poses the question whether it is possible to separate long and short range effects, for instance by a factorization of the cross sections. In order to obtain such a factorization we assume the secondaries to be "soft", i.e. we neglect energy-momentum-, spin- and unitary spin recoils associated with the emission of the secondaries.

Before discussing the factorization implied by this assumption, let us look at the experimental situation in order to see how realistic this third hypothesis is : Many of the observed cosmic ray events seem to be quasielastic, i.e. only a relatively small fraction [8] (about 30 - 40 %) of the initial energy is converted into secondaries, and the primary particles emerge with a relatively small energy loss after the scattering ("leading particles"). However, the analysis of accelerator experiments [10] suggests that large momentum transfers in inelastic nucleon-nucleon collisions are strongly suppressed and that most of the observed cosmic ray events are very likely associated with small momentum transfers of the primary particles. This conclusion is not completely convincing because present accelerator energies are quite small in comparison to the interesting cosmic ray energies ($E_{lab} > 10^2$GeV), and it may be that large momentum transfer reactions are more frequent at very high energies. A detailed analysis of this question in cosmic ray physics would be very interesting.

As it stands, the simple assumption of "soft" secondaries may be too unrealistic considering the present state of our knowledge although the soft pion hypothesis has met with considerable success at low energies in the framework of current algebra [11][12]. However, there may be at least two ways out of this dilemma: First, the collision cross sections may be "soft" or quasielastic in certain variables but not in others. Perhaps, if one is lucky, one will be able to pick the right variables in which the cross sections are insensitive to our "softness" approximation. It seems, for instance, that large momentum transfer elastic and inelastic nucleon-nucleon cross sections depend in a first approximation only on the transverse momentum transfer $p_\perp$ of the outgoing nucleons, not on their longitudinal momentum [4],[10],[13]. It is therefore tempting to assume these collisions to be quasielastic with respect to $p_\perp$. The assumption is supported by measurements of the pion c.m. $p_\perp$- and $p_\parallel$-spectra [14] in inelastic proton-proton collisions at 12.5 GeV. The $p_\perp$ - spectra have a Gaussian distribution around $p_\perp = 0$, but the maximum of the $p_{\parallel}$-spectra is around 0.5 GeV/c. Thus the pions seem indeed to be soft with respect to their transverse momentum transfer. Unfortunately,the result is inconclusive for us because the kinematics of the outgoing protons were not measured simultaneously, and so we do not know whether the soft mesons are associated with small or large momentum transfers of the primary nucleons. What is needed is an experiment which measures the kinematics of the primary and secondary particles at the same time.

We shall discuss this first refined possibility of a quasi-elastic approximation in detail in parts 2 and 3 of this section.

Secondly, we can weaken the "softness" - hypothesis by keeping it for the amplitudes but dropping it for the phase spaces in question. This is, of course, more realistic and the results are quite encouraging [15]-[17]. We shall deal with this approach in detail in part 5.

Let us briefly discuss now the factorization implied by the assumption of negligible energy-momentum and spin recoil. The unitary spin problem will be treated in part 4. Here we assume that there is only one type of primary particles and one type of secondaries. Furthermore we assume spin and parity effects to be negligible. There is practically no information as to whether this last assumption is justified for large momentum transfer collisions or not.

With these simplifications we get, as the cross section for the quasielastic scattering of two primary particles accompanied by the emission of n soft secondaries, the expression

$$\frac{d\sigma^{(n)}}{d\Omega} = \bar{\sigma}(E,\Theta)w_n(E,\Theta) \quad . \tag{1}$$

Here E is the c.m. energy of one of the primary particles and Θ their c.m. scattering angle.(In the following, we always mean c. m. quantities if not stated otherwise.)

The factorization (1) is nothing other than the multiplication law of probability theory with $w_n(E,\Theta)$ as the conditional probability for the emission of n soft secondaries, provided the primary particles are being scattered at energy E with scattering angles Θ and $\pi-\Theta$ respectively.

Negligible recoil means that the secondaries are emitted statistically independently, and therefore we have for w_n the Poisson probability

$$w_n(E,\Theta) = \exp\left[-\bar{n}(E,\Theta)\right] . \frac{1}{n!} . \left[\bar{n}(E,\Theta)\right]^n . \qquad (2)$$

As the second factor in eq. (1) describes the emission of secondaries, which according to our assumption b) is a consequence of the rearrangement of the long-range parts of the hadronic fields, this factor $w_n(E,\Theta)$ is characteristic for the long-range effects. Therefore the first factor, the so-called "potential" cross section $\bar{\sigma}(E,\Theta)$, should be characteristic for the short-range properties. Thus eq. (1) provides us with a separation of long- and short-range effects, valid, of course, at present only in the framework of our approximations.

The quantity $\bar{n}(E,\Theta)$ is the average number of secondaries, provided the primary particles have the energy E and scattering angles Θ and $\pi - \Theta$ respectively. This "differential" multiplicity is of considerable importance for our model.

We further notice that the factor $\exp[-\bar{n}]$ is contained in elastic and inelastic cross sections. It represents the overall consequence of all the inelastic

S-matrix unitarity restrictions for each individual channel, including the elastic one.

Summing over n in eq. (1) gives

$$\frac{d\sigma}{d\Omega}^{tot}(E,\Theta) = \sum_{n=o}^{\infty} \frac{d\sigma}{d\Omega}^{(n)} = \bar{\sigma}(E,\Theta) \quad . \qquad (3)$$

From this we conclude that the differential total cross section gives the cleanest information about the short-range interactions, not the individual ones! This interesting feature is a consequence of our approximations, but its realm of validity may go beyond them.

d) Our last main assumption concerns the form of the potential cross section at large momentum transfers, or, equivalently, the structure of the short-range interactions: We assume the short - range interactions of the "bare" primary hadrons - i.e. without their long - range meson cloud - to be pointlike.

In other words, the short-range interactions do not have any fixed hard core, all rest masses are negligible, and all relevant coupling constants do not contain a fixed length.

From this it follows that

$$\bar{\sigma}(E,\Theta) = E^{-2} A(\Theta); \quad \Theta \neq 0,\pi \quad ; E>>M \quad , \qquad (4)$$

because E is the only quantity left which can provide us with a dimension of length.

Another way of describing the hypothesis (4) is that we assume the short-range interactions to become invariant under dilatations, i.e. all fixed lengths like rest masses, etc., become negligible at very high energies and large momentum transfers as far as the potential cross section is concerned.

The assumption (4) certainly cannot be applied to

the second factor in eq. (1), because the range of the hadronic fields is determined by rest masses and the factor $w_n(E,\Theta)$ depends very much on this range. Therefore, we cannot neglect all rest masses in $w_n(E,\Theta)$. Thus, the naive assumption that all rest masses become negligible for large s, -t and -u obviously cannot be true in general!

2. Comparison of the model with experiments

In this part we shall compare some of the predictions of that version of the model described in part 1 which supposes the soft meson hypothesis to be a good approximation for some suitably chosen variables. The refined version which takes realistic phase spaces into account will be discussed in part 5.

2.1. Elastic scattering

For the elastic cross section $d\sigma^{el}/d\Omega \equiv d\sigma^{(o)}/d\Omega$ we get from eqs. (1), (2) and (4):

$$\frac{d\sigma^{el}}{d\Omega} = E^{-2} A(\Theta)\, e^{-\bar{n}(E,\Theta)} \quad ; \Theta \neq 0,\pi \quad . \tag{5}$$

We see that - at a fixed Θ - the only unknown energy dependence of the cross section is contained in the differential multiplicity $\bar{n}(E,\Theta)$. From accelerator and cosmic ray experiments we know only something about the integral multiplicity

$$\bar{n}_{tot}(E) = \int \frac{d\sigma^{tot}}{d\Omega}\, \bar{n}(E,\Theta)\, d\Omega/\sigma^{tot}(E) \quad . \tag{6}$$

Experimentally [8] we have $\bar{n}_{tot}(E) \sim E^{\alpha}$ in nucleon-nucleon collisions, where α varies roughly between 0.5

and 1. Under the assumption that the angle integration (6) does not change the energy dependence of $\bar{n}(E,\Theta)$ very much, we expect the elastic p-p cross section to drop exponentially with increasing energy E for fixed scattering angles $\Theta \neq 0,\pi$. This feature is borne out experimentally: Orear fitted [18] the earlier data [4] by the following expression :

$$\frac{d\sigma^{el}}{d\Omega} = \text{const.}\ E^{-2}\ e^{-ap_\perp} \quad , \tag{7}$$

where $a \simeq 5(\text{GeV}/c)^{-1}$ and $p_\perp = p \sin \Theta$. Later measurements [19] showed that the fit (7) is only a first approximation and that there is some fine structure. But as a first approximation the expression (7) is quite useful for us. It obviously has the same structure as the formula (5). It is therefore tempting to identify the two exponentials:

$$\bar{n}(E,\Theta) = a\ p_\perp \quad . \tag{8}$$

However, the identification (8) may be wishful thinking, because it is not a priori clear which percentage of the empirical exponential $\exp(-ap_\perp)$ has to be attributed to the inelastic channels characterized by $\exp(-\bar{n})$ in our model. If the fraction is β then we have $\bar{n}(E,\Theta) = ap_\perp + \ln\beta$. If β is energy- and angle-independent, we can neglect the constant $\ln\beta$ at large enough energies. But it is not clear whether $\beta=1$ is a realistic choice for the present accelerator energies. Nevertheless, let us stick to the relation (8) for the moment - $\beta = 1$ - and let us look at its implications:

Eq. (8) means that the differential multiplicity $\bar{n}(E,\Theta)$ in p-p scattering should depend in a first approximation only on the transverse momentum transfer of the nucleons, not on their longitudinal one. This will have to be tested by future experiments. Numerically we get $\bar{n} \simeq 15$ for $p_\perp = 3$ GeV/c. This seems to be a rather large number, which will probably be cut down by realistic phase space limitations (see part 5 of this section).

Furthermore, eqs. (7) and (8) combined imply $d\sigma^{tot}/d\Omega = \text{const.}\ E^{-2}$, i.e. the differential total cross section becomes independent of Θ if we use the fit (7). This contradicts the results [10] of Anderson et al. which show clearly that $d\sigma^{tot}/d\Omega$ decreases with increasing Θ. (see also ref. [17]). A large part of this discrepancy is probably due to the fact that the fit (7) describes only the gross features of the data.

2.2. Inelastic collisions

By inserting the relation (8) into eqs. (1) and (2) we get the prediction that the inelastic cross sections should depend dominantly on the nucleon transverse momentum transfer. Unfortunately the quantities $d\sigma^{(n)}/d\Omega$ or $d^2\sigma^{(n)}/d\Omega dp$ have not yet been measured for large momentum transfer p-p collisions. What has been measured, however, is the sum

$$\frac{d^2\sigma}{d\Omega dp} = \sum_{n=0}^{\infty} \frac{d^2\sigma^{(n)}}{d\Omega\, dp} .$$

Anderson et al. fit [10] their data for a given initial energy by the expression

$$\frac{d^2\sigma}{dp_\parallel dp_\perp} \sim p_\perp^2\, e^{-ap_\perp} \tag{9}$$

We see that this cross section indeed depends dominantly on the transverse momentum transfer. Moreover, we have the same exponential $\exp(-ap_\perp)$ with the same constant a as in Orear's fit (7) for the elastic data. The factor $p_\perp^2$ in front of the exponential in eq. (9) may be interpreted as the approximate replacement of the sum $\sum_n 1/n!\ (ap_\perp)^n$ by one of its terms.

We have to stress, however, that the comparison of the fit (9) with eq. (1) cannot be done without raising serious questions: In deriving eq. (1) we assumed the scattering to be quasielastic. On the other hand the protons suffer a considerable energy loss in the experiments of Anderson et al. As already pointed out earlier, one way out of this dilemma seems to be the feature [10] that the dynamics does not seem to care much about losses of longitudinal momentum. If, therefore, the reactions are somewhat quasielastic with respect to the transverse momentum transfer, our first version of the bremsstrahlung model may still be reasonable.

2.3. Differential total cross section

Eq. (4) predicts the differential total cross section

$$\frac{d\sigma}{d\Omega}^{tot} = \int dp \frac{d^2\sigma}{d\Omega dp}$$

to go down like $E_{c.m}^{-2}$ for fixed $\Theta_{c.m.}$. The date of ref. [10] allow for a certain comparison of this prediction at one single angle $\Theta_{c.m.} = 29^\circ$. A numerical graphical integration by hand yields

$$d\sigma^{tot}/d\Omega(E_{lab} = 10\ \text{GeV}) \simeq 3.0 \pm 0.5\ \text{mb/ sr}\ .$$

$$d\sigma^{tot}/d\Omega(E_{lab} = 30\ \text{GeV}) \simeq 2.2 \pm 0.5\ \text{mb/ sr}.$$

The theoretical prediction is 3:1 instead of 3:2.2. However, this comparison cannot claim to be a serious one for several reasons: At E_{lab} = 10 GeV we certainly cannot neglect all hadron rest masses in the c.m. system as has been supposed in eq. (4). In addition the momentum transfers are not large compared to the nucleon rest mass at $\Theta = 29^{o}$ and E_{lab} = 10 GeV. Finally the values of $d^2\sigma/dpd\Omega$ are not known experimentally for $p \leq 1$ GeV/c, although the maxima of the curves lie in this region. Qualitatively we can say that $d\sigma^{tot}/d\Omega$ seems to drop slowly with increasing energy, in agreement with the hypothesis (4).

2.4. Ericson fluctuations

If the high energy large momentum transfer scattering of nucleons is dominated by some compound scattering mechanism like the collisions in the region of overlapping resonances in nuclear physics, one would expect Ericson fluctuactions [20] in the elastic p-p channel. Such fluctuations have not been found [21]. This is nice for us, because in our model we do not assume the primary particles to participate in any thermodynamical equilibrium,and therefore we do not expect such Ericson fluctuations to occur.

2.5. Difference between proton- and pion-spectra

There is a striking qualitative difference between the proton- and pion-spectra $d^2\sigma/dp_\perp\, dp_\parallel$ in the work [10] of Anderson et al.: Whereas the proton spectra do not depend on $p_\parallel$ (proton) the pion spectra drop approximately exponentially with increasing $p_\parallel$ (pion). Such a feature is hard to understand on the basis of a thermodynamical model which assumes the outgoing protons and pions to be produced by the same mechanism. In our model, however, the emission of pions and protons during the scattering is considered to be qualitatively different, although we are not able at the moment to account quantitatively for the observed difference. In order to do this, more dynamical assumptions than those discussed here will be necessary.

2.6. Forward-backward asymmetry

Intuitively we expect $\bar{n}(E,\pi-\Theta) > \bar{n}(E,\Theta)$, $\Theta < \pi/2$, because the rearrangement of the long-range parts of the hadronic fields is stronger for larger angles, implying the emission of more secondaries. Thus we expect a forward-backward asymmetry in those reactions where we do not have to symmetrize (or antisymmetrize) the scattering amplitudes or cross sections with respect to Θ and $\pi-\Theta$. Such a forward - backward asymmetry is indicated by the present p-$\bar{p}$- and p-n-scattering data [22]. A more quantitative analysis will be possible in the near future when more precise data become available.

3. Predictions for future experiments

We shall list a number of predictions implied by the first version of our model. These predictions can be tested by future experiments. Here we assume only one type of primary and one type of secondary (= produced) particles. Predictions which result from the inclusion of unitary spins and more realistic phase spaces for the secondaries will be discussed in part 4 and 5.

3.1. Correlations between elastic and inelastic kinematics

One of the basic features of the model under consideration is the assumption of correlations between the kinematics of the primary particles in elastic and inelastic collisions, i.e. we assume the final state momenta of the two primary particles in inelastic reactions not to be completely distorted compared to their elastic scattering. The best we can hope for is a relation like

$$\vec{p}\,'_1 \simeq - \vec{p}\,'_2 \quad , \tag{10}$$

even in multiparticle final states, where $\vec{p}\,'_1$ and $\vec{p}\,'_2$ are the c. m. momenta of the primary nucleons in the final state. Even if the relation (10) would not hold, the model might still work if, for instance,

$$\vec{p}\,'_{1\perp} \simeq \vec{p}\,'_{2\perp} \quad , \tag{11}$$

because most of the known nucleon-nucleon cross sections seem to depend - as discussed above - in a first approximation on the nucleon transverse momentum transfer only.

3.2. Differential multiplicities

A very important quantity in the framework of our model is the differential multiplicity

$$\bar{n} = \bar{n}\,(E; \vec{p}\,'_1, \vec{p}\,'_2) \quad , \tag{12}$$

where E is the initial c.m. energy. The expression (12) is written in a more general form than previously.

Of course, if the assumptions (10) or (11) hold or even the prediction (8), the number of independent variables is reduced drastically. However, the quantity (12) may be of considerable interest beyond the scope of our specific model, because it expresses rather simple correlations between primary and secondary particles. The decomposition of the differential multiplicity (12) with respect to different types of secondaries will be discussed in part 4.

3.3. Fluctuations

A specific test of the assumed Poisson distribution (2) consists in measuring the r.m.s. of $\bar{n}$. In the case of the Poisson distribution these fluctuations are given by $\bar{n}^{1/2}$.

3.4. Differential total cross sections

It was already pointed out before that the differential total cross sections $d\sigma^{tot}/d\Omega(E,\Theta)$ seem to be particularly appropriate in order to test short-range properties. Again we assume that the primary particles go approximately in opposite directions in the c.m. system after the collision, even in inelastic events. If only the relation (11) holds, then

$$\frac{d\sigma^{tot}}{dp_\perp}(E,p_\perp) = \int dp \frac{d^2\sigma}{dp_\perp dp}$$

may be the better quantity to consider.

3.5. Isobars

In many cases it will probably happen that one - or both - of the primary nucleons leaves the collision region in an excited state (isobar) and then goes back to its ground state (nucleon). If this happens one will not describe the pion from the isobar decay by a factor w_n in eq. (1), because this factor is supposed to describe secondaries produced by a nonresonant mechanism (they may be meson resonances, however). In such a case the isobar would have to be dealt with in the potential cross section $\bar{\sigma}$. If the momentum transfers and energies are much larger than the rest masses of the produced isobars, not much should change in $\bar{\sigma}$ compared to the nucleon case. If not, one has to incorporate at least a Breit-Wigner formula in $\bar{\sigma}$ in order to account for the final state interaction between the nucleon and the pion. In that case the assumption of negligible rest masses is, of course, no longer applicable to $\bar{\sigma}$.

4. Incorporation of unitary symmetries

Up to now we have considered only one type of primary and one type of secondary particles. In this part we shall deal with the case where the primary and secondary particles separately belong to a "primary" and "secondary" multiplet of a certain unitary symmetry group SU_n.

In order to include some features of a unitary symmetry, we shall make several crude approximations: We ignore spin and parity of all particles. Consequently, we ignore the usually necessary symmetrizations or antisymmetrizations due to Bose- or Fermi-statistics. This means, for instance, that we neglect the interference terms between amplitudes at Θ and $\pi-\Theta$ in elastic p-p scattering. We merely add the cross sections.

We further assume the total unitary spins of the primary and secondary particles to be decoupled, i.e. we ignore any unitary spin exchange between primaries and secondaries for the time being. Thus the total unitary spin of all the secondaries will be assumed to be zero. Such an assumption is evidently wrong if, for instance, one additional π^+ is produced in p-p collisions, but it might not be too bad an approximation in the case of multi- particle final states.

The hypotheses just stated again imply a factorization of S-matrix elements into a factor $\langle out|in\rangle_I$ describing only the primary particles and a factor

$$\langle out|0;\, in\rangle_{II} = \langle in'|S_{II}|0;\, in\rangle_{II}$$

for the secondaries which depends on the kinematics of $\langle out|in\rangle_I$ (conditional probability amplitude!). The state $|0;in\rangle$ does not contain any secondary particles. The unitary spin dependence can be exhibited in detail by writing down the field equations

$$(\Box + m^2)A_\beta(x) = j_\beta(x), \qquad \beta=1,\ldots,r\,, \tag{13}$$

for the secondaries. The index β is a unitary spin index. Our assumptions from above are equivalent to approximating the sources $j_\beta(x)$ by c-numbers. In that case, we can solve the eqs. (13) for the S-matrix explicitly in a standard manner [23]:

$$S_{II} = \exp(-\tfrac{1}{2}b) \times \exp\Big[i\sum_{\beta=1}^{r}\int\frac{d^3\vec{k}}{\sqrt{2\omega}}\, j_\beta(k)\, a^+_{in}(k;\beta)\Big] \times$$
$$\times \exp\Big[i\sum_{\beta=1}^{r}\int\frac{d^3\vec{k}}{\sqrt{2\omega}}\, j^*_\beta(k)\, a_{in}(k;\beta)\Big], \tag{14}$$

where

$$\omega = +(\vec{k}^2+m^2)^{1/2};$$

$$b = \sum_{\beta=1}^{r} g_\beta, g_\beta(E,\Theta) = \int \frac{d^3\vec{k}}{2\omega} |j_\beta(k)|^2 ,$$

$$j_\beta(x) = \int_0^\infty \frac{d\omega}{2\pi} \int \frac{d^3\vec{k}}{(2\pi)^{3/2}} \left[j_\beta(k)\, e^{-ikx} + j_\beta^*(k)\, e^{ikx} \right] \tag{15}$$

The $a_{in}^{+}(k;\beta)$ and $a_{in}(k;\beta)$ are the creation and annihilation operators of the "in"-fields associated with the fields $A_\beta(x)$. The index II is to indicate that the S-matrix in eq. (14) describes only the secondary particles.

Before discussing some applications of the above formulae we notice that the S_{II}-matrix contains the overall SU_n - invariant factor exp(-1/2 b). According to our previous discussions, this factor dominates all cross sections for large momentum transfers of the primary particles. Thus, we expect a universal exponential decrease with increasing energy of all large momentum transfer cross sections for those particles which belong to unitary spin invariant couplings of "primary" and "secondary" multiplets. The universality may, of course, be distorted by symmetry breaking.

I would like to stress the following: Since the sources are c-numbers in eq. (13), the transformations $j_\beta \rightarrow V_{\beta\beta'}(SU_n)\, j_{\beta'}$ are not implementable by a unitary transformation. They are merely automorphisms in the vector spaces spanned by the fields $A_\beta(x)$ and the sources $j_\beta(x)$.

As a first application we discuss the isospin group SU(2). The primary particles, the nucleons, form an isodoublet, the secondaries, the pions, an isotriplet. We consider the following cases:

4.1. Nucleon-nucleon two-body collisions

Ignoring spins we have two independent amplitudes F_o and F_1 corresponding to the total isospins 0 and 1. Proton-proton and neutron-neutron cross sections are given by the same expression

$$\frac{d\sigma_{pp}}{d\Omega} = E^{-2}\left[|F_1(E,\Theta)|^2 e^{-b(E,\Theta)} + \right.$$

$$\left. + |F_1(E,\pi-\Theta)|^2 e^{-b(E,\pi-\Theta)}\right] = \frac{d\sigma_{nn}}{d\Omega} . \qquad (16)$$

We have already discussed the fact that $b(E,\pi-\Theta) = = \bar{n} = \bar{n}_+ + \bar{n}_- + \bar{n}_o$ is expected to be large compared to $b(E,\Theta)$ for $\Theta < \frac{\pi}{2}$, and therefore the second term in eq. (16) may be neglected for $\Theta < \frac{\pi}{2}$. For $\Theta > \frac{\pi}{2}$ the first term will become negligible.

In the case of proton-neutron collisions we have to differentiate between elastic and charge exchange scattering. Here we have

$$\frac{d\sigma}{d\Omega}\left(p + n \to \begin{matrix} p+n \\ n+p \end{matrix}\right) = \frac{1}{4E^2}|F_o \pm F_1|^2 e^{-b} . \qquad (17)$$

Experimentally, one cannot discriminate between elastic p-n scattering at the angle Θ and charge exchange scattering at $\pi-\Theta$. The observed cross section is therefore

$$\frac{d\sigma_{pn}}{d\Omega} = \frac{1}{4E^2}\left[|F_o+F_1|^2_\Theta \; e^{-b(\Theta)} + |F_o-F_1|^2_{\pi-\Theta} \; e^{-b(\pi-\Theta)}\right].$$

(18)

Here we did not write down the energy dependence explicitly. At the moment we do not know the relative phase between F_o and F_1 nor the ratio of their moduli.

Therefore, we cannot say whether one of the two terms in eq. (18) is negligible for certain values of Θ.

4.2. Two - pion production

Let us consider the process [24]

p + p → p + p + 2 pions,

where the protons are scattered with large momentum transfers, the total energy of the pions in the c.m. system is small compared to the nucleon energies, and the pions are "nonresonant" (no decay products of isobars or meson resonances). From eqs. (14) and (15) we get

$$\frac{d\sigma_{pp}^{(\pi^+\pi^-)}}{d\Omega} = \frac{d\sigma_{pp}^{el}}{d\Omega} \left(g_+^{(1)} \; g_-^{(2)} + g_+^{(2)} \; g_-^{(1)}\right) \tag{19a}$$

and

$$\frac{d\sigma_{pp}^{(\pi^o,\pi^o)}}{d\Omega} = \frac{d\sigma_{pp}^{el}}{d\Omega} \, g_o^{(1)} \; g_o^{(2)} \quad , \tag{19b}$$

where the $g_\beta^{(1)}$ etc. are the quantities defined in eq. (15). Now we make the plausible assumption that all the g_β, $\beta=+,-,o$, are equal :

$$g_+ = g_- = g_o = \frac{1}{3}\, b \quad . \tag{20}$$

Taking our b from eq. (8) we can predict the absolute values of the cross sections (19a) and (19b) by inserting the known values of the elastic cross section. A comparison of these predicted values with experiments has not yet been done.

4.3. Ratios of multiplicities for different types of secondaries

We now discuss some estimates for the ratios of differential multiplicities $\bar{n}(E,\Theta)$ for different types β of secondaries like pions, kaons, vector mesons etc. We can do this in the framework of SU(3) or probably even SU(6), because we assume the secondaries to "see" the primary particles in their nonrelativistic limit in the sense of negligible recoils. However, the notion of unitary symmetries is not necessary for the following.

We crudely approximate the integrand $|j_\beta(k)|^2$ in eq. (15) by the constant $g^2_{\beta BB}\, m^{-2}$, where $g_{\beta BB}$ is the (Yukawa) baryon coupling constant of the particle of type β . The arbitrary rest mass factor is introduced for dimensional reasons. With this approximation we have

$$\bar{n}_\beta(E,\Theta) = g^2_{\beta BB}\, m^{-2}\, \Phi_\beta(E,\Theta) \ , \qquad (21)$$

where $\Phi_\beta(E,\Theta) = \int \frac{d^3k}{2\omega}$ is the effective phase space of the particle of type β . This phase space is a function of the variables E and Θ of the primary particles. From eq. (21) we get

$$\bar{n}_{\beta_1}(E,\Theta) : \bar{n}_{\beta_2}(E,\Theta) = g^2_{\beta_1}\Phi_{\beta_1} : g^2_{\beta_2}\Phi_{\beta_2} \ . \qquad (22)$$

If the mass differences between the particles β_1 and β_2 are not too large one may be optimistic and assume $\Phi_{\beta_1} \simeq \Phi_{\beta_2}$.

Then we have

$$\bar{n}_{\beta_1} : \bar{n}_{\beta_2} = g^2_{\beta_1} : g^2_{\beta_2} \ . \qquad (23)$$

Thus the ratio $\bar{n}_{\beta_1} : \bar{n}_{\beta_2}$ becomes energy independent and rather simple in the crude approximation considered here.

Example:

Using the empirical values [25] for the nucleon coupling constants g_π, g_ρ, g_ω and g_ϕ for the pions, ρ-, ω- and ϕ- mesons, eq. (23) predicts

$$\overline{n}_\pi : \overline{n}_\rho : \overline{n}_\omega : \overline{n}_\phi \simeq 15 : 1 : 5 : 0 . \qquad (24)$$

The ratios apply to the charge state 0. In the case of other charge states, (π,ρ) , one has to multiply the above values by the number of charge states.

The predicted rather small multiplicities of ρ-mesons by eq. (24) may be misleading, because we have ignored the rather large magnetic coupling $f_\rho \simeq 4g_\rho$ of the ρ-meson. This coupling, which seems to be rather small for the ω and ϕ mesons, may enhance the ρ-value in eq. (24).

5. Bremsstrahlung model with realistic phase spaces

Up to now we have optimistically more or less ignored the amount of energy and momentum carried away by the secondaries, assuming this amount to be small or at least ignorable for a clever choice of the variables. Such an attitude is in many cases too optimistic and one has to weaken the assumptions. One interesting possibility is to keep the "softness"- or "smoothness" - hypothesis for the collision amplitudes but to use realistic phase spaces for the secondaries.

Such a procedure was first proposed by Anderson and Collins for inelastic p-p collisions [15]. A similar analysis was applied by Sosnowski and coworkers [16] to inelastic pion-nucleon scattering. Here we shall discuss the method of Mack which is most closely related to the model described in the previous parts of this section [17]. We shall not outline all the details of the calculations which may be taken from Mack's paper, but rather mention the general features. Only pions are being considered as secondaries.

According to eq. (15) we have for the quantity $\bar{n}$ in eq. (2) the expression

$$\bar{n}(E,\Theta) = \int \frac{d^3k}{2\omega} \, |j(k)|^2 .$$

We approximate this by

$$\bar{n}(E,\Theta) \simeq J(E,\Theta) \quad \Phi_1(E,\Theta) ,$$

$$J(E,\Theta) = |j(k \simeq o)|^2 ,$$

$$\Phi_1 = \int \frac{d^3k}{2\omega} .$$

Φ_1 is the effective phase space for one soft secondary. Eq. (2) then means that we have approximated the actual phase space of n pions by the n-th power of the one-pion phase space. In order to be more realistic one should use the actual phase space instead of the product Φ_1^n. If we keep the assumption that the two nucleons go into opposite directions after the scattering, even in inelastic collisions, and if only one proton is observed in the final state, we get for the reaction

$$p + p \rightarrow p + \text{nucleon} + n \text{ pions}$$

the cross section

$$\frac{d^2\sigma^{(n)}}{d\,\Omega dp}(E,p,\Theta) = C_n \, B\sigma^{el}(E,\Theta) \, \frac{1}{n!} \, J^n(E,\Theta)\Phi_n(W) . \tag{25}$$

Here $\Phi_n(W)$ is the invariant phase space for one nucleon and n pions with the invariant mass W(p), where p is the momentum of the observed proton after the scattering. C_n is a factor taking the different pion charge states into account and B is a kinematical factor. Both factors are defined in ref. 17 . The phase space can be calculated by the convenient method of Mazur and Lurçat [26] . The normalization in eq. (25) is such that

$$\int \frac{d^2\sigma^{(o)}}{dp\ d\Omega}\, dp = \sigma^{el} .$$

In order to exploit eq. (25) Mack's procedure is as follows: Experimentally, one knows [10] some values of

$$\frac{d^2\sigma}{dp\ d\Omega} = \sum_{n=o}^{\infty} \frac{d^2\sigma^{(n)}}{dp d\Omega}$$

for certain initial energies E and scattering angles Θ as a function of p. By a least square fit one determines σ^{el} and J as a function of E and Θ. As a result of this fitting one can say the following:

5.1. The experimental curves of $d^2\sigma/d\Omega dp$ can be accounted for very well by this 2-parameter fit (for fixed E and Θ).

5.2. The values of $\sigma^{el}(E,\Theta)$ so obtained are compatible with the experimentally measured ones. In particular one can predict the strong decrease of the elastic cross section with increasing energy E at fixed Θ.

5.3. Predictions can be made for the following quantities at the energies, angles and momenta dealt with in ref. 10):

a) The differential multiplicity $\bar{n}(E,\Theta,p)$,

b) the fluctuations $(\overline{n^2}-\bar{n}^2)(E,\Theta,p)$,

c) the differential total cross section $\bar{\sigma}(E,\Theta) =$

$$= \int dp \frac{d^2\sigma}{dp d\Omega}$$

d) and the multiplicity

$$\bar{n}_{av}(E,\Theta) = \frac{1}{\bar{\sigma}} \int dp\ \bar{n}(E,p,\Theta) \frac{d^2\sigma}{dp d\Omega} .$$

The numerical values of these predictions are contained in ref. 17 . One has to wait for future experiments in order to test them.

III. Electron-Nucleon Collisions

We shall now apply the general assumptions of section II to elastic and inelastic electron-nucleon scattering. Unfortunately, there is at the moment almost no experimental information about electroproduction cross sections where more than the electron and the recoil nucleon have been observed in multi-particle final states with large momentum transfers of the electron [27]. Thus one can only make predictions. In order to get something to compare with experiments we follow Mack [28],[29] and add some new dynamical assumptions in part 2 which allow us to calculate an explicit expression for the nucleon magnetic form factor at large momentum transfers.

1. A Model for the Electroproduction of Soft Mesons at Large Electron Momentum Transfer

We again start by assuming only one type of secondaries. Extensions to different types along the lines discussed for p-p collisions in section II are obvious. Therefore, we shall have only a few remarks about the role of unitary symmetries in this section.

Furthermore, we shall ignore all electromagnetic radiative corrections.

1.1. Kinematics

We use the variables

$$s=(p_{el} + p_{nucl})^2 \ , \ t = (p'_{el} - p_{el})^2, \ q^o = (p_{el}-p'_{el})^o \ ,$$

where the dashed variables correspond to the final state.

For s, $-t >> 4 M^2$ and $G_E(t) \leq G_M(t) \equiv G(t)$, where M is the nucleon mass and G_E and G_M the usual electric and magnetic form factors, the Rosenbluth formula [30] has the simple form

$$\frac{d\sigma^{el}}{dt} = \sigma_B \ (s,t) \ G^2(t) \quad , \tag{26}$$

where

$$\sigma_B = \frac{2\pi\alpha^2}{t^2 s^2} \ (t^2+2st+2s^2)$$

is the Born approximation which goes to

$$\sigma_B \xrightarrow{s\to\infty} \frac{4\pi\alpha^2}{t^2} \tag{27}$$

for $s \to \infty$ and fixed t. For details of the inelastic kinematics - which we do not need here - see ref. 31).

1.2. Elastic scattering

Comparing the eqs. (1) and (26) gives

$$\sigma_B(s,t)\, G^2(t) = \bar{\sigma}(s,t)\, e^{-\bar{n}(s,t)}, \quad -t \gg 4M^2 \quad . (28)$$

In order to extract more information from this equation we follow Wu and Yang [9] and assume the elastic form factor for large $-t$ to be completely determined by the inelastic channels. Since the inelastic channels in our model are accounted for by the exponential on the right hand side of eq. (28), it is tempting to put

$$\bar{n}(s,t) = -\ln G^2(t) \; . \qquad (29)$$

As in the case of p-p scattering (II,2.1.) we have the normalization problem here, too. Small deviations of the normalization constant $\beta = G^2(t)/\exp(-\bar{n})$ from the value 1 do not affect the relation (29) for large momentum transfers. Despite some doubts, let us stick to the normalization $\beta = 1$ in the following. At the end we shall use the slightly different normalization $\beta = \mu_p^2$, where μ_p is the magnetic moment of the proton (or neutron). This latter normalization seems to be more appropriate if one wants to extrapolate the relation (29) to small values of $-t$. It then seems reasonable to have $\bar{n} = 0$ for $t = 0$.

Eq. (29) gives us an interesting relation between the elastic magnetic form factor $G(t)$ and the differential multiplicity $\bar{n}(s,t)$ in quasielastic electron-nucleon scattering. Unfortunately, there are no data in order to test this relation. If we insert the experimental value [32] of $G(t)$ at $-t = 10(\mathrm{GeV}/c)^2$, we

get $\bar{n} \simeq 8-9$. This seems to be a rather large number, but one has to wait for the experiments and see!
One characteristic feature of the relation (29) is that it predicts that the differential multiplicity will become independent of the initial energy $\sqrt{s}$ for $s,-t \gg 4M^2$.

If eq. (29) is at least approximately reasonable, it may be useful for the determination of G(t) at very large momentum transfers, for the only quantities to be measured in order to determine G(t) are t and the average number of secondaries. Since the inelastic cross sections are expected to become larger than the elastic one for very large -t (see below), the application of eq. (29) may be helpful for the determination of G(t).

1.3. Inelastic collisions

From eqs. (1) and (29) we get

$$\frac{d\sigma^{(n)}}{dt} = \sigma_B(s,t)\ G^2(t)\ \frac{1}{n!}\ \left[-\ln G^2(t)\right]^n \ . \qquad (30)$$

Thus we have definite predictions for the cross sections of quasielastic electron-nucleon scattering with the additional emission of soft nonresonant secondaries. For small n one probably can neglect recoil effects associated with the emission of secondaries. For larger n one has to be more careful. There are several possibilities as to how one actually should compare eq. (30) with future data:

a) If $|t|^{1/2} \gg q^o_{c.m.}$ it may be that

$$\frac{d^2\sigma^{(n)}}{dt\,dq^o_{c.m.}}(s,t;q^o_{c.m.}) \simeq \frac{1}{q^o_{c.m.}}\frac{d\sigma^{(n)}}{dt}(s,t), \qquad (31)$$

where $d\sigma^{(n)}/dt$ is given by eq. (30).

b) Another possibility is

$$\int dq^{o}_{c.m.} \frac{d^2\sigma^{(n)}}{dt\; dq^{o}_{c.m.}} (s,t;q^{o}_{c.m.}) \simeq \frac{d\sigma^{(n)}}{dt} (s,t) \; , \tag{32}$$

where again the right hand side is given by eq. (30).

c) One makes similar phase space corrections as in the p-p case (see II,5.). So far this has not been done.

1.4. Short-range interactions and sum rule

Summing eq. (30) over n gives

$$\frac{d\sigma^{tot}}{dt} = \bar{\sigma}(s,t) = \sigma_B(s,t); \; s,-t >> 4M^2 \quad . \tag{33}$$

Since σ_B does not contain any rest masses (see eq. (26)), and since the fine structure constant α is dimensionless, we see that our fourth postulate from part 1 of section II, namely dilatation invariance for the short-range interactions, is fulfilled automatically here. In addition the sum rule (33) is compatible with a very similar inequality derived by Bjorken [33] for $d\sigma^{tot}/dt$ at fixed t in the limit $s \to \infty$ from chiral current algebra:

$$\frac{d\sigma_p^{tot}}{dt} + \frac{d\sigma_n^{tot}}{dt} \geq \frac{2\pi\alpha^2}{t^2} \quad . \tag{34}$$

The indices "p" and "n" refer to inelastic electron scattering off protons and neutrons respectively. If we assume the above proton and neutron cross sections to be approximately equal, we get from eq. (33) four times the value of the lower limit required by eq. (34). At the moment there are no data in order to test the relations (33) and (34) for $-t > 4M^2$. At $t \simeq -1(\mathrm{GeV}/c)^2$ the inequality (34) seems to be roughly fulfilled [31].

We should not forget, however, that the numerical value of eq. (33) depends on the normalization β . With our second normalization $\beta = \mu_p^2$ the expression (33) would be larger by a factor μ_p^2 .

1.5. Equality of proton and neutron magnetic form factors as a consequence of SU(2)

It was already pointed out in section II that the exponential $\exp(-\bar{n})$ is a unitary spin invariant. If we consider in particular the isospin group SU(2) and apply the above result to the proton and neutron magnetic form factors G_p and G_n by using the eq. (29), we obtain

$$\ln G_p(t) = \ln G_n(t) . \tag{35}$$

This relation is fulfilled experimentally up to the logarithm of the ratio of the proton and neutron magnetic moments squared, a number which is negligible compared to $\ln G^2$ for large $-t$. If we chose our second normalization, the result (35) would be the same as the usual scaling law [34].

It is interesting that we already get the relation (35) from SU(2). It has previously been derived from higher symmetries [35].

2. A calculation of the nucleon magnetic form factor for large momentum transfers

A characteristic feature of the elastic cross section (26) and of our model in particular is a factorization into a dilatation invariant (better: covariant) factor $\bar{\sigma}$ and a second one which does not have this invariance, for otherwise it would be a constant because it is dimensionless. This second factor is therefore in some sense associated with the breaking of dilatation invariance at very large momentum transfers. Thus, it seems likely that more detailed information about the structure of this symmetry breaking will yield some information about the large momentum transfer behaviour of form factors.

In order to describe this symmetry breaking it is very useful to employ the methods developed in connection with the hypothesis of a partially conserved axial vector current. This has been done by Mack [29]. We shall not repeat the details of calculations here but shall briefly sketch the ideas and one result which is interesting for us.

2.1. Partially conserved dilatation current

The action integrals of the coupling terms in many interesting Lagrangian field theories are invariant under the transformations

$$x^{\mu} \to x^{\mu\prime} = \rho x^{\mu}, \qquad \mu = 0,1,2,3 \quad ;$$

$$\phi(x) \to \phi'(x') = \rho^{\ell}\, \phi(x) \,, \tag{36}$$

where $\phi(x)$ is a certain spinor- or tensor-field with respect to the homogeneous Lorentz group. We have $\ell = -1$ for scalar, pseudoscalar and vector fields $A(x)$ and $A_{\mu}(x)$, $\mu = 0,1,2,3$, and $\ell = -3/2$ for spin 1/2 fields $\psi(x)$. Invariant are all interactions with a dimensionless coupling constant like $\bar{\psi}\psi\ A$, $\bar{\psi}\gamma_5\psi A$, $\bar{\psi}\gamma_{\mu}\psi A^{\mu}$ etc. The kinetic terms are invariant, too, except for the mass part: the rest masses destroy the dilatation invariance.

One can define a dilatation current $\mathcal{D}_\mu(x)$ in the usual way à la Noether and find its divergence $\partial^\mu \mathcal{D}_\mu$ proportional to the kinetic mass terms, for instance $m^2 A^+(x)A(x)$ for a free pseudoscalar field and $\alpha_o u_o + \alpha_8 u_8$ for the quark model, where $u_o = \bar{t}t$ breaks the chiral invariance and $u_8 = \bar{t}\,\lambda_8\, t$ breaks the SU(3) symmetry.

In the same way as in the algebra of currents one can define a time dependent "charge"

$$D(t) = \int d^3\vec{x}\ \mathcal{D}_o(x)$$

which generates the infinitesimal transformations (36):

$$\left[D(t)\ ,\phi(x)\right]_{x^o=t} = i^{-1}(-\ell+x^\nu\partial_\nu)\phi(x)\ . \qquad (37)$$

The important point is now that the divergence $U(x) \equiv \partial^\mu\ \mathcal{D}_\mu(x)$ is a local field with quantum numbers $J^P = 0^+$, $I=0$. This can be seen, for instance, from the examples given above. One now makes the "PCAC"-assumption that the matrix elements $\langle p'|U(o)|p\rangle$ are small for large momentum transfers $(p'-p)^2$, but relatively large for small $(p'-p)^2$. In addition one assumes $\langle 0^+,I=0|U(o)|0\rangle \neq 0$, i.e. $U(x)$ is an interpolating field for a 0^+, $I=0$ asymptotic free state. Such a state has the same quantum numbers as the more or less hypothetic σ-particle. But we do not require that such a particle really exists. The asymptotic state may be a 2-pion s-state, etc.

With the above assumptions one can play the usual soft - meson [36] game: Write down a generalized Ward-Takahashi identity, use the equal time commutation relations (37), take the Fourier transform etc. Having done all this one can express the τ-function for the process $A \to B + 0^+$(soft) in terms of the τ-function for the process $A \to B$:

$$\tau_{A\to B+0^+}[k(0^+) \simeq 0;\, p_2,\ldots,p_n] =$$

$$= -i \sum_{j=2}^{n} (\ell_j + 4 + p_j^\nu \frac{\partial}{\partial p_j^\nu})\, \tau_{A\to B}(p_2,\ldots,p_n) \ . \quad (38)$$

Application of this relation to the electromagnetic form factor $F_+(t)$ of a scalar particle gives the amplitude $T^\mu(p,p',k\simeq 0)$ for the corresponding process where one additional soft 0^+-particle is emitted:

$$T^\mu(p,p',k) = -\,2M^2 \Big[\frac{1}{-2p.k+k^2}(p'+p-k)^\mu +$$

$$+ \frac{1}{2p!k+k^2}(p'+p+k)^\mu\Big]\, F_+(t) + \quad (39)$$

$$+ 2(p'+p)^\mu\, t\frac{\partial}{\partial t} F_+(t) + O(k) \ .$$

In eq. (39) we have $p^2=p'^2= M^2$. The two first terms of the order k^{-1} describe those processes where the soft meson is emitted from the external lines, i.e. "before"and "after" the interaction of the primary hadron. The third term in eq. (39) describes the emission "during" the interaction.

2.2. Discussion of some approximations

In order to apply the relation (39) to the nucleon form factor G(t) we have to make some questionable approximations: First we shall ignore the spin of the nucleon. This does not seem to be very critical. For in the region $-t \gg 4M^2$ and $s \to \infty$ which we consider here, the electron-nucleon elastic cross section in eqs. (26) and (27) has the same form as the cross section for the electron scattering off spin zero particles. Furthermore, in one approach we shall treat the pions which we expect to form the bulk of the secondaries in inelastic electron-nucleon collisions as scalar particles rather than pseudoscalar ones. Two heuristic arguments for such a crude-looking approximation may be given: First we have seen that we can neglect the spin terms in the Rosenbluth formula in the ultrarelativistic region. Thus spin, and therefore parity, too, do not look so important any more. Secondly, if we look at the first order $\bar{u}(1')\gamma_5 u(1)\ A(2)$ of the pseudoscalar coupling, we have $\gamma_5 u(1) \simeq \pm\ u(1)$ in the extreme relativistic limit for the nucleons. Thus the pseudoscalar vertex looks very similar to the scalar one in this limit, and it may be that we can ignore parity altogether. Obviously, all these arguments are not reliable.

If we keep the small momenta k fixed in eq. (39), then the third term contains one more power of the energy in the numerator than the two first terms. Thus we expect it to dominate in the high energy region, although it is of a higher order in k than the first two terms. The trouble here is that one has to deal with two limits at the same time: $p \to \infty$, $k \to 0$, a problem which does not arise in low energy soft meson theorems. Since the physical mesons do have a finite rest mass after all, it seems reasonable to keep $k \neq 0$ fixed and perform the limit $p \to \infty$ first. Intuitively, this means that we assume the emission of secondary mesons at high energies and large momentum transfers to be more likely during the primary interaction than before and after.

It would be interesting to see how a "Low"-type formula [37] like eq. (39) looks for soft pseudoscalar pions. To the best of my knowledge, only those terms corresponding to the emission of soft pions from the external lines have been calculated [38]. Their application to high energy processes does not look very encouraging [12].

2.3. A differential equation for the form factor

Keeping only the third term in eq. (39) we get as the differential cross section for the electroproduction of one soft meson:

$$\frac{d^2\sigma^{(1)}}{dt\,d\Phi}(s,t;\,k\simeq 0) = \frac{4g^2}{M^2}\left[t\,\frac{\partial}{\partial t}\,G(t)\right]^2\sigma_B(s,t)\ . \quad (40)$$

Here Φ is the meson phase space and g the meson-nucleon coupling constant. Since we are not interested in the kinematics of the emitted meson, we rewrite eq. (40) in the following way

$$\frac{d\sigma^{(1)}}{dt}(s,t) = \frac{4g^2}{M^2}\,\sigma_B(s,t)\,\left[t\,\frac{\partial}{\partial t}\,G(t)\right]^2\,\Phi_{eff}\ , \quad (41)$$

where Φ_{eff} is some effective unknown phase space for the meson. Since we have assumed the mesons to be soft, their effective phase space should not depend too violently on s and t. In the following, we shall make the crude approximation Φ_{eff} = const.

The crucial step now is this: We assume that the cross section $d\sigma^{(1)}/dt$ in eq. (41) may also be expressed by the formula (30) with n = 1. This hypothesis leads to the consistency condition

$$[t \frac{\partial}{\partial t} G(t)]^2 \frac{4g^2}{M^2} \Phi_{eff} = - G^2(t) \ln G^2(t) . \qquad (42)$$

The solution of this differ ntial equation is

$$G(t) = \exp [- A \ln^2(-at)] , \qquad (43)$$

with $A = M^2/(8g^2\Phi_{eff})$ and a as a constant of integration. The above procedure leading to the form factor (43) may be interpreted in different physical terms:

a) We ignore the intrinsic parity of pions and identify the scalar mesons with them. The problems associated with this interpretation have already been discussed. If we insert for g^2 the value of the pion-nucleon coupling constant squared and for Φ_{eff} the estimated value m_π^2, we obtain an A which is of the order of magnitude 10^{-1}. The value of the constant a is, of course, not known. A rough guess is that it may be of the order m_π^{-2} . These estimates of A and a give indeed a reasonable G(t) as can be seen by a comparison of the expression (43) with the latest Stanford data [39] up to $t = -25(GeV/c)^2$.

b) We interpret the scalar meson as a 2-pion s-state (nonresonant or resonant). The questionable assumption then is that we can describe this pair by the formula (30) with n=1, although we are essentially dealing with two secondaries. Furthermore, it seems doubtful that the main bulk of secondaries in multiparticle final states is produced in such pairs. However, nothing is known experimentally about this. In this second interpretation g is an effective 2-pion-nucleon coupling constant and Φ_{eff} a two-pion effective phase space. It is hard to give estimates for these quantities, although one may take for g the effective g_σ-coupling constant from dispersion theoretical analysis [25],[40].

c) A third possibility is to keep the 2-pion interpretation of the scalar meson but to use the formula (30) with n=2. The right hand side of eq. (42) is then slightly changed, and we get as a solution of this new differential equation

$$G(t) = \exp\left(-b|t|^{B}\right), \qquad B = \frac{M}{2g\sqrt{\Phi_{eff}}} . \tag{44}$$

The quantity b is a constant of integration which should be positive if the expression (44) is to make any sense.

2.4. Ad hoc extension of the calculated G(t) to small t.

Despite the fact that the assumptions leading to the expression (43) are only valid for large -t, one may nevertheless ask how it behaves around t=0. It obviously has the wrong threshold properties at this point. The cut should start at $t = 4m_\pi^2$, not at $t = 0$. If one makes an ad hoc correction in order to have the cut started at the right t-value - even if the type of branch point remains wrong -, one gets

$$G(t) = \exp\{-A[\ln^2(-a(t-4m_\pi^2)) - \ln^2(a4m_\pi^2)]\}. \tag{45}$$

In ref. [28],[39] the normalization on the left hand side of eq. (45) is $G(t)/\mu_p$ where μ_p is the total magnetic moment of the proton. As already mentioned before, a comparison with experiments [39] shows the expression (45) to be quite reasonable.

Footnotes and References

1. We use the units h = 1 = c and the metric $a^2 = (a^o)^2 - (\vec{a})^2$.
2. Surveys of the latest developments are contained in the Proceedings of the Heidelberg International Conference, ed. H. Filthuth (North-Holland Co., Amsterdam, 1968); see also ref. 3 .
3. As to the present status of theories and experiments see the Proceedings of the CERN Topical Conference on High-Energy-Collisions of Hadrons, ed. L. Van Hove (CERN, 1968).
4. G. Cocconi et al., Phys. Rev. 138, B165 (1965).
5. For a review see the talk of H. Epstein in ref. 3.
6. Only two examples shall be listed here: T. T. Chou and C. N. Yang, Proceedings of the Conference on High-Energy Physics and Nuclear Structure, ed. G. Alexander (North-Holland Publ. Comp., Amsterdam, 1967), p. 348; R. Hagedorn, Nuovo Cim. Suppl. 3, 147 (1965). Further papers by Hagedorn and Hagedorn and Ranft are to be published in Nuovo Cim.
7. H. A. Kastrup, Phys. Rev. 147, 1130 (1966); this paper contains a long list of earlier work on the bremsstrahlung model; the following paper should be added to the list: R. C. Arnold and P. E. Heckman, Phys.Rev.,164,1822(1967).Additional aspects of the model are discussed by H. A. Kastrup, Nucl. Phys. B1, 309 (1967) and in a paper contained in ref. 3.
8. See,for instance, V. S. Barashenkov et al., Fortschr. d. Phys. 14, 357 (1966) and 15, 435 (1967).
9. T. T. Wu and C. N. Yang,Phys. Rev. 137 , B708 (1965).
10. E. W. Anderson et al., Phys. Rev. Letters 19, 198 (1967).
11. For a review, see,for example, the Proceedings of

the 1967 International Conference on Particles and Fields, ed. C. R. Hagen, G. Guralnik and V. A. Mathur, (Interscience Publ., New York, London, Sydney, 1967).

12. The application of the usual soft-pion techniques to high-energy collisions has met with some difficulties as discussed in the following papers: H. D. I. Abarbanel and S. Nussinov, Ann. Physics (N . Y.) 42, 467 (1967); R. Perrin , Phys. Rev. 162, 1343 (1967).
13. A model in order to explain the dominant $p_\perp$ - dependence was discussed by K. Huang, Phys. Rev. 146, 1075 (1966).
14. L. G. Ratner et al., Phys. Rev. Letters 18, 1218 (1967).
15. E. W. Anderson and G. B. Collins, Phys. Rev. Lett. 19, 198 (1967).
16. G. Bialkowski and R. Sosnowski, Phys. Letters 25B, 519 (1967).
17. G. Mack, Phys. Letters 26B, 515 (1968).
18. J. Orear, Phys. Letters 13, 190 (1964); practically the same expression had been proposed earlier by D. S. Narayan and K. V. L. Sarma, Phys. Letters 5, 365 (1963).
19. For details see ref. 3 . A compilation of all data is contained in the recent paper by J. A. Ållaby et al., to be publ.
20. T. Ericson and T. Mayer-Kuckuck, Ann. Rev. Nucl. Science 11, 183 (1966).
21. J. V. Allaby et al., Phys. Letters 23, 389 (1966).
22. A. Bialas and O. Czyzewski, Nuovo Cim. 49A, 273 (1967).
23. See, for example, J. D. Bjorken and S. D. Drell, Relativistic Quantum Fields (McGraw-Hill Book Comp., New York, 1965), p. 202.

24. The production of single vector mesons in the framework of the bremsstrahlung model has been discussed by Arnold and Heckman. ref. 7 .
25. G. Köpp and D. Söding, Phys. Letters 23, 494 (1966). The numerical values for some g_β vary with the analyzing authors; For a recent compilation and a new analysis see D. V. Bugg, Nucl. Phys. B5, 29 (1968). As to the strange particles see J. K. Kim, Phys. Rev. Letters, 19, 1079 (1967).
26. F. Lurçat and P. Mazur, Nuovo Cim. 31, 140 (1964).
27. F. W. Brasse, J. Engler, E. Ganßauge and M. Schweizer, DESY-Report 67/34.
28. G. Mack, Phys. Rev. 154, 1617 (1967).
29. G. Mack, Nucl. Physics B5, 499 (1968).
30. See, for instance, G. Källén, Elementary Particle Physics (Addison-Wesley Publ. Comp., Inc., Reading, Mass., 1964), p. 224.
31. F. J. Gilman, SLAC-PUB-357, to be publ. in Phys.Rev.
32. W. Albrecht et al,Phys.Rev.Letters 18,1014(1967).
33. J.D.Bjorken, Phys.Rev.Letters 16,408 (1966).
34. As to the present experimental status see the report by W. K. H. Panofsky in ref. 2.
35. See, for example, R. Delbourgo and R. White, Nuovo Cim.40A, 1228(1965); this paper contains further references.
36. See,for instance, ref. 11.
37. F. E. Low, Phys. Rev. 110, 974 (1958).
38. S. Weinberg, Phys. Rev. Letters 16,879 (1966); see also S. L. Adler, Phys. Rev. 139, B1638(1965).
39. D. H. Coward et al., Phys. Rev. Lett.20, 292 (1968).
40. C. Lovelace, talk given at the Heidelberg International Conference, ref. 2.

GAUGE PROPERTIES OF THE MINKOWSKI SPACE†

By

H. A. KASTRUP

Universität München, Sektion Physik, München,
Germany

I. Introduction

Gauge groups have been playing an important role in particle physics. Some examples are:

1. Phase transformations of states ($\phi \to e^{i\alpha}\phi$) or of non-observable fields ($\psi(x) \to e^{i\alpha}\psi(x)$). Such symmetry transformations are associated with the conservation of charge-like quantities such as, for instance, the baryon number.

2. Local gauge transformations $A_\mu(x) \to A_\mu(x) + \partial_\mu f(x)$, $\psi(x) \to e^{if(x)}\psi(x)$ in quantum electrodynamics which imply the conservation of the electric charge and which are considered to be intimately connected with the vanishing rest mass of the photon. These position dependent transformations further have the im-

† Seminar given at the VII. Internationale Universitätswochen f.Kernphysik, Schladming,February 26-March 9,1968.

portant consequence that all charged particles are coupled in a universal way [1].

3. Generalizations of these local gauge transformations to strong interactions (Yang-Mills-groups [2]) played an important role in the creation of the Eightfold Way [3]. There is, however, the problem that the associated vector particles, which take the place of the photon, have finite rest masses. Perhaps for this reason, and certainly because there are mass differences between members of the same multiplets, these symmetries are only approximately valid.

4. As a last example we mention the chiral group $\psi(x) \to e^{i\alpha\gamma_5}\,\psi(x)$, acting on spinor fields and being an exact symmetry only in the limit of (at least) vanishing pion masses [4].

I mention these different well-known examples of gauge transformations for the following reasons: They show that even approximate gauge groups which are broken by rest masses can be very useful for particle physics. We further notice that all the groups listed above transform only the state vectors or fields but not the coordinates x^i. One may, therefore, ask whether there are useful gauge groups, perhaps only approximate ones, which transform the coordinates $x = (x^0,\vec{x})$ of the Minkowski space, too.

By a gauge group of the Minkowski space we mean the following: We assume distances ds to be defined by the differential quadratic form

$$ds^2 = (dx^0)^2 - (dx^1)^2 - (dx^2)^2 - (dx^3)^2 \quad . \qquad (1)$$

We take the differential form (1) because we want to allow for nonlinear transformations, in particular local ones. The structure of the Minkowski space is preserved by requiring the form (1) for all space-time points.

As a gauge transformation of the Minkowski space we denote any mapping $x^i \to x^{i\prime} = f^i(x)$, $i=0,1,2,3$, which multiplies the form (1) by a - in general position dependent - factor.

The continuous and discrete groups which induce such transformations in x-space were already determined in the last century. Liouville did this analysis for three [5] and S. Lie for arbitrary [6] dimensions [2]. We shall call the largest group of the Minkowski space, which has the property in question, the "full Liouville group of the Minkowski space". Its orthochronous proper part, i.e. that part of the group which is continuously connected with the identity transformation, consists of the following subgroups:

The orthochronous proper Poincaré group $D^{\uparrow}_{10}(a;\Lambda)_+$:

$$x^j \to \Lambda^j{}_i\, x^i + a^j, \quad j = 0,1,2,3,$$

$$ds^2 \to ds^2 \quad ; \tag{2}$$

the dilatations $D_1(\alpha)$:

$$x^j \to e^{\alpha}\, x^j, \quad j = 0,1,2,3,$$

$$ds^2 \to e^{2\alpha}\, ds^2 \quad , \tag{3}$$

and the special Liouville group $SL_4(c)$:

$$x^j \to RT(c)\, Rx^j = \frac{1}{\sigma(x)}\,(x^j - c^j x^2), \quad j=0,1,2,3,$$

$$ds^2 \to \frac{1}{\sigma^2(x)}\, ds^2 \quad , \tag{4}$$

where

$$Rx^j = -\frac{x^j}{x^2} \; ; \; T(c)\, x^j = x^j + c^j \; ;$$

$$\sigma(x) = 1-2c.x + c^2x^2 \quad .$$

The full 15-parameter Liouville group L_{15} may be obtained from the proper transformations (2) - (4) by adjoining the usual parity transformation P and the time inversion T or, equivalently, the "length inversion" R and the discrete group $\tilde{R}\, x^i = x^i/x^2$.

The group L_{15} is isomorphic to the group O(2,4). We notice that the special Liouville group SL_4 and the length inversion R are not well-defined in x-space, because some light cones($(x^j-c^j/c^2)(x_j-c_j/c^2)c^2 = \sigma(x)=0$ or $x^2 = 0$ respectively) are mapped into infinity. We shall deal with this problem later on, when we discuss the geometrical meaning of the transformations (3) and (4) in detail.

II. The Classical Free Relativistic Particle.

In order to illustrate the "conserved" quantities associated with the dilatations and the special Liouville group, we consider the very simple case of a classical relativistic free particle characterized by the relations

$$p^o = + (\vec{p}^2 + m^2)^{1/2} \quad , \qquad \vec{r} = \frac{\vec{p}}{p^o}\, x_o + \vec{a} \quad . \tag{5}$$

The constants of motion associated with the homogeneous Lorentz group are

$$\vec{m} \equiv \vec{r} \times \vec{p} = \vec{a} \times \vec{p} \quad , \vec{n} \equiv \vec{r}\, p_o - x_o\, \vec{p} = p_o\, \vec{a} \quad . \tag{6}$$

The "dilatation momentum" is [7]

$$s \equiv p^o x^o - \vec{r}.\vec{p} = \frac{m^2}{p_o} x_o - \vec{a}.\vec{p} \quad , \tag{7}$$

and the corresponding "Bessel-Hagen momenta" of the special Liouville group are [7]

$$h_o \equiv 2\, x_o\, x^j p_j - x^2 p_o = \frac{m^2}{p_o} x_o^2 + \vec{a}^2\, p_o \; ,$$

$$\vec{h} \equiv 2\vec{x}\, x^j\, p_j - x^2 \vec{p} =$$

$$= \frac{m^2}{p_o} \left(\frac{\vec{p}}{p_o} x_o^2 + 2\vec{a}\, x_o\right) + \vec{a}^2\vec{p} - 2(\vec{a}.\vec{p})\, \vec{a} \quad . \tag{8}$$

We see that the gauge groups $D_1(\alpha)$ and $SL_4(c)$ lead to conserved quantities only in the limits $m \to 0$ or $p_o \to \infty$. Therefore the situation here is very similar to those of the last two examples discussed in the Introduction.

It is interesting to notice that in the limits $m \to 0$ or $p \to \infty$ all of the 15 conserved quantities listed above may be expressed by the two independent vectors $\vec{p}$ and $\vec{h}$:

$$p_o = + (\vec{p}^2)^{1/2} \quad , \; h_o = + (\vec{h}^2)^{1/2} \quad , \quad 2s^2 = p.h \quad ,$$

$$\vec{h} \times \vec{p} = 2s\, \vec{m} \; , \qquad p_o \vec{h} - h_o \vec{p} = 2\, s\, \vec{n} \; . \tag{9}$$

Thus the quantities s, $\vec{m}$ and $\vec{n}$ are determined by $\vec{h}$ and $\vec{p}$ up to a relative sign.

The relations (9) bear a close resemblance to the following commutator of the generators P_j and K_ℓ of the groups $T_4(a)$ and $SL_4(c)$:

$$[P_j, K_\ell] = 2i(g_{j\ell} D - M_{j\ell}) \; , \tag{10}$$

where D and $M_{j\ell}$ are the generators of the groups $D_1(\alpha)$ and $L_+^\uparrow$. Because of $K_j = RP_jR$ one can generate the full Liouville group by the translations T_4, the length inversion R, and the time inversion T.

Under the discrete groups P, T and R the above constants of motion transform as follows:

$$P : s \to s\ , \quad \vec{h} \to -\vec{h}\ , h_o \to h_o\ ;$$

$$T : s \to -s\ , \quad \vec{h} \to -\vec{h}\ , h_o \to h_o\ ;$$

$$R : \vec{m} \to \vec{m}\ , \vec{n} \to \vec{n}\ , s \to -s\ , h_j \to p_j,\ \ p_j \to h_j\ ,$$

$$j = 0,1,2,3,$$

In addition we have $\tilde{R} = PTR$.
We see that the transformation R leaves the energy p_o positive because $h_o \geq 0$. The same is true for the group SL_4: the quantity h_o stays positive under translations (see eq. (8)). Because of eq. (4) this means that p_o stays positive under special Liouville transformations.
Remark:

For any free elementary excitation with a dispersion law

$$E = A\ p^\alpha\ , \qquad (11)$$

where A and α are constants, the Bessel-Hagen momentum $\vec{h}$, and because of the eqs. (9) the quantities h_o, s, $\vec{m}$ and $\vec{n}$, too, are constants of the free motion [8]. This can be seen by replacing $(x^o, \vec{x})$ in the above formulae by $(vt, \vec{x})$, where $\vec{v} = \partial E/\partial \vec{p}$. In particular we have all these constants of motion for the nonrelativistic particle with $E^2 = p^2/2m$. Thus, all these free motions are invariant under the group [9] O(2,4).

However, the situation is different, if there is an energy gap in the dispersion law: $E = A(p^{\alpha}+B)^{\beta}$. The "gap" parameter B introduces a fixed length into the theory, which breaks the dilatation and Liouville symmetries.

III. Geometrical Interpretation

The geometrical interpretation of the dilatations is relatively easy: they map a given length ds onto another length $e^{\alpha}ds$. Therefore, physical systems which contain a fixed length can only be approximately invariant under the group [10].

I think that the interpretation of the special Liouville group is very similar: it induces local geometrical gauge transformations instead of the global dilatations (the gauge factor $\sigma(x)$ in eq. (4) is position dependent but e^{α} is not!). There are, however, some rather complicated problems associated with the special Liouville group in x-space. I shall discuss three of them:

1. It was already pointed out in the Introduction that the Liouville group is not well-defined in x-space because the mapping (4) is not one-to-one. This feature can be taken care of quite easily by separating a Poincaré invariant unit of length κ^{-1} (scale factor) from the position coordinates x^j. One defines

$$x^j = \frac{\eta^j}{\kappa} , \quad j = 0,1,2,3, \tag{12}$$

and the spurious coordinate λ by

$$\kappa \lambda = \eta^j \eta_j \quad . \tag{13}$$

For each point in space-time the four numbers η^j characterize its position and κ the unit of length employed at this point [7].

The coordinates η^j transform like the x^j under the Poincaré group ($\eta^j \to \eta^j + a^j\kappa$, $\eta^j \to \Lambda^j_i \eta^i$). Under dilatations we have

$$\eta^j \to \eta^j , \kappa \to e^{-\alpha}\kappa, \lambda \to e^{\alpha}\lambda \quad , \qquad (14)$$

and the length inversion R gives

$$\eta^j \to \eta^j , \quad \kappa \to -\lambda \quad , \quad \lambda \to -\kappa \quad . \qquad (15)$$

Since the group R generates the special Liouville group, the relations (15), which leave the position unchanged, make it even more evident that we are dealing with gauge transformations.

The full 15-parameter Liouville group leaves the quadratic form $\eta_j \eta^j - \kappa\lambda$ invariant, which shows its isomorphy to the group O(2,4).

The length ds can be expressed in terms of the new coordinates by

$$ds^2 = dx^j dx_j = \frac{1}{\kappa^2} (d\eta^j d\eta_j - d\kappa\, d\lambda) \ . \qquad (16)$$

with the subsidiary condition (13).

2. In the framework of the Poincaré group plus dilatations we can define space-like and time-like separations by the sign of the form $(x_1 - x_2)^2$. This is no longer true if we include the special Liouville group. The reasons are the following [7]:

a) The transformations (4) are nonlinear in x-space. From all that we know from differential geometry and general relativity we then have to use differential forms in order to define distances, not global ones like $(x_1 - x_2)^2$.

b) Because the transformations (4) are not well-defined in x-space, it does not make much sense to use the x-coordinates themselves in order to define space- and time-like separations.

c) The sign of the global form $(x_1 - x_2)^2$ is not invariant [7] under the transformations (4).

For all these reasons we have to define space- and time-like separations in the space of the coordinates η^j, κ and λ. Eq. (14) suggests how to do this: Locally we still can define space- and time-like separations in a Liouville invariant way by the sign of ds^2 - see eq. (4) - , or because $\kappa^2 \geq 0$ by the sign of $d\eta_j\, d\eta^j - d\kappa\, d\lambda$. Since all the transformations induced by the Liouville group are linear in η^j, κ and λ, we can define space- and time-like separations by the sign of the invariant form

$$(\eta_1 - \eta_2)^j (\eta_1 - \eta_2)_j - (\kappa_1 - \kappa_2)(\lambda_1 - \lambda_2) . \qquad (17)$$

This definition seems to be the natural generalization of the usual one by means of the sign of $(x_1 - x_2)^2$.

Remark: Because of the property c) the transformations (4) seem to violate a theorem by Zeeman [11] on the causal automorphisms of the Minkowski space. However, this is not the case, because Zeeman assumes the mappings of the Minkowski space he is dealing with to be one-to-one. The transformations (4) do not fulfill this assumption.

3. The conventional interpretation of the special Liouville group in x-space is that it represents a transformation from a system at rest to a relativistic uniformly accelerated one [12] (hyperbolic motion). I do not consider this interpretation to be a fruitful or even a right one for the following reasons:

a) The interpretation applies only to the spatial parameters c^j, j=1,2,3, not to c^o.

b) The group is not well-defined in x-space.

c) Invariance under translations and the special Liouville group implies [13] invariance under dilatations (eq. (10)). This means that uniformly accelerated systems would have continuous rest masses and continuous energy spectra. However, there is no evidence that the discrete energy spectra of atoms or the rest masses of elementary particles become continuous under uniform accelerations.

d) The group velocity of the wave packets formed by the eigenfunctions of the generators K_j of the special Liouville group describes linear motions [14] , whereas the phase velocity describes hyperbolic motions. From all that we know from quantum mechanics, the group velocity describes the motions of particles, not the phase velocity.

e) We have seen that the Liouville symmetry is not confined to the relativistic case, but that it occurs in any situation where there are elementary excitations of the form (11). Whereas the "gauge"-interpretation can be applied to all these situations, the "acceleration"-interpretation cannot.

IV. The Nonrelativistic Case

It has already been mentioned that the nonrelativistic free motion is invariant under the full Liouville group O(2,4). The crucial difference between the relativistic and nonrelativistic case is in the way the symmetry is broken: in the relativistic case many interesting interaction Lagrangians are dilatation and Liouville invariant [15], but almost all nonrelativistic potentials are not [9]. Relativistically, these gauge symmetries are in general broken by the kinetic mass terms, not by the interaction Lagrangians, but nonrelativistically, the symmetries are broken by the potentials, not by the kinetic terms.

Several applications of the broken dilatation and Liouville symmetries for relativistic systems have been discussed elsewhere [16]. In the following we shall discuss some features of the nonrelativistic case [9],[17]:

1. The conserved quantities associated with the O(2,4)-invariance of the free particle may be obtained by the replacement $x^o \to y^o = vt$, for instance

$$s \equiv 2Et - \vec{r} \cdot \vec{p} \quad , \quad \vec{h} = 2\,\vec{x}\,s - (v^2t^2 - \vec{x}^2)\,\vec{p} \; ,$$

where $E = 1/2\ \vec{v}\cdot\vec{p}$. The energy E itself is not among the 15 constants of motion associated with the group O(2,4). It is a function of $p^o = p$. The same is true for the "Galilei"-momentum $\vec{g} = v^{-1}\,\vec{n}$, where $\vec{n}$ is the "Lorentz" - momentum $\vec{n} = \vec{x}\,p - vt\,\vec{p}$. More details are contained in ref. 9 .

2. It is very illustrative to analyze the additional constraints imposed on elastic scattering by the assumption of conserved dilatation- and Bessel-Hagen momenta. We consider the scattering of two particles with equal masses in the c.m. system with the c.m.coordinate $\vec{R} = \vec{r}_1 + \vec{r}_2 = 0$.
At large negative times t we have

$$\vec{r}_i = \pm \frac{\vec{p}}{m}\,t \pm \vec{a} \; , \qquad i=1,2 \; .$$

For large positive times we get - as a consequence of

energy, momentum, and Galilei-momentum conservation- :

$$\vec{r}_i' = \pm \frac{\vec{p}'}{m} t \pm \vec{a}' , \quad i=1,2; \quad |\vec{p}'| = |\vec{p}| .$$

The conservation of angular momentum implies that the perpendicular projection of $\vec{a}'$ on $\vec{p}'$ is the same as that of $\vec{a}$ on $\vec{p}$:

$$\vec{a}' \times \vec{p}' = \vec{a} \times \vec{p} .$$

There are no constraints on the parallel projection $\vec{a}' \cdot \vec{p}'$ so far, and this freedom accounts for the possibility of positive and negative time delays during the interaction. However, if we require the conservation of the total dilatation momentum s_1+s_2 we get

$$\vec{a}' \cdot \vec{p}' = \vec{a} \cdot \vec{p} . \qquad (18)$$

Eq. (18) means that there can be no time delays if the total dilatation momentum is conserved. This is, of course, a severe restriction which is not valid for most of the nonrelativistic interactions. In quantum mechanical language the vanishing time delay means [18] that $d\delta_\ell / dp = 0$, where δ_ℓ is the usual phase shift. Thus only constant phase shifts are allowed by the conservation of the total dilatation momentum. The phase shifts do not have to vanish, however.

One can check that the conservation of the total Bessel-Hagen momenta is fulfilled automatically in our example of elastic scattering if all the other total momenta are conserved.

3. It is interesting to see how the dilatation- and Liouville symmetries can be broken in the nonrelativi-

stic case in order to describe more realistic interacting systems. Because of the qualitative difference between relativistic and nonrelativistic systems the following procedure may not be very useful for the relativistic case [19].

We consider the simplified situation of a stationary system (t=0). Then we have only the Euklidian group, the dilatations and a 3-dimensional Liouville group. A free nonrelativistic spinless quantum mechanical system is described by states $\phi(\vec{p})$ which form a Hilbert space with the scalar product

$$(\phi_1, \phi_2) = \int d^3\vec{p}\ \phi_1^*(\vec{p})\ \phi_2(\vec{p}) \ . \tag{19}$$

In the space of these states we have a unitary representation of our groups. The Hermitian generators of this representation are given by

$$P^j = p^j, \quad M^{jk} = i^{-1}(p^j \partial_k - p^k \partial_j) \ ,$$

$$\partial_k = \frac{\partial}{\partial p^k} \ ; j,k = 1,2,3 \tag{20a}$$

$$K^j = 3\partial_j + 2p^k \partial_k \partial_j - p^j \Delta \quad , \ D = i(\tfrac{3}{2} + p^j \partial_j) \ . \tag{20b}$$

It is important for later to mention that the first term in K^j and D is typical for the behaviour of the functions $\phi(\vec{p})$ themselves under finite transformations, for inst. $\phi'(p) = e^{3/2\ \alpha}\ \phi(e^{\alpha}\ \vec{p})$, - "intrinsic" parts of D and K^j -, whereas the rest is a consequence of the transformations of their argument $\vec{p}$ - "orbital" parts - or of the argument of their Fourier transform $\tilde{\phi}(\vec{x})$. We write down only one commutation relation:

$$[P^j \ , K^{\ell}] = 2i\ (\delta^{j\ell}\ D - M^{j\ell}) \ . \tag{21}$$

The operator $H_o = 1/2m\ p^2$ for the free particle energy belongs to the enveloping algebra of the above Lie-algebra. Because of

$$e^{i\alpha D}\ H_o\ e^{-i\alpha D} = e^{-2\alpha}\ H_o$$

the spectrum of H_o is continuous.

Let us consider a system with interactions now. We assume the interaction to be described by a potential V in the energy operator $H = H_o + V$, where V is a function of the generators P^k and K^ℓ (because of eq. (21) all the other generators may be expressed by these). If we confine ourselves to velocity independent and rotational invariant potentials we have $V = V(\vec{K}^2)$. Now, if the dilatations are to be a symmetry group, then V has to transform in the same way as H_o. This implies $V \sim (\vec{K}^2)^{-1}$. Nature obviously is much richer than this single potential would allow her to be. More generally, invariance under finite dilatations implies a continuous energy spectrum which, however, is not observed in most of nonrelativistic atomic physics.

Therefore, we have to break the dilatation symmetry. We notice that the above constraints on the spectrum of H are no longer valid if D would commute with H. This happens if we drop the second (orbital) term of D in eq. (20b). In order to describe [20] this procedure mathematically we write $D_\varepsilon = i(3/2 + \varepsilon p^j \partial_j)$, where $0 \le \varepsilon \le 1$. $D_{\varepsilon=o}$ is a multiple of the unity operator and therefore commutes with H.

However, because of the closure relation (21) we cannot keep K^j as it is in eq. (20b) - we want to keep the momenta $\vec{p}$. We have to drop the orbital part of K^j, too, in order to get $D_{\varepsilon=o}$ on the right hand side of eq. (21). With

$$K^j_\varepsilon = 3\,\partial_j + \varepsilon(2p^k\partial_k\partial_j - p^j\,\Delta)$$

we get instead of eq. (21):

$$[P^j, K^\ell_\varepsilon] = 2i(\delta^{j\ell}\,D_\varepsilon - \varepsilon\,M^{j\ell}) \quad . \tag{22}$$

In the limit $\varepsilon=0$ we have $Q^j = \frac{1}{3i}K^j_{\varepsilon=o}$ and $I = \frac{2}{3i}D_{\varepsilon=o}$, where Q^j is the usual quantum mechanical position operator and I the identity operator. Thus, the occurrence of the position operator is a consequence of our breaking of the dilatations,and quantum mechanics appears as a broken Liouville symmetry [17]. The fundamental relation (21) of the group goes over into the usual canonical commutation relation.

We still have to deal with the following problem: the quantities $D_{\varepsilon=o}$ and $K^j_{\varepsilon=o}$ are both purely imaginary. This means that

$$u(\alpha) = e^{+i\alpha D_{\varepsilon=o}} \quad , \quad u(\vec{c}) = e^{ic_j K^j_{\varepsilon=o}}$$

are no longer unitary if α and c_j are kept real. We can avoid this by considering $\alpha = \alpha(\varepsilon)$ and $c_j = c_j(\varepsilon)$ as complex-valued functions of ε which are real for $\varepsilon=1$ and imaginary for $\varepsilon=0$. Then $u[\alpha(\varepsilon=o)]$ and $u[\vec{c}(\varepsilon=o)]$ are unitary. In this way the dilatations degenerate into a phase transformation.

Footnotes and References

1. E. P. Wigner, Proc. Nat. Acad. Sciences (USA) 38, 449 (1952).
2. C. N. Yang and R. L. Mills, Phys. Rev. 96, 191 (1956).

3. M. Gell-Mann and Y. Ne'eman, The Eightfold Way (W. A. Benjamin, Inc., New York, 1966).
4. See, for instance, the lecture of F. Gürsey during the school.
5. J. Liouville, editorial note VI in G. Monge's book Application de l'Analyse à la Geométrie (Bachelier, Paris, 1850).
6. S. Lie and F. Engel, Theorie der Transformations-gruppen (B. G. Teubner, Leipzig, 1893) Vol.III, Chaps. 17 and 18.
7. H. A. Kastrup, Phys. Rev. 150, 1183 (1966).
8. H. A. Kastrup, Nuovo Cim. 48, 271 (1967).
9. H. A. Kastrup, Gauge Properties of the Galilei Space, to be published.
10. H. A. Kastrup, Nucl. Phys. 58, 561 (1964).
11. E. C. Zeeman, Journ. Math. Phys. 5, 490 (1964).
12. See, for instance, T. Fulton, F. Rohrlich and L. Witten, Nuovo Cim. 26, 652 (1962).
13. We always assume the generators we are dealing with to be represented by self-adjoint operators which can be integrated to continuous finite unitary representations. See, however, M. Flato, and D. Sternheimer, Phys. Rev. Letters 16, 1185 (1966).
14. H. A. Kastrup, Phys. Rev. 143, 1021 (1966).
15. H. A. Kastrup, Phys. Letters 3, 78 (1962).
16. H. A. Kastrup, lectures given at this school.
17. H. A. Kastrup, Quantum Mechanics as a Broken Symmetry, to be published.
18. E. P. Wigner, Phys. Rev. 98, 145 (1955); see also ref. 10.
19. Contractions of the relativistic Liouville group were first introduced by I. E. Segal, Duke Math. J. 18, 221 (1951). Their physical significance was

20. Technically we are dealing with contractions of Lie-algebras: I. E. Segal, ref. 19; E. Inönü and E. P. Wigner, Proc. Nat. Acad. Sciences (USA) 39, 510 (1953) and 40, 119 (1954).

SOMETHING NEW IN COSMIC RAYS †

By

S. L. GLASHOW

Department of Physics, Harvard University, USA

The remarks I shall make are inspired by the beautiful recent cosmic ray experiment done in Utah by Keuffel et al. reported in the Christmas 1967 issue of Physical Review Letters. The "sec Θ law" which they observe to fail was deduced by Barrett et al. in Review of Modern Physics 1952. Whatever I say in the way of controvertial explanation of the experiment is due to Curtis Callan and myself.

The Experiment

Energetic cosmic radiation found in deep mines is usually assumed to consist of muons, for no other particle of sufficient penetrating power is known. The vertical intensity of this radiation as a function of rock depth has been very well studied. Muons of given energy have a more or less well defined range in

† Seminar given at the VII. Internationale Universitätswochen für Kernphysik, Schladming, February 26 - March 9, 1968.

rock, which may be approximately computed quantum-electrodynamically. Thus, from a knowledge of vertical intensity as a function of depth we may establish the vertical flux of muons at sea level as a function of energy. The Utah experiment attempts to measure the muon intensity as a function both of rock slant depth and zenith angle. In order to decouple these variables, the mine is situated under a mountainous overburden, so that contours of constant slant depth correspond to a wide range of zenith angles:

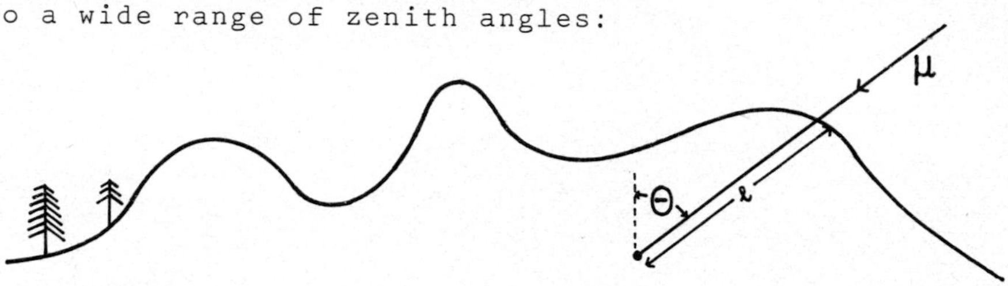

In this experiment, slant depths of 2000 - 8000 feet are studied corresponding to muon energies of 10^3 - 10^4 GeV. According to theory, the Utah counting rate should have been

$$F(\Theta,\ell) = F(\ell) \cdot \sec\Theta$$

where $F(\ell)$ is the well measured vertical cosmic ray intensity. In fact, what was observed was isotropy at fixed ℓ:

$$F(\Theta,\ell) = F(\ell)$$

If this experiment is to be believed (and, it appears to be very well done) we must abandon the hypothesis that leads to the sec Θ-law. More precisely, we must conclude that most of the penetrating radiation at depths of 2000 - 8000 feet of rock does <u>not</u> consist of muons that are the progeny of pions and kaons which were themselves produced by primary cosmic rays.

The Sec Θ-Law

Let us imagine that the high energy cosmic ray muons are produced by the chain

primary nucleon —(collision)→ energetic pion —(decay)→ energetic muon
energetic pion —(collision)→ no muons

Assume that the flux of primaries is isotropic at high energies. We are interested in muon (and hence pion) energies of 10^3 - 10^4 GeV. At these energies, the pion travels 30 - 300 km before it decays. Because most primary interactions take place at altitudes ∿ 10km, most of these pions will suffer nuclear interactions before they have time to decay: only a small fraction of the pions will succeed in producing muons.(Of course, pions which interact will produce more pions. However, these "secondary" pions will not produce an appreciable muon flux for two reasons: they are of lower energy, and are negligible in number compared to "primary" pions produced by the more copious primary cosmic ray. Moreover, they are produced in denser atmosphere where their chance of decay is even smaller than that of their ancestors). We attempt to estimate the distance a "primary" pion travels, on the average, before it interacts. Let y denote altitude in units of the earths atmospheric scale depth. Thus, density is given by $N\ e^{-y}$ where N is the sealevel density. Let σ_p be the cross section for a primary proton off atmospheric nuclei, and

σ_π that of a pion. Let Θ be the zenith angle of the primary, which is also that of the energetic pion to a good approximation. On the average, the primary proton will interact at altitude y_o where

$$\int_{y_o}^{\infty} \frac{dy\ N\ e^{-y}}{\cos\Theta} = 1/\sigma_p \quad , \; e^{-y_o} = \frac{\cos\Theta}{N\sigma_p} \quad .$$

The energetic pion produced in the collision will proceed to the lower altitude y_1, where

$$\int_{y_1}^{y_o} \frac{dy\ Ne^{-y}}{\cos\Theta} = 1/\sigma_\pi \quad , \; e^{-y_1} = \frac{\cos\Theta}{N}\left(\frac{1}{\sigma_\pi} + \frac{1}{\sigma_p}\right)$$

We may conclude that

$$y_o - y_1 = \log\left[\frac{\cos\Theta}{N}\left(\frac{1}{\sigma_\pi} + \frac{1}{\sigma_p}\right)\right] - \log\left(\frac{\cos\Theta}{N}\,\frac{1}{\sigma_p}\right)$$

$$y_o - y_1 = \log\left(1 + \frac{\sigma_p}{\sigma_\pi}\right)$$

and that the distance transversed by pions of given energy is proportional to sec Θ . But,this means that the probability that a given pion will decay is also given by sec Θ . Thus, at energies sufficiently great that pion decay is small compared to pion interaction, the angular distribution at sea-level of these muons be given by

$$F(E)\ \sec\Theta$$

where F(E) is the vertical muon flux. A more careful calculation yields an angular distribution like $(\cos\Theta + E_o/E)^{-1}$ where $E_o \sim 100$ GeV. A similar calculation may be performed for muons which are the progeny of kaons.

Because the kaon lifetime is half that of the pion, and because the kaon is heavier than the pion, those kaons producing muons of 10^3 - 10^4 GeV have a shorter decay length by a factor of ~ 4 than corresponding pions. For the kaon progeny, the sec Θ law is only partially effective at 10^3 GeV, but it should be fully realized by 10^4 GeV. The isotropy observed in Utah is quite incompatible even with the extreme assumption that all the muons arise from kaon decay.

Some Explanations

1. Suppose that at high energies, muons somehow develop strong interactions and are produced directly in pairs. Thus, we avoid the conflict between decay and interaction which yields the sec Θ law. However, could a muon with such strong interactions be expected to penetrate 8000 feet of rock? No!

2. Suppose there is a flux of primary cosmic photons of very high energies which is 10^4 as copious as primary nucleons. (It is somewhat difficult to guess their source). Though they would make the required muons, they would lead to far more energetic air showers than are observed. No!

3. Suppose that somehow, at high energies, the weak interactions become strong and W's are copiously produced by "weak interactions". Rapidly decaying into leptons these W's could produce enough isotropic muons if the cross section for energetic W production were comparable to that of energetic pions. However, since weak interactions have become strong, we are again faced with the problem that the μ's cannot be expected to penetrate rock as if they had <u>no</u> strong interactions.

4. We can explain the paradox of copious production, but no strong interaction by the following type of artifice, analogous to associated production. Let there exist a fermion X which enjoys conventional weak interactions like those of leptons, i.e.

$$X \begin{array}{l} \rightarrow \mu + \nu + \bar{\nu} \\ \rightarrow e + \nu + \bar{\nu} \end{array}$$

For a X mass of 1 GeV we anticipate a lifetime of 10^{-11} sec, so that all X's succeed in decaying before reaching earth, and so that heavy lepton searches that have been performed at accelerators would have failed. Suppose furthermore, that X has strong quadratic couplings with hadrons, so that $X\bar{X}$ pairs are copiously produced by primary cosmic rays. Artificial though this explanation is, it works.

5. Another way to obtain enough isotropic muons is to give strong quadratic couplings to the W's. Thus they may be copiously produced, and may decay semi-weakly into muon plus neutrino (perhaps one third of the time). This is a possible explanation of the Utah experiment, but strong W couplings do present some new problems. They will in general induce neutral lepton currents, and would lead to weak elastic scattering of neutrinos from protons:

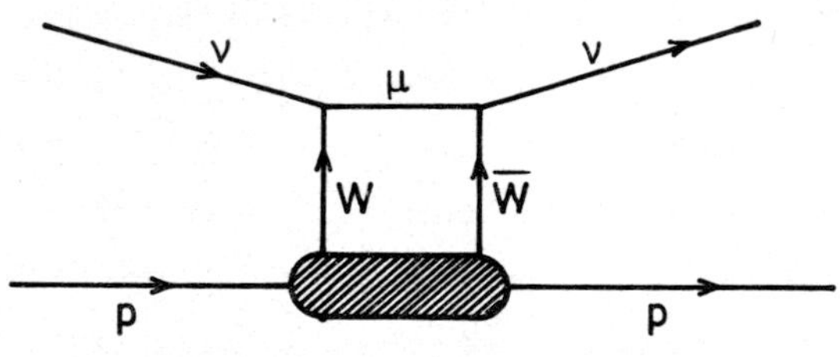

However, this process may be suppressed if the strong W couplings are invariant under the interchange of W and $\bar{W}$, for then

$$M_E(\nu p \to \nu p) = M_E(\bar{\nu} p \to \bar{\nu} p)$$

This is because crossing symmetry tells us that

$$M_E(\nu p \to \nu p) = -M_{-E}(\bar{\nu} p \to \bar{\nu} p)$$

and $M_E(\nu p \to \nu p)$ is then an odd function of energy vanishing at E = 0. Another potential difficulty is that strong W couplings can disrupt the universality of $G_\mu = G_V$ because of the diagram

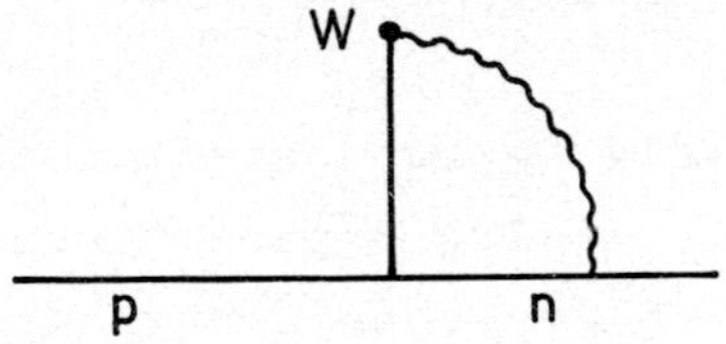

where the wavy line denotes the strong bilinear coupling of hadrons to $\bar{W}W$. Again, this diagram may be made to vanish at zero momentum transfer if the strong $\bar{W}W$ coupling is made gauge invariant, $\bar{W}_{\mu\nu} W_{\mu\nu} \times$ hadron - fields. But, the problem remains that weak magnetism effects will still be induced by the new hypothetical coupling of W's. There is another difficulty with any model involving the relatively copious production of heavy particles, like W, which decay into muons. This is the fact that high energy jets only very rarely involve particles with large transverse momenta. Decays of W's would certainly produce such particles.

6. Finally, we consider the radical hypothesis that the "muons" seen in deep mines are not really

muons. Imagine that there are in cosmic primaries new particles, U, with the following properties: they are massive ($\geq$ 2,5 GeV), singly charged, and they lack strong interactions and are stable. Such particles will penetrate to deep mines even more effectively than muons. We allege that most of the particles seen at depths greater than 2000 feet are U-particles. They are isotropic simply because the incident primary flux is isotropic. In order to explain the Utah experiment, it is necessary to suppose that $\sim 10^{-3}$ of the primary flux at energies of 10^{3} - 10^{4} GeV consist of U particles. This hypothesis has the virtue that it may be easily tested: it is straightforward to measure the mass of cosmic rays in deep mines to verify whether or not they are really muons.

It seems reasonably to suppose that if the Utah experiment is really seeing U particles, that, in fact, 10^{-3} of all primary cosmic rays are U's. In that case, a rather substantial number of U's should have accumulated on earth. Positive U's will form a new heavy "isotope" of hydrogen, which should have a natural abundance, in seawater, of $\sim 10^{-14}$, which is well within the range of experimental detectability. Negative U's will bind to prevalent atomic nuclei making new heavy "isotopes" of nitrogen and of carbon

$$U^{-} + N \rightarrow \text{heavy } C$$
$$U^{-} + O \rightarrow \text{heavy } N$$

Possibly, these "isotopes" are concentrated by some biological or geological processes, and they may occur with detectable abundances. Great scientific and technological vistas would be revealed, should U-matter really exist!

η PRODUCTION IN $K^- + p \to \Lambda + \eta$ AND SOME NEW RESULTS ON THE η DECAY MODES†

By

B. B. COX, L. R. FORTNEY, D. W. CARPENTER, C. M. ROSE, W. SMITH and E. C. FOWLER

Starting in 1965, the high energy physics group at Duke University, Durham, North Carolina obtained about 500.000 stereophotographs of the Columbia-Brookhaven National Laboratory 30-inch hydrogen bubble chamber exposed to an electrostatically separated 750 MeV/c K^- beam emerging from BNL's Alternating Gradient Synchrotron. The major purpose of the experiment was to collect a sample of about 1200 η mesons decaying via the $\pi^+ \pi^- \pi^0$ mode in order to test the possibility that the electromagnetic interaction of hadrons fails to some extent to be invariant under the charge conjugation operation C [1]. Experiments prior to that time had indicated that for this decay there might be an asymmetry between the π^+ and the π^- of as much as 10 %. This experiment, as the Columbia University experiment [2] with η's produced in π^+ + deuterium, would have given a definitive result for an asymmetry of 10 %. The Duke result for a little less than 1000 events stands now at

---

†Talk given by E. C. Fowler at the VII. Internationale Universitätswochen für Kernphysik, Schladming, February 26 - March 9, 1968.

1.3 standard deviations from perfect symmetry. A very good spark chamber experiment [3] from the European Organization for Nuclear Research (CERN Laboratory) gave 0.3 % ± 1.0 % showing that any possible effect was small enough that it could probably not be found with present bubble chamber techniques. The latest result which I have heard about is that of a new spark chamber experiment by W. Y. Lee of Columbia University at the AGS. My understanding is that he is getting 1.0 % ± 1/2 %. Barring the remote chance of something truly unusual (like there being two states instead of one that the η mass), there is little likelihood that the Duke bubble chamber work will yield new information of importance to this aspect of η decay. We are now concentrating on the question of establishing the existence of a resonance at threshold for Λ , η production in K^-p interaction and on measuring the branching ratios of η decay.

Table I shows the well-established static properties of the η state, which may conveniently be thought

Table I

Properties of State

Mass	548.8	MeV/c^2
Spin	0	
Parity	-	
I-Spin	0	
G-Parity	+	

of as a bound state of four pions. Because it has positive G-parity, decay via the strong interaction to three pions is forbidden, but it can decay to three pions via the electromagnetic interaction.

Table II furnishes the latest compilation concerning the major observed decay modes of the η , and rather than hold anything back, the results of the Duke experiment are presented here. It is seen that the current situation is unsatisfactory as regards the $\pi^0\gamma\gamma$ mode.

Table II

Comparison of Duke Results with Latest Summary on the Decay of the State (Latest Rosenfeld, Roos et al., 1.Jan.,1968)

η Decay Mode	"New" Data	"Old" Data	Duke with $\pi^0\gamma\gamma$	Duke without $\pi^0\gamma\gamma$
$\gamma\gamma$	0.42±.02	0.34±.02	0.40±.05	0.42±.05
$\pi^0\gamma\gamma$	0.01±.02	0.19±.03	0.17± ?	------
$3\pi^0$	0.28±.03	0.18±.03	0.17± ?	0.33± ?
$\pi\ \pi\ \pi^0$	0.24±.01	0.23±.01	0.20±.02	0.20±.02
$\pi\ \pi\ \gamma$	0.06±.005	0.05±.005	0.07±.01	0.07±.01
$\pi^0e^+e^-$	<0.01			

In 1962 and 1963, shortly after the discovery of the η state, there was not enough experimental information to judge whether or not any decay to neutrals beyond two γ's and three π^0's was present. An experiment at Frascati [4] gave the first indication (Spring 1966) of $\pi^0\gamma\gamma$ and then others also showed evidence of this mode. Without going into detail, more recent experiments (most of them reported at the Heidelberg conference last September) have shown no strong evidence for this mode; hence the division into "new" and "old" data. The statement made by the people who made this compilation is that if all the errors are taken at face value, the probability that the various experiments in the "new" group were drawn from the same original data as those of the "old" group is less than 1 in 10.000. I would like to be able to tell you that the Duke experiment is adequate to settle the question, but it is clear that it is not. Our data certainly do not demand the $\pi^0\gamma\gamma$ decay, but if we make the hypothesis that it is there, we get almost as good a fit to the experimental histogram (see fig. 9) which currently has only about 130 events in it. There is still a reasonable chance that by the time we have completed the analysis of all the Duke photographs we shall obtain a more stringent separation between hypotheses.

No one has yet published any strong theoretical reason why there should be a large amount of $\pi^0\gamma\gamma$ decays relative to the others. Of course the simplest way of thinking about η decay does not seem to work. One of the early mysteries of η decay remains unanswered: How can the decay to three pions be considerably more probable than the decay to two pions and a γ when the latter has one less power of α,the fine structure constant, "against" it? In a similar crude way, one would certainly expect the decay to two γ's to dominate $\pi^0\gamma\gamma$. There is another problem which is closely related to the way the experiments are done. Even from the early analysis of Gell-Mann and Rosenfeld [5] it is clear that for η decaying to three pions in an I=1 state (which is consistent with the distribution on the Dalitz plot for π^+, π^-, π^0) the ratio of three π^0's to the charged pion decay should be close to the limit indicated by Clebsch-Gordan coefficients for an almost 100% symmetric final state wave function for the three pions; namely,

$$R_1 = \frac{\Gamma(\pi^o \ \pi^o \ \pi^o)}{\Gamma(\pi^+ \ \pi^- \ \pi^o)} = 1.73$$

(This would be 3/2 except for the fact that the π^o is about 5 MeV/c^2 "lighter" than the charged pions). Indeed there is a small admixture of the mixed symmetry states needed to account for the decrease in population of the Dalitz plot as the π^o energy increases, but this turns out to be sufficiently small that a final value for R_1 of about 1.3 or 1.4 would be quite satisfactory. Now many experiments give a definite limit on the ratio

$$3 : 1 \doteq R_2 = \frac{\Gamma(\eta \to \text{neutrals})}{\Gamma(\eta \to \pi^+ \ \pi^- \ \pi^o)} \quad ;$$

moreover, the decay mode $\eta \to \gamma + \gamma$ usually can be identified unambiguously. In this situation if one introduces, or is forced to introduce, the possibility of a decay mode such as $\pi^o\gamma\gamma$ it is inevitable that the ratio R_1 will be reduced, because the "new" decay mode must be "stolen" from events previously assigned to $3\pi^o$ decays - they could seldom be confused with the $\eta \to \gamma\gamma$. From this standpoint, as illustrated in Table II, the conclusion that about 20% of the η's decay to π^o and 2 γ's is a complication, and the recent indications that there is only a very small percentage of the all neutral decays to be assigned to that mode represents a desirable trend towards simplicity. Finally, there is a rather definite prediction [6] by Gell-Mann and collaborators of the ratio,

$$R_3 = \frac{\Gamma(\pi^+ \ \pi^- \ \gamma)}{\Gamma(\gamma\gamma)} \quad \text{between} \ \frac{1}{4} \ \text{and} \ \frac{1}{8}$$

based on a dominant, intermediate vector meson model which seems to be borne out by the data. A number of attempts have been made to explain a large $\pi^o\gamma\gamma$ decay mode and the consequent low value of R_1, but in view of the uncertain experimental situation, I shall not refer to them here.

Turning to the question of production, the Duke data thus far is based on the η's decaying to $\pi^+\pi^-\pi^0$. Figure 1 shows a typical event in the 30" chamber.

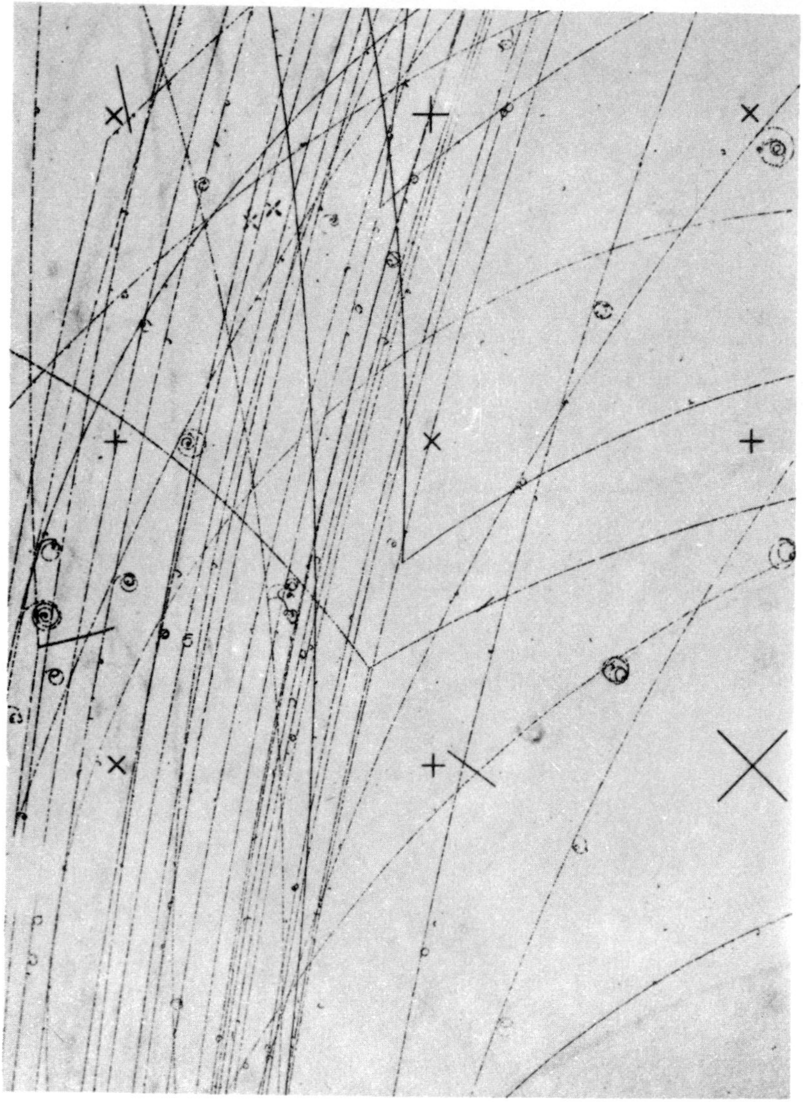

Figure 1. Photographic print of View 1 of a typical η production event with subsequent decay to $\pi^+\pi^-\pi^0$. The negatively charged K mesons enter from the bottom of the photograph.

Note that the beam curvature corresponds to its entering from the bottom of the page and being negatively charged. Both the positive pion from the primary vertex and the positive arm of the "vee" (corresponding to the proton in $\Lambda \to p + \pi^-$) appear dense as a result of a larger bubble density owing to their lower velocity in the laboratory compared to the other particles of the event. The π^o is, of course, "missing", but after measurement can be traced by using conservation of energy and momentum in the event. The momentum of the Λ itself, transformed into the center of mass of the original collision between K^- and target proton, gives a direct indication of the fact that the three pions were produced initially as an η, with the unique mass of 548.8 MeV/c^2. In fact, if we first eliminate the approximately 6.000 events in which the two pions in the primary and the decaying Λ fit with nothing "missing" and then accept all of the remaining events which fit the hypothesis of a single missing π^o (approximately 700), we are left with a 95 % pure sample of η's as indicated by the plot of the raw data for Λ momentum shown in Figure 2 (it is given in terms of the raw "missing mass", which is completely equivalent just to the Λ momentum in CMS). Note that the intervals in the plot of Figure 2 are 1/2 MeV/c^2. The rather sharp mass resolution displayed here is a result of the fact that the η's are produced very near threshold. Even at 100 MeV/c above threshold, the uncertainty of mass shown by such a plot would be more than five times that displayed here. As had been found previously by Berley and collaborators [7] the sample of η's found in this way is very nearly free of background and the general level of the cross section for η production is between 0.5 and 1.0 millibarns. Thus far, we have used only the events found in this

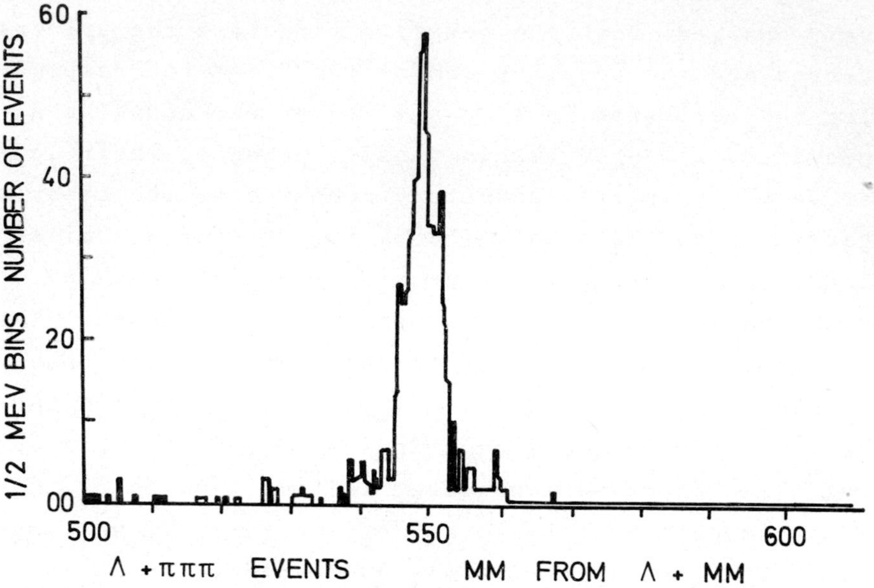

FIGURE 2

The raw measurement of the Λ momentum and knowledge of the beam momentum is used to evaluate the overall invariant mass for the remainder of the reaction products for those two-pronged plus Λ events which unambiguously fit $\Lambda\pi^+\pi^-\pi^0$ in SQUAW. It is seen that the selection of a missing π^0 yields an almost unambiguous selection of η's in this experiment.

---

way (hence only η's decaying via $\pi^+\pi^-\pi^0$) to construct a plot of production cross section versus bombarding energy. These data are shown in Figure 3. In the running of the experiment, every effort had been made to collect the largest possible sample of η's for a given amount of AGS time, consequently the Duke data do not cover a very wide interval of bombarding momenta. Figure 3 however includes the data of Berley et al. and also some early results of the Alvarez group at

Berkeley, so that a reasonable complete curve can be constructed under the hypothesis of

a) an $S^{1/2}$ scattering length at threshold,

b) a $P^{1/2}$ scattering length at threshold, or

c) an $S^{1/2}$ resonance just above threshold.

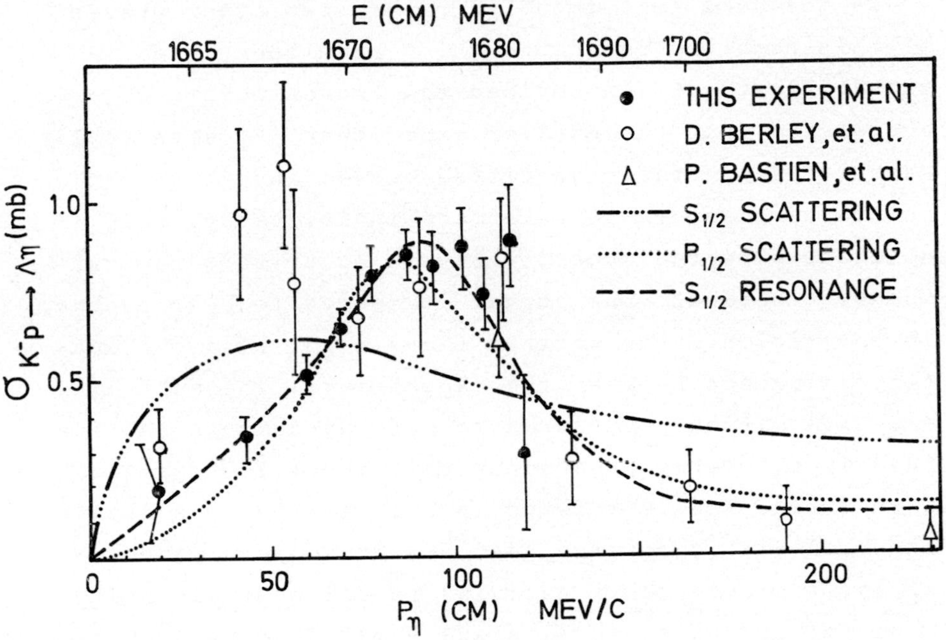

FIGURE 3

The total cross section for production of $\Lambda+\eta$ in K^-+p interactions is plotted against the final state momentum of the η in the center of mass. Only the $\pi^+\pi^-\pi^0$ decays have been used to construct this curve. The overall factor to obtain the total cross section is 4.0 (see Table II). Data from Berley et al. and from Bastien et al. (see Ref. 7) have been included.

Though the results are not completely conclusive, the best fit to all the experimental points is the $S^{1/2}$ resonance. The parameters for this best fit are very similar to the earlier ones found by Berley et al;

namely E_0^* = 1675 MeV and width = 15 MeV. Recently, a similar sharp rise in the production of η's just above the threshold for $K^- + p \rightarrow \Sigma^0 + \eta$ has been observed. It is interesting to speculate that there might be an $S^{1/2}$, parity minus, octet analogous to the baryon octet [8].

In the Duke experiment there are two other phases of this possible resonance which have been investigated. First, it was noticed that three points in succession from the earlier experiment [7] seem to lie rather far off the best fitted curve. Though these points have rather large errors individually, and could be expected occasionally to occur this far off the hypothetical true curve, it seemed to bear further investigation as to whether there might be a systematic difference between the experiments. The most likely explanation considered to date depends on the fact that in the early experiment only about 190 examples of η production were found in all, so that η's decaying via all modes were used and consequently most of the η's were those decaying to all neutrals. As a matter of convenience, since a much larger sample, was available, all of the 700 Duke events used in Figure 3 are η's decaying via $\pi^+\pi^-\pi^0$. We have not completed a systematic analysis comparable to the first sample, but we have recently collected a sample of about 800 η's decaying to all neutrals. This group shows a dependence on bombarding momentum indistinguishable from the Duke points in Figure 3. We conclude that the apparent wide swing in the earlier data was just statistical.

A Λ-η resonance must have isotopic spin zero. The most likely competitive channel is $\Sigma^0 - \pi^0$, which cannot be detected unambiguously by fitting event by event, but which can be detected "in the large" through analysis of missing mass spectra. Recently, we have begun to obtain some interesting data on the dependence of the total cross section for $K^- + p \to \Sigma^0 + \pi^0$. The measuring machines at Duke are of the conventional Franckenstein variety, but they are connected to a relatively sophisticated control computer (DDP24). It has been possible to make an important new use of

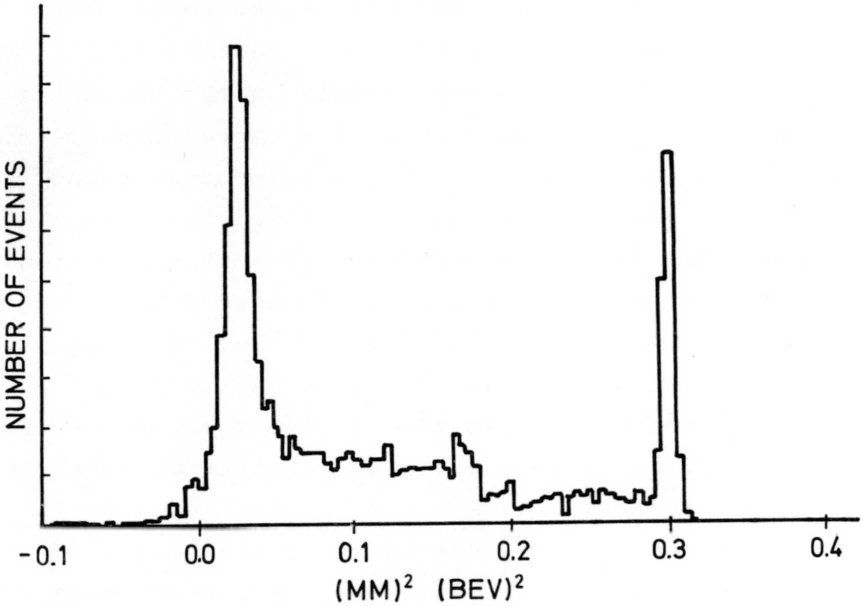

FIGURE 4

For all zero-pronged events which have an identified Λ decay associated, the number of events is plotted against the invariant "missing mass". The peak at the left corresponds to $\Lambda+\pi^0$, that at the right to $\Lambda+\eta$, the step function near the center is at the upper limit for this calculation mass when $\Sigma^0 + \pi^0$ is produced.

---

the computer in the part of the Duke experiment having to do with finding and measuring "zero-pronged" events with a "vee". The "vee" is usually a decaying Λ and is found by area scanning. No search for a corresponding disappearing beam track (zero-prong) is made in scanning; instead, every vee is set up for measurement, one fiducial mark is measured and the two tracks from the decay of the Λ are measured in a single view. The computer takes in these measured points and by fitting circles determines the projected momentum of each track, from these it reconstructs the line of flight of the Λ and drives the cross-hair on the measuring machine to a point which shows the operator where to look for a possible associated zero-pronged event. If found, the operator proceeds to record and measure the entire event in all three views. If nothing is seen in the first view, the process is repeated in a second view. If no zero prong is found there, the search is abandoned. This process, even with pictures full of beam tracks made it possible to find the zero-pronged events with about 95 % efficiency. After measurement, and space reconstruction the first test applied is to fit whether or not the vee is truly the result of a Λ decay (some 10 or 15 % of the vees are the result of the decay of neutral K-mesons). For all zero-pronged events with an identified Λ decay, the invariant missing mass has been calculated. The results are shown in Figure 4. The peak at the left corresponds to the single missing π^o or to the process $K^-+p\rightarrow\Lambda+\pi^o$, the peak at the right corresponds to the η's discussed earlier. Near the middle of the abscissa there is a sudden drop in the number of events as one reads on to the right (toward higher and higher missing mass). The "edge" corresponds to the limit of missing mass for the process $K^-+p\rightarrow\Sigma^o+\pi^o$ where the missing mass has been cal-

$\rightarrow\Lambda+\gamma$

culated from the detected Λ alone. Figure 5 shows the results of a Monte Carlo calculation of the missing mass spectrum for this experiment taking into account the π^0 and η production, the $\Sigma^0 - \pi^0$ production and a background of $\Sigma^0\pi^0\pi^0$ production. Having worked out these shapes, the data have been divided according to bombarding K^- momentum.

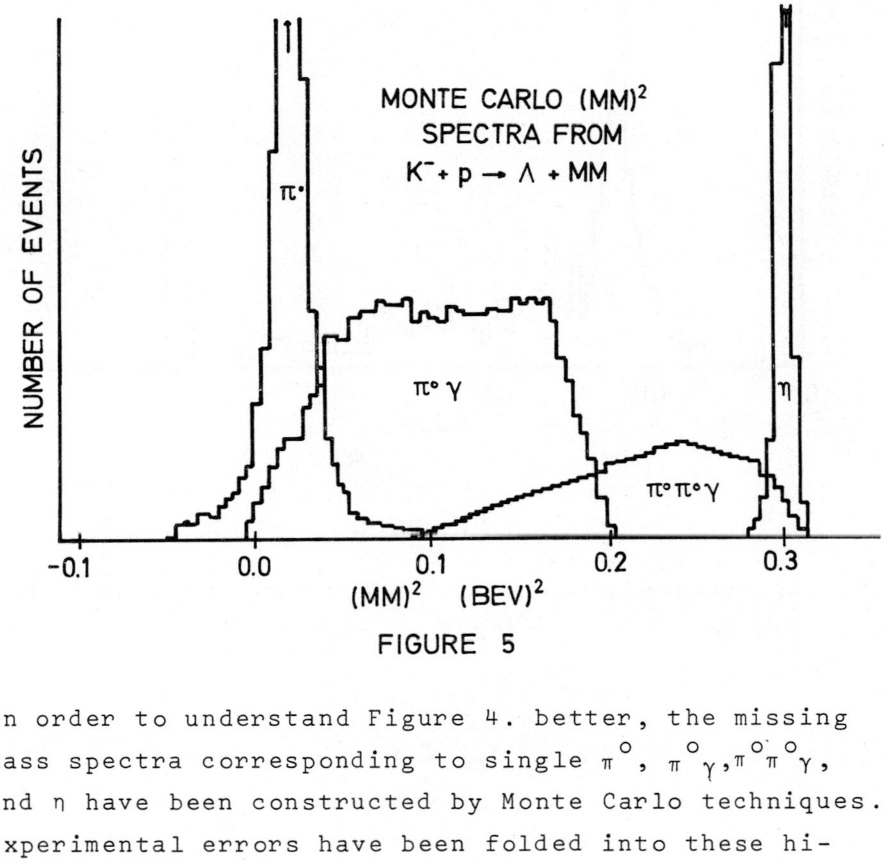

FIGURE 5

In order to understand Figure 4. better, the missing mass spectra corresponding to single π^0, $\pi^0\gamma$, $\pi^0\pi^0\gamma$, and η have been constructed by Monte Carlo techniques. Experimental errors have been folded into these histograms.

Figures 6, 7 and 8 show; respectively, 852 events for momentum less than 735 MeV/c (E^* = 1669 MeV), 1851 events for the interval 735 to 745 MeV/c (E^* 1669 to

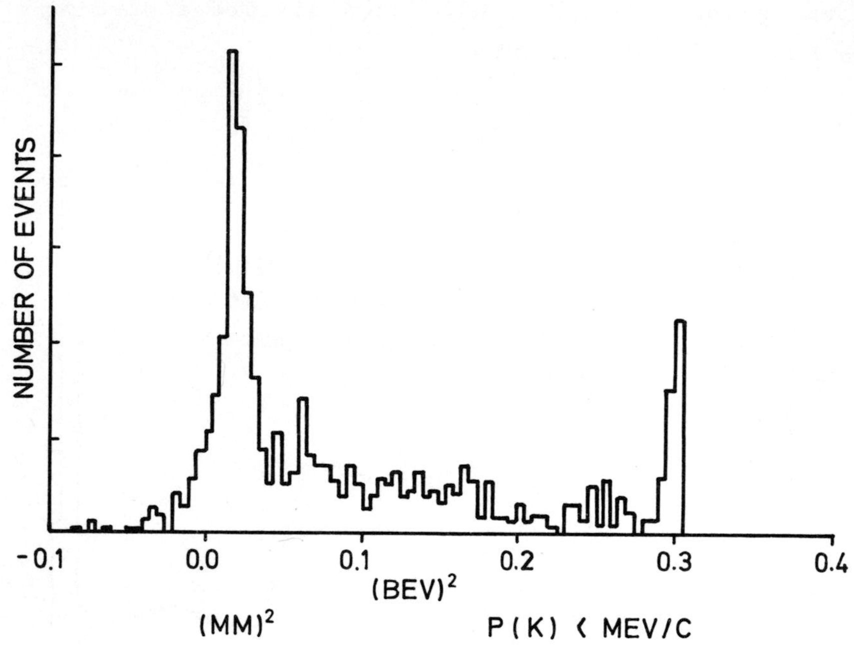

FIGURE 6

852 zero-pronged events with bombarding momentum less than 735 MeV/c (E^* = 1669 MeV) are shown plotted against the missing mass. Compare with Figures 7 and 8.

1674), and 1301 events for all momenta above 745 MeV/c. The Duke experiment contains scarcely any data for momenta above 760 MeV/c as can be seen from the points in Figure 3. This sequence of graphs shows clearly the rise in η production with increasing momentum and it gives a strong indication of a similar rise in the production of $\Sigma^0 + \pi^0$. I do not have the corresponding graphs for our latest results, but my coworkers at Duke have recently increased the total sample of zero-pronged events with identified Λ decay to 6.000

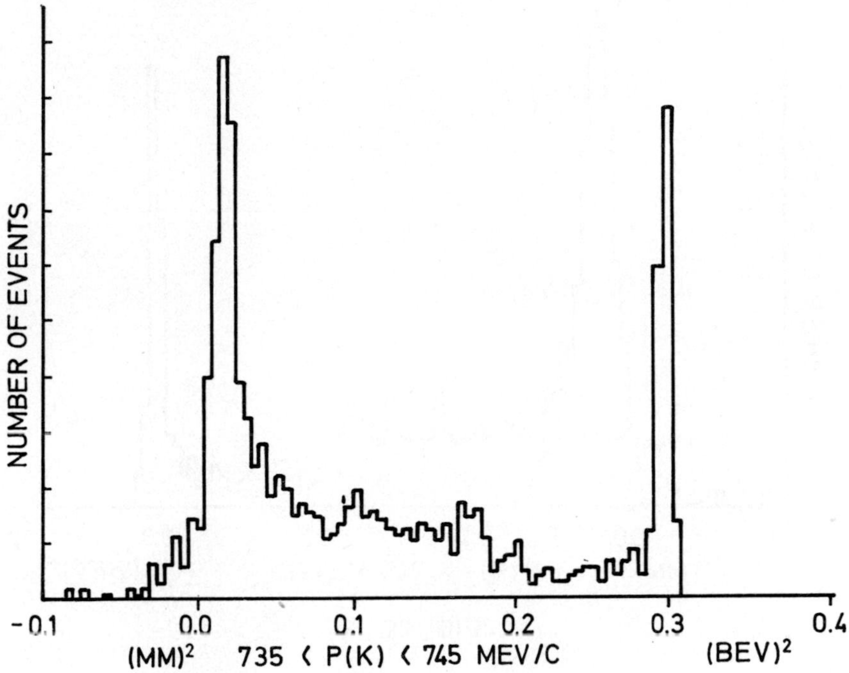

FIGURE 7

1851 zero-pronged events with bombarding energy between 735 and 745 MeV/c (E^* between 1669 and 1674 MeV) are shown plotted against missing mass. Compare with Figures 6 and 8.

---

and have been able to break these into five groups corresponding to E^*'s of 1668, 1670, 1673, 1675 and 1679 MeV and fit by maximum likelihood technique to find the best mixture of the four possibilities (π^0, $\pi^0\gamma$, $\pi^0\pi^0\gamma$ and η) shown in Figure 5. These latest results show a bump of 3.0 ± 0.3 millibarns at 1675 or

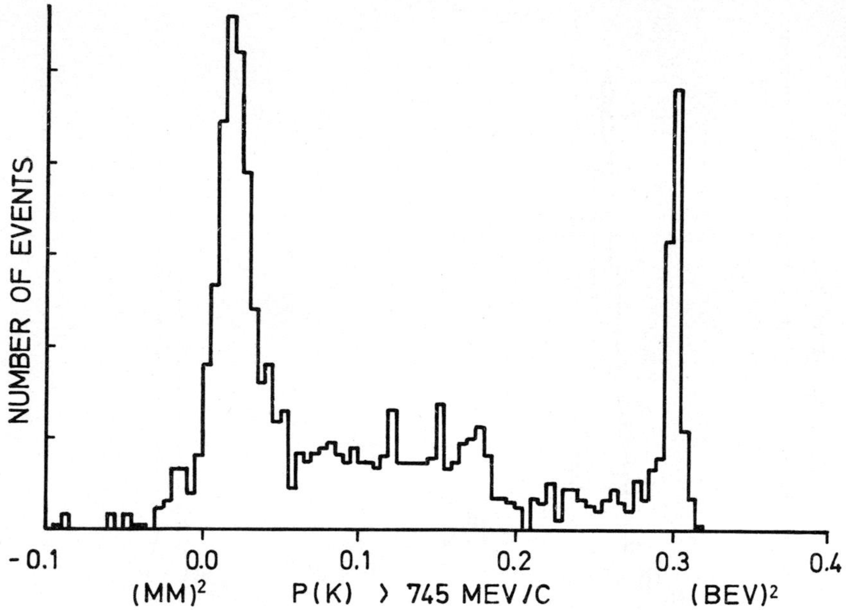

FIGURE 8

1301 zero-pronged events with bombarding energy greater than 745 MeV/c are shown plotted against missing mass. Compare with Figures 6 and 7.

---

746 MeV/c. If this new result is combined with the older experiment [7] there is a clear indication of a bump in the $\Sigma^0\pi^0$ production (see Table III). The significance of this result is that it eliminates the possibility of the "single strong scattering length" as

the primary explanation of the large cross section at threshold for Λ-η since the Σ^{o} - π^{o} channel is far

Table III

Cross Section for Production of Σ^{o} π^{o} in $K^{-}p$ Interactions as a Function of Bombarding Momentum*

Momentum (MeV/c)	Cross section (millibarns)	Comment
705	2.0 ± 0.7	
725	1.9 ± 0.7	
741	1.7 ± 0.7	
746	3.0 ± 0.3	New Duke result
768	2.4 ± 0.5	
802	1.0 ± 0.9	
820	1.3 ± 1.0	

above threshold at 746 MeV/c. The interpretation of the effects as a resonance has been strengthened. We are studying the angular distribution of the production, but must be satisfied with an indirect observation. Individual events cannot be fit to the hypothesis of a Σ^{o} and a π^{o} with consequent definite directions in space, but a large sample of events near the center portion of the plots shown in Figures 6 through 8 consists mostly of such events. For this sample we observe the angular distribution of the Λ with respect to the beam. With reasonable assumptions, one can infer the angular distribution of the sigmas.

---

* See footnote 7., and private communication from Dr. E. Hart.

Decay Modes of the η

The best measurement concerning the decay modes of the η to be obtained from this experiment is the ratio we have called

$$R_3 = \frac{\Gamma(\eta \to \text{all neutrals})}{\Gamma(\eta \to \text{charged})} .$$

This experiment yields extremely clean "signals" for the η mass, with relatively smooth backgrounds in the general vicinity of 5 % of the signal. It has already been mentioned that special precautions were taken to assure a high scanning efficiency for the zero-pronged events. After finding a Λ by area scanning, a scanner seldom misses an associated two-pronged event corresponding to the $\eta \to$ charged decays. Our results for R_3 is 2.7 ± 0.4, to be compared with the compilation in Table II which gives 2.37 from the "new" column.

Our analysis of the branching ratios within the all neutral modes is based on the observation of single electron-positron pairs either internally or externally converted. Though limited in terms of the total number of events because of the low probability of external conversion in hydrogen, each event corresponds to about 95 % certainty that an η was produced with a known momentum at a known position in the chamber. When the electron-positron pair is observed in hydrogen, the energy of the γ ray which produced it can normally be determined within an error of about 10 %. With reasonable assumptions about the decay distributions, one can predict the overall single γ ray spectrum from the observed η sample. This technique was first used for the η decays when I was working with Frank Crawford and others at the Lawrence Radiation Laboratory [9].

The final result is that of having a selection mechanism for η production with the addition of a γ ray spectrometer, the same in principle as the Frascati experiment [4]. The recent bubble chamber results from Columbia University [10] have been analyzed in this same way. Though there is not a specific slide showing the distributions, one can immediately see that there is a situation here similar to that for the various distributions going to make up the "missing mass" distribution of Figure 5. There is a single peak in γ ray energy corresponding to the $\gamma\gamma$ decay mode, a distribution starting at the energy corresponding to a single missing π^o from the $\pi^o\gamma\gamma$ decay when one of the two γ's is converted, a wide distribution corresponding to that same mode when a γ ray from π^o decay is converted (leaving three"missing" γ's), and finally a distribution starting at the energy corresponding to two missing π^o's for the $\eta \to \pi^o\pi^o\pi^o$ mode. There are, of course, different efficiencies for conversion in each of the modes. We have calculated the conversion efficiencies for γ rays of various energies, starting at various points within the chamber and checked these calculations with the γ's from single π^o production in this experiment. Also, we have made a subtraction of a few percent for the general background of $\Lambda\pi^o\pi^o$ production and $\Sigma\pi^o\pi^o$ production as determined from our fits to Figures 6,7 and 8. We are left with the best fit which can be made to the 130 events of Figure 9. The ratios are given in the two Duke columns of Table II, the first under the assumption that $\pi^o\gamma\gamma$ is an important decay mode; the second under the assumption that this decay mode is negligible small. The errors shown are mostly statistical at this level; however, since we cannot settle the question of the presence or absence of the $\pi^{0}\gamma\gamma$ mode, clearly the errors on the

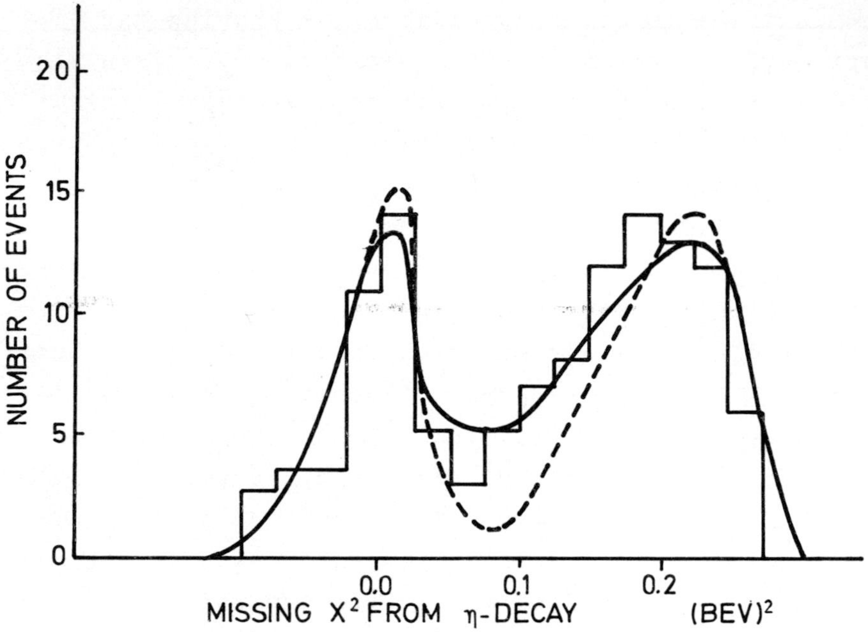

FIGURE 9

When a neutral decaying η accompanied by a single converted γ ray is observed, the overall invariant mass in other neutral particles is calculated (called "Missing X^2"). The histogram shows the distribution of 130 events with the value of Missing X^2 and two possible fits to the data. The solid line corresponds to the Duke fit with 17 % $\pi^{0}\gamma\gamma$ decays. The dashed line corresponds to the assumption of no $\pi^{0}\gamma\gamma$ decay. See Table II.

$3\pi^o$ decay are rather large. Our conclusion is that our experiment by no means demands the $\pi^o\gamma\gamma$ mode, though it can be brought into agreement with about 15 % with little difficulty. All experiments to test the existence of this mode are very sensitive to the efficiencies assigned to γ ray conversions and the dependence of those efficiencies on the energy of the γ rays. This is especially true of those spark chamber experiments which depend on counting 2,3,4,5 and 6 conversions in a single event. It seems that the best new idea settling whether or not the $\pi^o\gamma\gamma$ decay mode of the η is appreciable is to produce the η at a high enough energy that the resulting γ rays all have relatively high energies and relatively constant conversion efficiencies. I understand from Professor G. Salvini that such new experiments are planned.

Finally, we come to a part of the Duke experiment which is again relatively straightforward. This is the determination of the ratio of

$$\frac{\Gamma(\eta \to \pi^+\pi^-\gamma)}{\Gamma(\eta \to \pi^+\pi^-\pi^o)} .$$

Only the two-pronged events with a vee are involved, but the situation is not quite as simple as it would first appear because of the presence of the process $K^- p \to \pi^+\pi^-\Sigma^o_{\hookrightarrow\Lambda\gamma}$, the same overall final state as $K^{\mp}\ p \to \Lambda\ \eta_{\hookrightarrow\pi^+\pi^-\gamma}$. Returning to the complete sample of identified Λ's with a primary two pronged event (about 10.000), about 8.000 events survive a selection that requires the line of flight of the lambda to be between 0.5 and 10.0 cm in length and that the two charged tracks in the primary event each be at 10 cm in length. Almost 5000 events are removed as giving excellent fits to the hypothesis that nothing is missing (i.e. they fit $\Lambda^o\pi^+\pi^-$). This is a four constraint class fit, and it has been well established that events which really have a missing particle, even a γ ray, can seldom "fake" the "nothing-missing" type of event. All of the remaining 3.000 events yield a good fit to a single missing γ ray - an easy fit to achieve since the γ has zero rest mass. When the invariant mass of that γ combined with the Λ is plotted against the invariant mass of the same γ combined with the two pions in the same event, the scatter plot of Figure 10 is obtained. The broad horizontal bond just above $M(\Lambda\gamma) = 1200\ MeV/c^2$ corresponds to Σ production. The narrow vertical line of points corresponding to $M(\pi\pi\gamma) = 550\ MeV/c^2$ corresponds to Σ production. At this stage a cut on $M(\Lambda\gamma) \leq 1212.6\ MeV/c^2$ has been made to remove most of the Σ events. This cut, under the assumption that we know approximately the general population of the Dalitz plot for $\eta \rightarrow \pi^+\pi^-\gamma$ decay, removes about 40% of the true η events (see Figure 11). It leaves us, however, with the clean part of Figure 10, where the η identification on the basis of mass is almost certain. After the Σ's have been removed, the remaining $\sim$ 1000 events are fit to the hypothesis that an η-Λ combination was produced with the η decaying to $\pi^+ + \pi^-$ + "missing mass". Again the Σ test was applied, if $M(\Lambda;-MM) \leq 1212.6\ MeV/c^2$, the event was rejected (this removed 317 events, leaving 736. Figure 12 shows a plot of these 736 events; $(\text{missing mass})^2$ is shown along the abscissa, number of events on the ordinate. The peak at zero corresponds to the $\pi^+\pi^-\gamma$ mode, the peak at $0.020(GeV/c^2)^2$ corresponds to the $\pi^+\pi^-\pi^o$ mode. After applying the 40% correction for η's lost in the γ cut, the results displayed in Table II are obtained. A careful check has verified that the technique described (primarily the cut on $M(\Lambda\gamma)$) does not cause appreciable loss of η's decaying

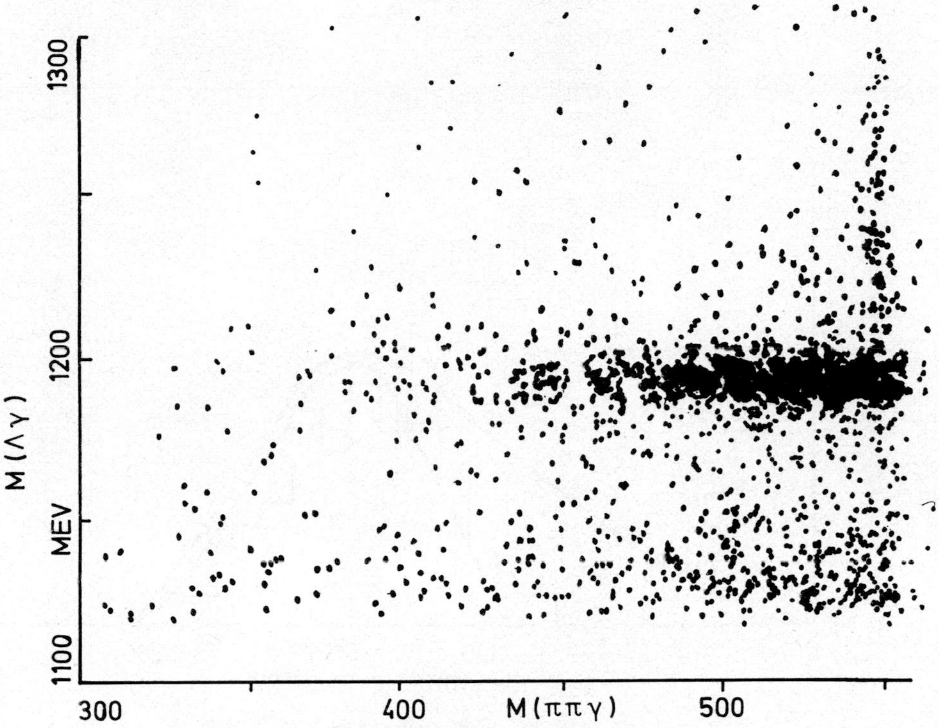

EVENTS WITH Λππ-FIT REMOVED

FIGURE 10

After eliminating events which fit $\Lambda\pi^+\pi^-$ production, twoprongs plus Λ are fit to the hypothesis of a single missing γ ray. This scatter plot shows the invariant mass of the Λ-γ combination versus that of the two pions combined with the γ ray. The broad horizontal band represents $\Sigma^0\pi^+\pi^-$ events and the narrow vertical band represents Λη production .

---

via $\pi^+\pi^-\pi^o$. No correction has been applied to the 642 which are seen in the peak shown in Figure 12.

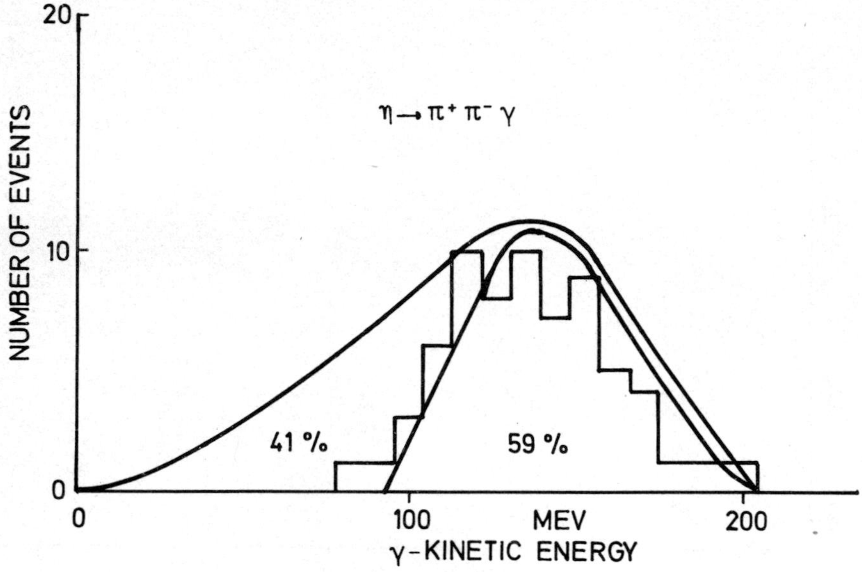

FIGURE 11

The simplest matrix element for the decay $\eta\rightarrow\pi^+\pi^-\gamma$ has been used to evaluate the effect of the $M(\Lambda\gamma)$ cut-off applied at the value 1212.6 MeV/c^2 in order to remove $\Sigma^o\pi^+\pi^-$ events from the η sample.

The heavier curve shows the dependence of number of η's on the γ ray energy. The lighter curve shows the way in which the effective mass cut-off would affect an ideal η sample. The histogram shows the identified η events after the cut-off was applied.

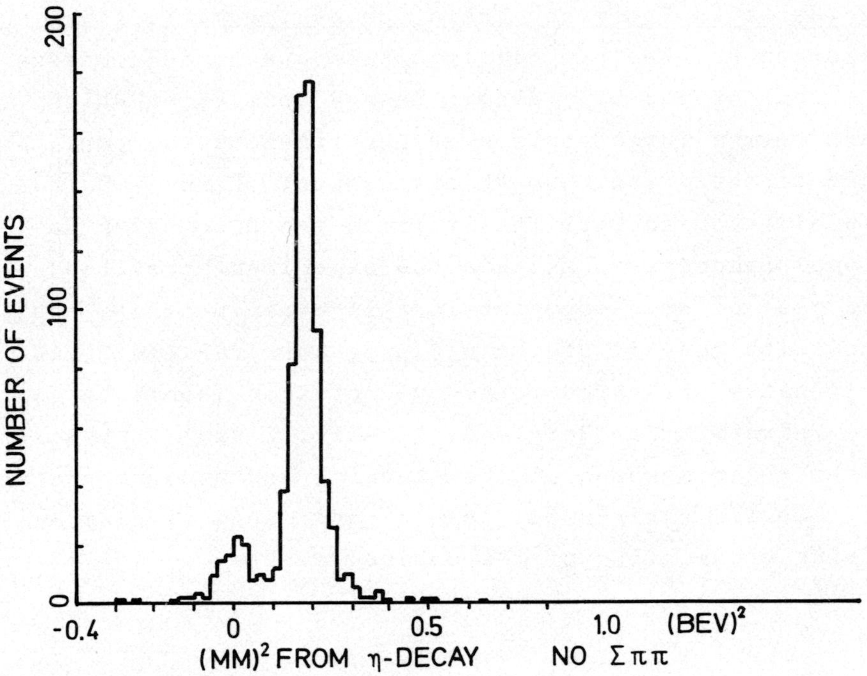

FIGURE 12

η production events which survived the 1212.6 MeV/c^2 cut-off made to remove $\Sigma^o\pi^+\pi^-$ events are tested for the mass of the unseen neutral particle. The distribution shows a peak at zero corresponding to the $\pi^+\pi^-\gamma$ decay mode and a second, larger peak at 0.02 $(BeV/c^2)^2$ corresponding to the $\pi^+\pi^-\pi^o$ decay mode. Based on the results illustrated in Figure 11., the smaller peak is corrected by almost a factor of 2 to obtain the final ratio.

We at Duke are continuing with this analysis, and we hope eventually to bring the total numbers of identified events to nearly double the number reported here.

I would like to take this opportunity to express my thanks to Professor Urban for asking me to give this report. Further, it is a pleasure to acknowledge the help of the U.S. Atomic Energy Commission which furnishes a large portion of our financial support, both directly and through its support of the U.S. National Laboratories. The staff of the Brookhaven National Laboratory has made the experiment possible, and most of our computing in this experiment has been done with the aid of the new programs TVGP and SQUAW originally presented to us by Professor Thomas Day of the University of Maryland. Finally, I want to thank the regular members of the scanning and measuring staff at Duke for their usual careful and rapid finding, measuring and sorting of the desired events.

References

1. J. Bernstein, G. Feinberg, and T. D. Lee, Phys. Rev. 139, B1650 (1965); see also the Proceedings of the XIII. International Conference on High Energy Physics, Lawrence Radiation Laboratory, Berkeley, California USA, 1966.
2. C. Baltay, P. Franzini, J. Kim, L. Kirsch, D. Zanello, J. Lee-Franzini, R. Loveless, J. McFadyen, and H. Yarger, Phys. Rev. Letters 16, 1224 (1966).
3. A. M. Cnops, J. C. Lassalle, G. Finocchiaro, P. Mittner, P. Zanella, J. P. Dufey, B. Gobbi, M. A. Pouchon, and A. Muller, Phys. Letters 22, 546 (1966).
4. G. Di Guigno, R. Querzoli, G. Troise, F. Vanoli, M. Giorgi, P. Schiavon, and V. Silvestrini, Phys.

Rev. Letters 16, 767 (1966).

5. M. Gell-Mann and A. H. Rosenfeld, pp. 407-478 Annual Reviews of Nuclear Science 7, Stanford Annual Reviews Inc. Stanford, Cal. (1957).
6. M. Gell-Mann, D. Sharp, and W. G. Wagner, Phys. Rev. Letters 8, 261 (1962).
7. D. Berley, P. L. Connolly, E. L. Hart, D. C. Rahm, D. L. Stonehill, B. Thevenet, W. J. Willis, and S. S. Yamamoto, Phys. Rev. Letters 15, 641 (1965).
8. J. Meyer, p. 152, Proceedings of the Heidelberg International Conference on Elementary Particles, ed. by H. Filthuth, North Holland Publ. Comp. Amsterdam (1968).
9. F. S. Crawford, Jr., L. J. Lloyd, and E. C. Fowler, Phys. Rev. Letters 10, 546 (1963).
10. C. Baltay, P. Franzini, J. Kim, R. Newman, N. Yeh, and L. Kirsch, Phys. Rev. Letters 19, 1495 (1967).

Comment by G. Källén:

I'll close this evening with a little anecdote: When I was a young student there was a meeting in the late 1940's in Copenhagen, and at the end of this meeting there was a joking summary made by Casimir - as you know this was in the days when everybody was very excited about the existence of two different kinds of mesons (π and μ), new counting techniques etc.- and in this summary Casimir was making fun of all the techniques, of course, and his biggest joke was the following: He showed an absolutely blank slide, and then he said:"Here you see a really exciting thing: one neutral particle decaying into two other neutrals". And of course, everybody was laughing very heartily in those days. I believe, if people had been able to look 20 years ahead and know that the experimentalists 20 years afterwards would have the impertinence not only to discuss the decay: one neutral particle into two neutrals, but actually to discuss the branching ratios between the three different neutral modes in the decay of one neutral particle, they would have been really impressed.

SUMMARY - First Week†

By

H. PIETSCHMANN

Physikalisches Institut, Universität Bonn
and
Institut für Theoretische Physik, Univ. Wien

Although I am a theoretician, I would like to begin the summary of the first of this years Schladming weeks by telling you the results of an experiment with has been run with several shifts in the surrounding peaks. The process under consideration is schion* - tadpole scattering. Just as the cardinal number in mathematics, a schion shall be denoted by a Hebrew letter,  , whereas the tadpole is as usual denoted by  . Hence the process reads

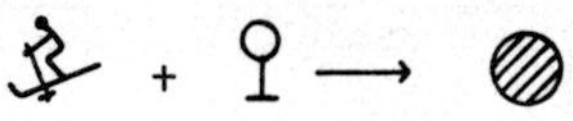 (-2)

where the black box on the right hand side of eq. (-2) originates, as usual, in our ignorance on the

† Summary given at the VII. Internationale Universitätswochen für Kernphysik, Schladming, February 26-March 9, 1968.

* Not to be confused with schizon although also a schion can sometimes carry zero momentum but ice-spin of several units.

form of the interaction. The energy of the process was about 5.10^{16} MeV. Very fast schions are called x-perts. In the approach of Prof. Julian Gradient, tadpoles are called "trees" and denoted by  .

As is well known, the S-operator consists of one part, the unit operator, that describes passage of the schion without an event. In graphical language, this is denoted by

(-1)

The remaining part is the transition operator which reads

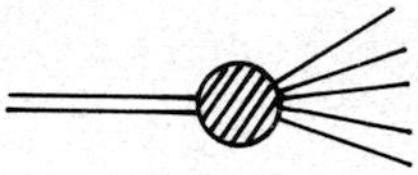

(o)

We are only concerned here with the transition part. The hugh majority of the events were of purely elastic origin and form the broad background. Only inelastic events were reported. Their classification is given in table 0 (further experiments are in progress).

Table 0: Inelastic schion events

Type	Number of events
Broken symmetry	1?
Random face scratches	2?

In addition, there was some violation of crossing symmetry. By this we mean that it happened that schi-on-legs were twisted and after twisting back they did not resume the original state any more. In the language of complex angular momenta, one would say, if schion poles conspire, a daughter trajectory may emerge and one is very lucky if he can avoid complex cuts. -

Let me now turn to the lectures of the first week. Jan Nilsson and the speaker gave two introductory courses into discrete symmetries and weak interactions. It has become standard practice to give introductory courses during the first week and I would like to emphasize once more the extreme value of this practice. Not only are there younger colleagues who benefit more from these courses than from more high-brow ones but they also serve the purpose to clarify many concepts and hence to minimize the confusion that necessarily propagates these days when everybody has to work hard to keep up to date even in a limited field. These lectures were complemented by a seminar of M. Nauenberg on "$\pi\pi$ - Interaction from $K_{\ell 4}$ Decay". Among other things, Nauenberg showed that in some cases conventional techniques for calculating spin-summed square of matrix elements may be intransparent to the way, a specific result is borne out by physical laws such as selection rules and so on.

Y. Dothan gave a review of recent developments in Regge-pole theory. To non-experts, this field has become almost a secret science, because of its closed set of terminology and phrasology that very often hides simple physical facts. It may therefore be advisable to have one review on Regge poles each year so that non-experts may bring their dictionary of technical terms up to date. It may be an unfortunate necessity to describe high energy scattering by this separated technique. However, research that tries to link the Regge-pole description with other theoretical areas should be strongly encouraged.

On different grounds, such a link has been found. The large number of exciting results that current algebra has produced can now also be borne out by the more conventional method of effective Lagrangians. F. Gürsey has actually shown how all the current algebra results are produced by this technique. It would be nice to have a friendly competition between these two fields. So far, it seems that the effective Lagrangian approach is a few steps behind in so far as it has not yet produced a result that was unknown before. A possible exception is the ratio G_A/G_V, given by

$$5/3\sqrt{2} = 1.17 \tag{1}$$

Since new measurements gave a larger value for G_A/G_V, I would like to anticipate further numerical research by stating

$$\sqrt{3}/\sqrt{2} = 1.22 \tag{2}$$

which can also be written as

$$(G_A/G_V)^2 = |\mu_p/\mu_n| \tag{3}$$

where μ_p and μ_n are the total magnetic momenta of proton and neutron.

Let us be serious again and return to chiral invariance. In an extremely transparent manner F. Gürsey has explained how all these different models of effective Lagrangians can be viewed from a unified point of view. He also pointed out the differences in numerical predictions. This is very valuable because it may serve as a means to rule out some of the alternatives by precise measurements. Connecting his approach to Yang-Mills type arguments, he showed that the theory does allow finite bare masses for the vector and axial vector mesons, provided the pion is massless.

This brings us to the problem of masslessness of the pion. Considering the above argument, it is certainly true that a zero mass pion with finite mass vector mesons is a much more comforting approximation than the opposite alternative. A similar approximation is included in the so-called "soft pion technique" of current algebra. However, I would like to point to a difference which should be kept at the back of one's mind. In the soft pion techniques the pion is off its mass-shell, continued to $q_\mu = 0$. If the mass-term in a Lagrangian is dropped, this means that the pion stays on its mass-shell but that it goes together with its mass-shell to the light-cone. The 2 alternatives are illustrated in fig. 1.

Fig. 1 : Two alternatives for "soft pions"

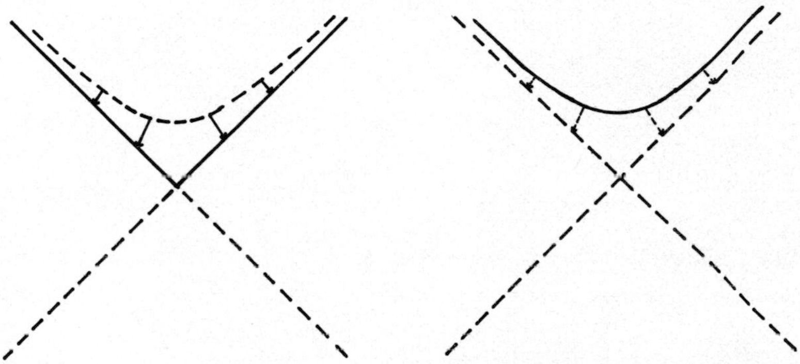

I would like to end the discussion of F. Gürseys very fine lectures by wishing him good luck for attempts of a synthesis of SU(6) with chiral groups.

Last not least let me mention two lectures of K. Koller on an attempt to obtain a quantum field theory with infinite dimensional representations of the Lorentz group. It should serve as a mathematical means to understand some of the concepts of generalized Regge poles. However, the mathematical difficulties are not yet completely solved.

In a wonderful school like this, the summarizer should not fail to notice that the weather has been at its very best throughout the whole of the first week.

SUMMARY - Second Week

By

G. KÄLLÉN
Department of Theoretical Physics
University of Lund
Sweden

To begin with I want to say a few words about the talk given by Rubbia on the decay of the longlived K-meson, i.e., about the reactions

$$K_L \to \begin{cases} \pi^+ \pi^- \\ \pi^o \pi^o \end{cases} .$$

This is something which we were excited about last year. Perhaps you might remember that at the time of the Schladming meeting in 1967 there were two absolutely conclusive pieces of evidence to the effect that the so-called "superweak theory" could not be correct. The relevant numbers are two complex parameters η_{+-} and η_{oo} which are related to the matrix elements for the two reactions mentioned above. We are particularly interested in the absolute values of these two complex numbers and a phase ϕ_{+-}

Summary given at the VII. Internationale Universitätswochen f.Kernphysik, Schladming, February 26 - March 9,1968.

quite important to calculate the radiative corrections to ordinary ß-decay if one wants to determine the Cabibbo angle from this process. There have been various suggestions in the literature this fall how to remove the divergence. Of course, this means that in some way or other one must change the basic assumptions of the previous calculation. All of the suggestions which have been made up to now seem to accept ordinary quantum electrodynamics and to assume that there is something which has to be done to the current algebra techniques. Perhaps it should be mentioned that the result quoted above, viz. that a divergence survives in the final answer holds only for the vector part of the weak interaction in ß-decay. It has been suggested by several groups that one should actually look for a compensation between the infinity from the vector part and the infinity from the axial vector part. However, it turns out that if you use one of the favourite models of ordinary current algebra, viz. the quark model, you will not get the desired cancellation between the vector and the axial vector terms. Consequently, several authors (Cabibbo et al., K. Johnson et al.) have suggested that one should abandon the ordinary quark model in favour of a theory where one has three basic triplets, denoted S, U, B by Cabibbo and coll. These basic triplets can be arranged in such a way that one actually gets a cancellation between the infinities from the vector and axial vector contributions to the radiative corrections. There is no question about the fact that this is a logically consistent scheme and if you believe in the existence of these basic triplets you get the desired result. However, my personal feeling is that you have to be rather brave to draw such far reaching conclusions about the basic structure of physics from the material at hand. From the more pragmatic point of view the argument is perhaps not very fruitful as it will be a long time before the existence or non-existence of the quarks, the SUB-particles or anything else on the same level can be demonstrated experimentally.

Another line of arguing which people have tried is to use the intermediate boson W for the weak interactions. Actually it has been shown, e.g., by Norton et al. that one can introduce this boson in such a way that the otherwise finite radiative corrections to μ-decay become divergent but the ratio between the μ-decay corrections and the corrections in β-decay becomes finite. By using available information about the Cabibbo angle, one has even been able to calculate that the mass of the W-boson should be about 10 GeV. This, of course, is a dangerous game. Essentially two things might happen. Either, the W will be experimentally discovered in the not too distant future and one might find that its mass is different from the value suggested on the basis of the radiative corrections to weak interactions. It is quite important to note that the mass of the W is predicted to have a very special value. The radiative correction estimate does not give only a lower or an upper bound for this mass. Alternatively, the W may not be discovered but sooner or later the experimental limit on its mass will be pushed above the value calculated from the radiative corrections effect. I don't know what the best bound on the mass of W is today but it is certainly less than 10 GeV. In summary, I believe the explanation using the SUB-particles will not be disproved for a long time while the explanation using the W stands a good chance of being proved wrong in a few years.

It is perhaps of interest to contrast this approach with the kind of calculation we discussed during the years of 1966 and 1967 in Schladming, Of course, everybody realizes that I can't pretend to give a completely unbiased point of view here but I cannot resist the temptation to express my personal feelings. In this context, I should like to say that while these more or less deep attempts to discuss what happens in the radiative corrections in terms of various high energy mechanisms are of interest, they really do not tackle the most burning issue for the moment. One thing is quite sure, viz.that the matrix elements which really appear in the radiative corrections include strong interaction effects. The high brow attempts try to invent models to calculate how the strong interactions actually do influence the matrix elements under consideration. What we did earlier was essentially much more modest. We did not make any attempt to calculate the matrix elements which appear inside the integrals but tried to take them from independent experimental data as far as available. Being a bit optimistic and using the material at hand rather liberally, one then ends up with a conclusion not only to the effect that the effects are finite (this is rather put in as an assumption) but one actually gets a numerical value for the radiative corrections in ordinary ß-decay. Therefore, the exact machinery that makes the radiative corrections finite is not at all discussed in these earlier attempts. In this respect, the calculations which were presented here during previous years are not in contradiction with the things Sirlin has been talking about. Everybody seems to agree that the final radiative corrections should be finite. However, the exact way in which they do become finite has been discussed very seriously this year but was essentially bypassed previously. In this respect, the discussion of this year has been more fundamental but perhaps also more dangerous than the earlier calculations.

I would now like to turn to the other talks of the second week and I take them essentially in random order. In this way I find that the next item on my list is the talk by Kummer. He demonstrated some calculations using current algebra. I do not want to go into too many details but let me try to emphasize one thing which I believe was the real essential point he was making. Kummer considered some process where a π-meson decays into a final state $|f\rangle$ and assumes that this decay is described by a certain current $j_b(o)$. In his first lecture he went through a rather complicated argument and ended up by assuming that the matrix element of interest fulfills a dispersion relation of the form

$$\langle f|j_b(o)|\pi\rangle = \frac{1}{f_\pi} T(o) - \frac{1}{\pi f_\pi} \int_{(3m_\pi)^2}^{\infty} \frac{\sigma(s')}{s'} ds' .$$

According to Kummer, standard current algebra using the PDDAC assumption means that one keeps just the pole term in this dispersion relation and drops the integral. As is rather well-known, this assumption leads to predictions for the decays $\pi^o \to 2\gamma$ and $\eta \to 2\gamma$ which are rather badly contradicted by the available data. Kummer suggests, if I understood him correctly, that one can improve upon this situation by considering also the integral. In particular, Kummer emphasized that when the pole term is small the integral cannot be neglected. This is all very nice, but the trouble starts when you have to evaluate the integral and find that you really don't know anything about the quantity which appears in the integrand. I believe, that Kummer then said that he hoped the weight $\sigma(s')$ can be approximated by a δ-function $\delta(s'-m^2)$ with a residue R.

Therefore, he effectively introduced two parameters R and m^2 instead of one parameter, viz. the value of the integral. He then tried to make plausibility guesses about what the residue R and the parameter m^2 could be for the decays and ended up with some predictions for the two γ-decays of π^o and η which, in his own opinion, did not look too good. Therefore, he remarked that as far as the η-decay goes one can always imagine that there is a mixing of η and η' (or X_o) and that this mixing can influence the ratio between the decay rates under consideration. Finally, he concluded very diplomatically that after all we probably have both these mechanisms and, for the moment, it is undecidable which of the two mechanisms is the most important one. I believe everybody will agree with this conclusion.

A somewhat related talk was given by Glashow. As a matter of fact, I believe his philosophy was rather different. Kummer said that the important term to consider is the integral while Glashow insisted on using only the pole term. However, and in contrast to Kummer, Glashow introduced an additional constant in the pole term, viz, a residue Z. Normally, one assumes that for a renormalized field the residue of the pole is equal to the coupling constant. Glashow said that he is not using this convention but some other normalization condition which I did not quite understand in detail (if he ever explained it). However, from the pragmatic point of view, what Glashow did was about the same as what Kummer did, viz. to introduce one extra parameter in the formalism. This, of course, makes the fitting of experimental data more complicated than what would otherwise be the case. Further, he considered not only the two simple decays that Kummer was discussing but had a whole set of other particles like ω, Φ, κ, K^* etc.

For each particle he had a decay parameter F_i, a mass m_i (which is known from experiments), the parameter Z_i from the pole term and form factors $f_i^{\pm}(o)$. By hammering a bit on this formalism and keeping only pole terms for the matrix elements, he found various relations between all his constants. It appeared to me that one of his most interesting predictions was the result

$$\frac{\Gamma(\omega \to e^+e^-)}{\Gamma(\Phi \to e^+e^-)} \simeq 6 \quad .$$

The number 6 here is, I believe, a rather new suggestion. Everybody else who has made similar calculations found a much smaller number. Glashow and all the rest of us are waiting for experimental data either to confirm this result or to prove that it is incorrect. As of today, no data are available for the ratio above.

As you have perhaps noted, Kummer and Glashow made rather opposite and contradictory assumptions about what the most important terms in the dispersion relation are. Nevertheless, it appeared to me that there was no very violent argument during the lectures of either person. I don't know if this means that none of them is so sure about his own grounds that he wanted to attack the other or if it is just a matter of general politeness.

Another somewhat related subject was discussed by Leutwyler. His problem was not quite so practical as the other speakers'. Leutwyler was interested in one special version of current algebra, viz. the technique where one lets the three-dimensional momentum of the particles go to infinity, in particular the z-component $p_3 \to \infty$. I believe his main point was that this is a somewhat peculiar mathematical scheme and that there should be simpler ways of doing it. Therefore, he introduced an average of the currents $j_i^\mu(x)$ which he called $\tilde{j}_i(x)$ and defined with the aid of a line integral

$$\tilde{j}_i(x) = \int_L j_i^\mu(x)\, n_\mu \, d\lambda \quad .$$

Here, n_μ is a light like vector with its space component along the z-direction. The line integral goes over the surface $x_o + x_3 = \sigma$ and the parameter λ is defined from

$$x_o = \frac{\sigma+\lambda}{2} \qquad x_3 = \frac{\sigma-\lambda}{2} \quad .$$

This new quantity $\tilde{j}_i(x)$ is then, by assumption, made to fulfill the standard current commutator relations with structure constants f

$$[\tilde{j},\tilde{j}] \sim f\, \tilde{j} \quad .$$

According to Leutwyler, this mathematical scheme is reasonably well defined. Further, it turns out that it gives the same result which one has previously obtained from the infinite momentum limit. Of course, this is a nice and interesting remark. Next, Leutwyler went on to saturate these commutators with one particle states. The same technique is normally also used by those who consider the infinite momentum limit. Then, he added various technical assumptions which, as usual, are quite important for the final conclusion but which I will not try to sketch here. However, at the end of his argument he was able to calculate a mass spectrum of all particles in the world. The result was reasonably realistic in so far as he found the masses of the particles should be increasing functions of the angular momentum quantum number. Unfortunately, there were also other parts of the mass spectrum which essentially corresponded to imaginary masses. Such states are sometimes called "ghost states". My personal impression is that this last part of the argument including the technical assumptions which lead to the mass spectrum should not be taken too seriously in the sense that they are not supposed to correspond very closely to the physical world. Rather, the whole argument should be considered as a model calculation showing how one can get a mass spectrum from this line of reasoning. I incidentally believe that the techniques Leutwyler used are reasonably similar to the methods of those people who work with infinite components wave functions and who also calculate a mass spectrum.

In contradistinction to the last few Schladming meetings there have not been very many discussions of group theory this year. In my own personal opinion this is a healthy trend in theoretical physics over the last years or so. The talk by Leutwyler had some relation to group theory but not very much.

Actually, Kastrup gave one talk where the title contained something about the conformal group but it appears to me that the main subject of his discussion was a consideration of inelastic processes where low energy particles are emitted. In particular, he discussed the emission of π-mesons and other particles in terms of a successive bremsstrahlung process. Using statistical arguments, he found that the cross section for an inelastic reaction with n particles in the final state could be written as follows

$$\frac{d\sigma^{(n)}}{d\Omega} = \sigma_B \frac{1}{n!} (\log G_M)^n G_M^2 \quad .$$

Here, σ_B is a Born approximation cross section and the last factor on the right-hand side is essentially a

Poisson-distribution, which appears because of the statistical nature of the argument. The interesting point of this formula is that the logarithm of the magnetic form factor (in the terminology of Sachs) appears as the average number of particles emitted in the reaction. This is a statement which, in principle, is possible to check experimentally and Kastrup showed a certain number of curves which are supposed to support this conclusion. On the basis of certain self consistency arguments he and collaborators have been able to go further and actually calculate an explicit expression for the form factor G_M which looked approximately like

$$G_M \sim \exp(-A \log^2(t-t_o)) \quad .$$

Kastrup also compared this form factor with the latest Stanford electron-proton scattering data, and the measurement appeared to agree reasonably well with the Kastrup formula for the form factor. It must, however, be mentioned that there are other formulae around in the literature, which are qualitatively different from Kastrup's suggestion and which also seem to fit the experiments reasonably well.

There was also a lecture by someone discussing the appearance or non-appearance of the certain gradient terms in the commutator of current operators. However, I don't think I would be able to add anything to this talk which has not already been said by the lecturer so I leave this subject out of my summary.

Finally, it should be also mentioned that Glashow gave a seminar talk where he tried very hard to convince us that certain modern cosmic-rays experiments offer strong evidence for the existence of a heavy lepton, which Glashow called the U-particle. I am not an expert on experimental data but it was my impression that you have to stretch your imagination rather much to be able to draw this conclusion from the material presented. However, I also had the impression that Glashow himself believes in the existence of these U-particles and also that he wants us to believe in them.

At the very end, I have the pleasant task to express my own and everybody else's sincere gratitude to the organizers of this meeting. In expressing these thanks let us start with the township of Schladming itself and its officials, especially its mayor. They certainly have done very much for our general entertainment here. The hard work and the main burden of the conference has been born by Miss Schmaldienst, Dr. Kühnelt and their collaborators. We are all very grateful to them for the excellent work they have done. Also, it is a great pleasure and honour to thank Professor Urban, the originator of these meetings. His kindness and friendliness is the same year after year and we are all very grateful to him that he gives us this opportunity of coming to Schladming, trying to learn various things like physics, skiing etc.